VOLKSWAGEN AIR-COOLED 1970-81 REPAIR MANUAL

CHILTON'S

Covers all U.S. and Canadian models of Beetle, Super Beetle, Karmann Ghia, Transporter, Vanagon, Fastback, Squareback, 411 and 412

by George B. Heinrich, A.S.E., S.A.E.

PUBLISHED BY **HAYNES NORTH AMERICA, Inc.**

Manufactured in USA
© 1997 Haynes North America, Inc.
ISBN 0-8019-8975-2
Library of Congress Catalog Card No. 97-65602
4567890123 9876543210

Haynes Publishing Group
Sparkford Nr Yeovil
Somerset BA22 7JJ England

Haynes North America, Inc
861 Lawrence Drive
Newbury Park
California 91320 USA

ABCDE
FGHIJ
KLMNO

Chilton is a registered trademark of W.G. Nichols, Inc., and has been licensed to Haynes North America, Inc.

Contents

1 GENERAL INFORMATION AND MAINTENANCE

- 1-2 HOW TO USE THIS BOOK
- 1-3 TOOLS AND EQUIPMENT
- 1-7 SERVICING YOUR VEHICLE SAFELY
- 1-8 FASTENERS, MEASUREMENTS AND CONVERSIONS
- 1-14 HISTORY
- 1-14 MODEL IDENTIFICATION
- 1-14 SERIAL NUMBER IDENTIFICATION
- 1-20 ROUTINE MAINTENANCE
- 1-43 FLUIDS AND LUCRICANTS
- 1-48 TRAILER TOWING
- 1-50 TOWING THE VEHICLE
- 1-50 JACKING AND HOISTING
- 1-51 JUMP STARTING A DEAD BATTERY
- 1-53 HOW TO BUY A USED VEHICLE

2 TUNE-UP

- 2-3 TUNE-UP PROCEDURES
- 2-8 FIRING ORDERS
- 2-8 POINT TYPE IGNITION
- 2-9 ELECTRONIC IGNITION
- 2-10 IGNITION TIMING
- 2-12 VALVE LASH
- 2-13 IDLE SPEED AND MIXTURE ADJUSTMENTS

3 ENGINE AND ENGINE REBUILDING

- 3-2 ENGINE ELECTRICAL
- 3-16 ENGINE MECHANICAL
- 3-36 ENGINE LUBRICATION
- 3-39 ENGINE COOLING
- 3-40 ENGINE REBUILDING

4 EMISSION CONTROLS

- 4-2 AIR POLLUTION
- 4-3 AUTOMOTIVE EMISSIONS
- 4-5 EMISSION CONTROLS
- 4-14 VACUUM DIAGRAMS

5 FUEL SYSTEM

- 5-2 BASIC FUEL SYSTEM DIAGNOSIS
- 5-2 CARBURETED FUEL SYSTEM
- 5-15 GASOLINE FUEL INJECTION SYSTEM
- 5-26 FUEL TANK

6 CHASSIS ELECTRICAL

- 6-2 UNDERSTANDING AND TROUBLESHOOT ELECTRICAL SYSTEMS
- 6-12 HEATER
- 6-14 WINDSHIELD WIPERS
- 6-15 INSTRUMENT CLUSTER
- 6-16 SEAT/BELT STARTER INTERLOCK
- 6-16 LIGHTING
- 6-20 FUSES
- 6-21 TRAILER WIRING
- 6-22 WIRING DIAGRAMS

Contents

7 DRIVE TRAIN

- 7-2 MANUAL TRANSAXLE
- 7-5 AUTOMATIC STICK SHIFT TRANSAXLE
- 7-8 DRIVESHAFT AND CONSTANT VELOCITY (CV) JOINT
- 7-10 CLUTCH
- 7-17 FULLY AUTOMATIC TRANSAXLE

8 SUSPENSION AND STEERING

- 8-2 TORSION BAR FRONT SUSPENSION
- 8-5 STRUT FRONT SUSPENSION
- 8-7 COIL SPRING FRONT SUSPENSION
- 8-10 DIAGONAL ARM REAR SUSPENSION
- 8-13 COIL SPRING/TRAILING ARM REAR SUSPENSION
- 8-15 STEERING

9 BRAKES

- 9-2 BRAKE SYSTEM
- 9-4 HYDRAULIC SYSTEM
- 9-6 FRONT DRUM BRAKES
- 9-10 FRONT DISC BRAKES
- 9-14 REAR DRUM BRAKES
- 9-19 PARKING BRAKE

10 BODY

- 10-2 EXTERIOR
- 10-14 AUTO BODY REPAIR
- 10-15 AUTO BODY CARE

GLOSSARY

- 10-37 GLOSSARY

MASTER INDEX

- 10-41 MASTER INDEX

SAFETY NOTICE

Proper service and repair procedures are vital to the safe, reliable operation of all motor vehicles, as well as the personal safety of those performing repairs. This manual outlines procedures for servicing and repairing vehicles using safe, effective methods. The procedures contain many NOTES, CAUTIONS and WARNINGS which should be followed, along with standard procedures to eliminate the possibility of personal injury or improper service which could damage the vehicle or compromise its safety.

It is important to note that repair procedures and techniques, tools and parts for servicing motor vehicles, as well as the skill and experience of the individual performing the work vary widely. It is not possible to anticipate all of the conceivable ways or conditions under which vehicles may be serviced, or to provide cautions as to all possible hazards that may result. Standard and accepted safety precautions and equipment should be used when handling toxic or flammable fluids, and safety goggles or other protection should be used during cutting, grinding, chiseling, prying, or any other process that can cause material removal or projectiles.

Some procedures require the use of tools specially designed for a specific purpose. Before substituting another tool or procedure, you must be completely satisfied that neither your personal safety, nor the performance of the vehicle will be endangered.

Although information in this manual is based on industry sources and is complete as possible at the time of publication, the possibility exists that some car manufacturers made later changes which could not be included here. While striving for total accuracy, the authors or publishers cannot assume responsibility for any errors, changes or omissions that may occur in the compilation of this data.

PART NUMBERS

Part numbers listed in this reference are not recommendations by Haynes North America, Inc. for any product brand name. They are references that can be used with interchange manuals and aftermarket supplier catalogs to locate each brand supplier's discrete part number.

SPECIAL TOOLS

Special tools are recommended by the vehicle manufacturer to perform their specific job. Use has been kept to a minimum, but where absolutely necessary, they are referred to in the text by the part number of the tool manufacturer. These tools can be purchased, under the appropriate part number, from your local dealer or regional distributor, or an equivalent tool can be purchased locally from a tool supplier or parts outlet. Before substituting any tool for the one recommended, read the SAFETY NOTICE at the top of this page.

ACKNOWLEDGMENTS

The publisher expresses appreciation to Volkswagen of America, Ltd, for their generous assistance.

All rights reserved. No part of this book may be reproduced or transmitted in any form or by any means, electronic or mechanical, including photocopying, recording or by any information storage or retrieval system, without permission in writing from the copyright holder.

While every attempt is made to ensure that the information in this manual is correct, no liability can be accepted by the authors or publishers for loss, damage or injury caused by any errors in, or omissions from, the information given.

HOW TO USE THIS BOOK 1-2
WHERE TO BEGIN 1-2
AVOIDING TROUBLE 1-2
MAINTENANCE OR REPAIR? 1-2
AVOIDING THE MOST COMMON MISTAKES 1-2
TOOLS AND EQUIPMENT 1-3
SPECIAL TOOLS 1-6
SERVICING YOUR VEHICLE SAFELY 1-7
DO'S 1-7
DON'TS 1-8
FASTENERS, MEASUREMENTS AND CONVERSIONS 1-8
BOLTS, NUTS AND OTHER THREADED RETAINERS 1-8
TORQUE 1-9
TORQUE WRENCHES 1-11
TORQUE ANGLE METERS 1-12
STANDARD AND METRIC MEASUREMENTS 1-12
HISTORY 1-14
MODEL IDENTIFICATION 1-14
SERIAL NUMBER IDENTIFICATION 1-14
VEHICLE (CHASSIS) 1-14
VEHICLE CERTIFICATION LABEL 1-14
ENGINE 1-14
TRANSAXLE 1-15
ROUTINE MAINTENANCE 1-20
AIR CLEANER 1-22
REMOVAL & INSTALLATION 1-22
FUEL FILTER 1-25
SERVICING 1-25
CRANKCASE VENTILATION 1-26
SERVICING 1-26
FUEL EVAPORATION CONTROL SYSTEM 1-26
SERVICING 1-26
BATTERY 1-26
GENERAL MAINTENANCE 1-26
BATTERY FLUID 1-27
CABLES 1-28
CHARGING 1-29
REPLACEMENT 1-30
BELTS 1-30
INSPECTION 1-30
ADJUSTMENT 1-31
AIR CONDITIONING
SAFETY PRECAUTIONS 1-32
GENERAL SERVICING PROCEDURES 1-33
SYSTEM INSPECTION 1-34
DISCHARGING, EVACUATING AND CHARGING 1-34
WINDSHIELD WIPERS 1-35
ELEMENT (REFILL) CARE AND REPLACEMENT 1-35
TIRES AND WHEELS 1-39
TIRE ROTATION 1-39
TIRE DESIGN 1-40
TIRE STORAGE 1-40
INFLATION & INSPECTION 1-40
CARE OF SPECIAL WHEELS 1-42
FLUIDS AND LUBRICANTS 1-43
FLUID DISPOSAL 1-43
FLUID LEVEL CHECKS 1-43
ENGINE OIL 1-43
TRANSAXLE 1-43
BRAKE FLUID 1-44
DIFFERENTIAL 1-44
STEERING GEAR (EXCEPT RACK AND PINION TYPE) 1-44
ENGINE OIL AND FUEL RECOMMENDATIONS 1-44
FLUID CHANGES 1-45
ENGINE 1-45
TRANSAXLE 1-47
FINAL DRIVE HOUSING 1-48
CHASSIS GREASING 1-48
WHEEL BEARINGS 1-48
REMOVAL, PACKING & INSTALLATION 1-48
TRAILER TOWING 1-48
GENERAL RECOMMENDATIONS 1-48
TRAILER WEIGHT 1-48
HITCH (TONGUE) WEIGHT 1-48
COOLING 1-49
ENGINE 1-49
TRANSMISSION 1-49
HANDLING A TRAILER 1-49
TOWING THE VEHICLE 1-50
JACKING AND HOISTING 1-50
JACKING PRECAUTIONS 1-51
JUMP STARTING A DEAD BATTERY 1-51
JUMP STARTING PRECAUTIONS 1-52
JUMP STARTING PROCEDURE 1-52
HOW TO BUY A USED VEHICLE 1-53
TIPS 1-53
USED VEHICLE CHECKLIST 1-53
ROAD TEST CHECKLIST 1-54
SPECIFICATION CHARTS
CHASSIS NUMBER CHART 1-16
ENGINE IDENTIFICATION CHART 1-19
OIL VISCOSITY SELECTION CHART 1-45
CAPACITIES 1-55
ENGLISH TO METRIC CONVERSION CHARTS 1-56

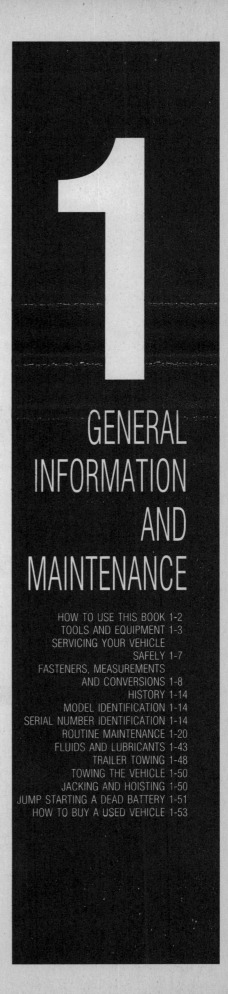

1

GENERAL INFORMATION AND MAINTENANCE

HOW TO USE THIS BOOK 1-2
TOOLS AND EQUIPMENT 1-3
SERVICING YOUR VEHICLE SAFELY 1-7
FASTENERS, MEASUREMENTS AND CONVERSIONS 1-8
HISTORY 1-14
MODEL IDENTIFICATION 1-14
SERIAL NUMBER IDENTIFICATION 1-14
ROUTINE MAINTENANCE 1-20
FLUIDS AND LUBRICANTS 1-43
TRAILER TOWING 1-48
TOWING THE VEHICLE 1-50
JACKING AND HOISTING 1-50
JUMP STARTING A DEAD BATTERY 1-51
HOW TO BUY A USED VEHICLE 1-53

1-2 GENERAL INFORMATION AND MAINTENANCE

HOW TO USE THIS BOOK

Chilton's Total Car Care manual is intended to help you learn more about the inner workings of your vehicle while saving you money on its upkeep and operation.

The beginning of the book will likely be referred to the most, since that is where you will find information for maintenance and tune-up. The other sections deal with the more complex systems of your vehicle. Operating systems from engine through brakes are covered to the extent that the average do-it-yourselfer becomes mechanically involved. This book will not explain such things as rebuilding a differential for the simple reason that the expertise required and the investment in special tools make this task uneconomical. It will, however, give you detailed instructions to help you change your own brake pads and shoes, replace spark plugs, and perform many more jobs that can save you money, give you personal satisfaction and help you avoid expensive problems.

A secondary purpose of this book is a reference for owners who want to understand their vehicle and/or their mechanics better. In this case, no tools at all are required.

Where to Begin

Before removing any bolts, read through the entire procedure. This will give you the overall view of what tools and supplies will be required. There is nothing more frustrating than having to walk to the bus stop on Monday morning because you were short one bolt on Sunday afternoon. So read ahead and plan ahead. Each operation should be approached logically and all procedures thoroughly understood before attempting any work.

All sections contain adjustments, maintenance, removal and installation procedures, and in some cases, repair or overhaul procedures. When repair is not considered practical, we tell you how to remove the part and then how to install the new or rebuilt replacement. In this way, you at least save the labor costs. Backyard repair of some components is just not practical.

Avoiding Trouble

Many procedures in this book require you to "label and disconnect . . ." a group of lines, hoses or wires. Don't be lulled into thinking you can remember where everything goes—you won't. If you hook up vacuum or fuel lines incorrectly, the vehicle will run poorly, if at all. If you hook up electrical wiring incorrectly, you may instantly learn a very expensive lesson.

You don't need to know the official or engineering name for each hose or line. A piece of masking tape on the hose and a piece on its fitting will allow you to assign your own label such as the letter A or a short name. As long as you remember your own code, the lines can be reconnected by matching similar letters or names. Do remember that tape will dissolve in gasoline or other fluids; if a component is to be washed or cleaned, use another method of identification. A permanent felt-tipped marker can be very handy for marking metal parts. Remove any tape or paper labels after assembly.

Maintenance or Repair?

It's necessary to mention the difference between maintenance and repair. Maintenance includes routine inspections, adjustments, and replacement of parts which show signs of normal wear. Maintenance compensates for wear or deterioration. Repair implies that something has broken or is not working. A need for repair is often caused by lack of maintenance. Example: draining and refilling the automatic transmission fluid is maintenance recommended by the manufacturer at specific mileage intervals. Failure to do this can ruin the transmission/transaxle, requiring very expensive repairs. While no maintenance program can prevent items from breaking or wearing out, a general rule can be stated: MAINTENANCE IS CHEAPER THAN REPAIR.

Two basic mechanic's rules should be mentioned here. First, whenever the left side of the vehicle or engine is referred to, it is meant to specify the driver's side. Conversely, the right side of the vehicle means the passenger's side. Second, most screws and bolts are removed by turning counterclockwise, and tightened by turning clockwise.

Safety is always the most important rule. Constantly be aware of the dangers involved in working on an automobile and take the proper precautions. See the information in this section regarding SERVICING YOUR VEHICLE SAFELY and the SAFETY NOTICE on the acknowledgment page.

Avoiding the Most Common Mistakes

Pay attention to the instructions provided. There are 3 common mistakes in mechanical work:

1. **Incorrect order of assembly, disassembly or adjustment.** When taking something apart or putting it together, performing steps in the wrong order usually just costs you extra time; however, it CAN break something. Read the entire procedure before beginning disassembly. Perform everything in the order in which the instructions say you should, even if you can't immediately see a reason for it. When you're taking apart something that is very intricate, you might want to draw a picture of how it looks when assembled at one point in order to make sure you get everything back in its proper position. We will supply exploded views whenever possible. When making adjustments, perform them in the proper order; often, one adjustment affects another, and you cannot expect even satisfactory results unless each adjustment is made only when it cannot be changed by any other.

2. **Overtorquing (or undertorquing).** While it is more common for overtorquing to cause damage, undertorquing may allow a fastener to vibrate loose causing serious damage. Especially when dealing with aluminum parts, pay attention to torque specifications and utilize a torque wrench in assembly. If a torque figure is not available, remember that if you are using the right tool to perform the job, you will probably not have to strain yourself to get a fastener tight enough. The pitch of most threads is so slight that the tension you put on the wrench will be multiplied many times in actual force on what you are tightening. A good example of how critical torque is can be seen in the case of spark plug in-

GENERAL INFORMATION AND MAINTENANCE 1-3

stallation, especially where you are putting the plug into an aluminum cylinder head. Too little torque can fail to crush the gasket, causing leakage of combustion gases and consequent overheating of the plug and engine parts. Too much torque can damage the threads or distort the plug, changing the spark gap.

There are many commercial products available for ensuring that fasteners won't come loose, even if they are not torqued just right (a very common brand is Loctite®). If you're worried about getting something together tight enough to hold, but loose enough to avoid mechanical damage during assembly, one of these products might offer substantial insurance. Before choosing a threadlocking compound, read the label on the package and make sure the product is compatible with the materials, fluids, etc. involved.

3. **Crossthreading.** This occurs when a part such as a bolt is screwed into a nut or casting at the wrong angle and forced. Crossthreading is more likely to occur if access is difficult. It helps to clean and lubricate fasteners, then to start threading with the part to be installed positioned straight in. Then, start the bolt, spark plug, etc. with your fingers. If you encounter resistance, unscrew the part and start over again at a different angle until it can be inserted and turned several times without much effort. Keep in mind that many parts, especially spark plugs, have tapered threads, so that gentle turning will automatically bring the part you're threading to the proper angle, but only if you don't force it or resist a change in angle. Don't put a wrench on the part until it's been tightened a couple of turns by hand. If you suddenly encounter resistance, and the part has not seated fully, don't force it. Pull it back out to make sure it's clean and threading properly.

Always take your time and be patient; once you have some experience, working on your vehicle may well become an enjoyable hobby.

TOOLS AND EQUIPMENT

Naturally, without the proper tools and equipment it is impossible to properly service your vehicle. It would also be virtually impossible to catalog every tool that you would need to perform all of the operations in this book. Of course, It would be unwise for the amateur to rush out and buy an expensive set of tools on the theory that he/she may need one or more of them at some time.

The best approach is to proceed slowly, gathering a good quality set of those tools that are used most frequently. Don't be misled by the low cost of bargain tools. It is far better to spend a little more for better quality. Forged wrenches, 6 or 12-point sockets and fine tooth ratchets are by far preferable to their less expensive counterparts. As any good mechanic can tell you, there are few worse experiences than trying to work on a vehicle with bad tools. Your monetary savings will be far outweighed by frustration and mangled knuckles.

Begin accumulating those tools that are used most frequently: those associated with routine maintenance and tune-up. In addition to the normal assortment of screwdrivers and pliers, you should have the following tools:

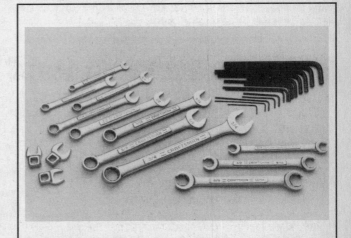

In addition to ratchets, a good set of wrenches and hex keys will be necessary

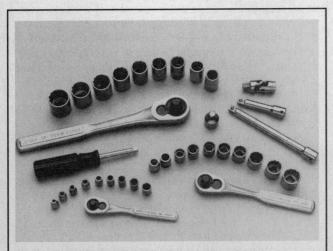

All but the most basic procedures will require an assortment of ratchets and sockets

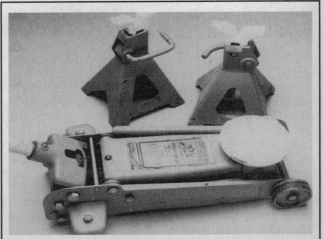

A hydraulic floor jack and a set of jackstands are essential for lifting and supporting the vehicle

1-4 GENERAL INFORMATION AND MAINTENANCE

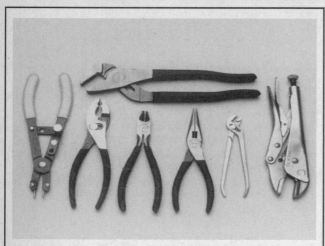

An assortment of pliers, grippers and cutters will be handy for old rusted parts and stripped bolt heads

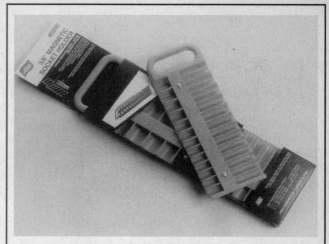

Tools from specialty manufacturers such as Lisle® are designed to make your job easier . . .

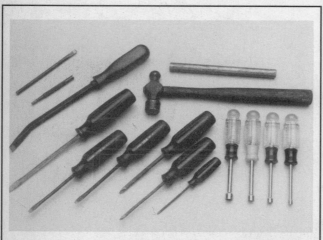

Various drivers, chisels and prybars are great tools to have in your toolbox

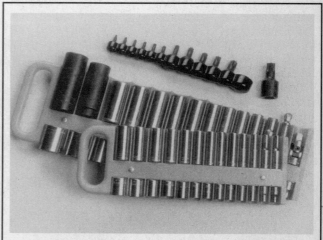

. . . these Torx® drivers and magnetic socket holders are just 2 examples of their handy products

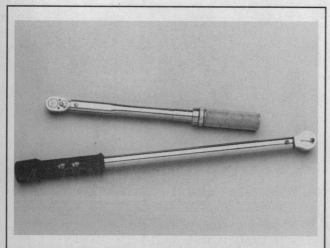

Many repairs will require the use of a torque wrench to assure the components are properly fastened

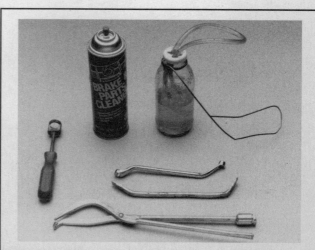

Although not always necessary, using specialized brake tools will save time

GENERAL INFORMATION AND MAINTENANCE 1-5

A few inexpensive lubrication tools will make maintenance easier

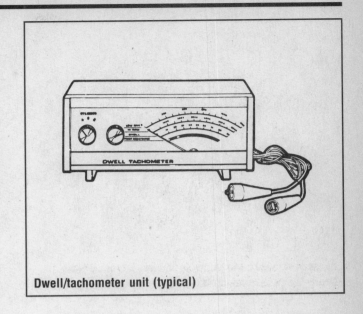
Dwell/tachometer unit (typical)

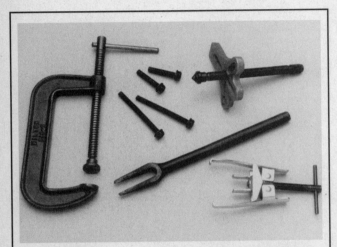
Various pullers, clamps and separator tools are needed for many larger, more complicated repairs

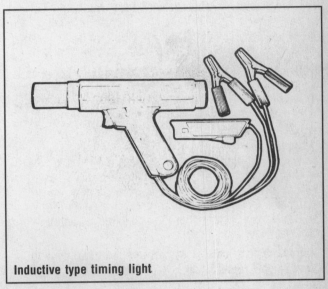
Inductive type timing light

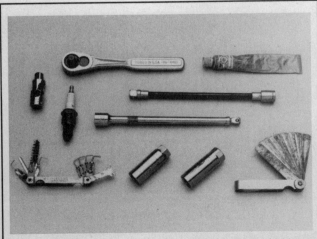
A variety of tools and gauges should be used for spark plug gapping and installation

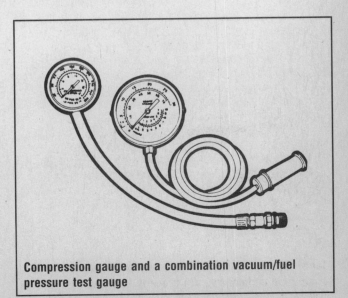
Compression gauge and a combination vacuum/fuel pressure test gauge

1-6 GENERAL INFORMATION AND MAINTENANCE

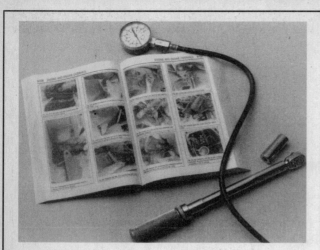

Proper information is vital, so always have a Chilton Total Car Care manual handy

- Wrenches/sockets and combination open end/box end wrenches in sizes from 1/8–3/4 in. or 3mm–19mm (depending on whether your vehicle uses standard or metric fasteners) and a 13/16 in. or 5/8 in. spark plug socket (depending on plug type).

➡ If possible, buy various length socket drive extensions. Universal-joint and wobble extensions can be extremely useful, but be careful when using them, as they can change the amount of torque applied to the socket.

- Jackstands for support.
- Oil filter wrench.
- Spout or funnel for pouring fluids.
- Grease gun for chassis lubrication (unless your vehicle is not equipped with any grease fittings—for details, please refer to information on Fluids and Lubricants found later in this section).
- Hydrometer for checking the battery (unless equipped with a sealed, maintenance-free battery).
- A container for draining oil and other fluids.
- Rags for wiping up the inevitable mess.

In addition to the above items there are several others that are not absolutely necessary, but handy to have around. These include Oil Dry® (or an equivalent oil absorbent gravel—such as cat litter) and the usual supply of lubricants, antifreeze and fluids, although these can be purchased as needed. This is a basic list for routine maintenance, but only your personal needs and desire can accurately determine your list of tools.

After performing a few projects on the vehicle, you'll be amazed at the other tools and non-tools on your workbench. Some useful household items are: a large turkey baster or siphon, empty coffee cans and ice trays (to store parts), ball of twine, electrical tape for wiring, small rolls of colored tape for tagging lines or hoses, markers and pens, a note pad, golf tees (for plugging vacuum lines), metal coat hangers or a roll of mechanics's wire (to hold things out of the way), dental pick or similar long, pointed probe, a strong magnet, and a small mirror (to see into recesses and under manifolds).

A more advanced set of tools, suitable for tune-up work, can be drawn up easily. While the tools are slightly more sophisticated, they need not be outrageously expensive. There are several inexpensive tach/dwell meters on the market that are every bit as good for the average mechanic as a professional model. Just be sure that it goes to a least 1200–1500 rpm on the tach scale and that it works on 4, 6 and 8-cylinder engines. (If you own one or more vehicles with a diesel engine, a special tachometer is required since diesels don't use spark plug ignition systems). The key to these purchases is to make them with an eye towards adaptability and wide range. A basic list of tune-up tools could include:

- Tach/dwell meter.
- Spark plug wrench and gapping tool.
- Feeler gauges for valve or point adjustment. (Even if your vehicle does not use points or require valve adjustments, a feeler gauge is helpful for many repair/overhaul procedures).

A tachometer/dwell meter will ensure accurate tune-up work on vehicles without electronic ignition. The choice of a timing light should be made carefully. A light which works on the DC current supplied by the vehicle's battery is the best choice; it should have a xenon tube for brightness. On any vehicle with an electronic ignition system, a timing light with an inductive pickup that clamps around the No. 1 spark plug cable is preferred.

In addition to these basic tools, there are several other tools and gauges you may find useful. These include:

- Compression gauge. The screw-in type is slower to use, but eliminates the possibility of a faulty reading due to escaping pressure.
- Manifold vacuum gauge.
- 12V test light.
- A combination volt/ohmmeter
- Induction Ammeter. This is used for determining whether or not there is current in a wire. These are handy for use if a wire is broken somewhere in a wiring harness.

As a final note, you will probably find a torque wrench necessary for all but the most basic work. The beam type models are perfectly adequate, although the newer click types (breakaway) are easier to use. The click type torque wrenches tend to be more expensive. Also keep in mind that all types of torque wrenches should be periodically checked and/or recalibrated. You will have to decide for yourself which better fits your purpose.

Special Tools

Normally, the use of special factory tools is avoided for repair procedures, since these are not readily available for the do-it-yourself mechanic. When it is possible to perform the job with more commonly available tools, it will be pointed out, but occasionally, a special tool was designed to perform a specific function and should be used. Before substituting another tool, you should be convinced that neither your safety nor the performance of the vehicle will be compromised.

Special tools can usually be purchased from an automotive parts store or from your dealer. In some cases special tools may be available directly from the tool manufacturer.

GENERAL INFORMATION AND MAINTENANCE 1-7

SERVICING YOUR VEHICLE SAFELY

It is virtually impossible to anticipate all of the hazards involved with automotive maintenance and service, but care and common sense will prevent most accidents.

The rules of safety for mechanics range from "don't smoke around gasoline," to "use the proper tool(s) for the job." The trick to avoiding injuries is to develop safe work habits and to take every possible precaution.

Do's

• Do keep a fire extinguisher and first aid kit handy.
• Do wear safety glasses or goggles when cutting, drilling, grinding or prying, even if you have 20–20 vision. If you wear glasses for the sake of vision, wear safety goggles over your regular glasses.
• Do shield your eyes whenever you work around the battery. Batteries contain sulfuric acid. In case of contact with the eyes or

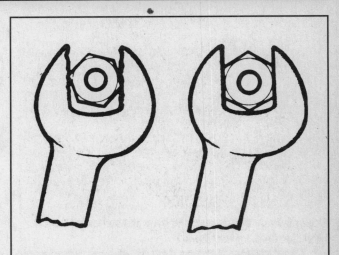

Using the correct size wrench will help prevent the possibility of rounding off a nut

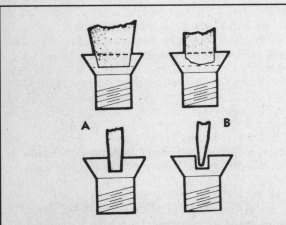

Screwdrivers should be kept in good condition to prevent injury or damage which could result if the blade slips from the screw

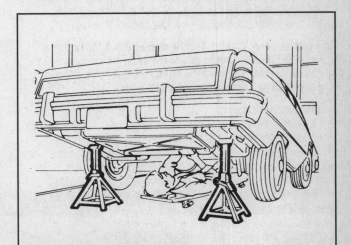

NEVER work under a vehicle unless it is supported using safety stands (jackstands)

skin, flush the area with water or a mixture of water and baking soda, then seek immediate medical attention.
• Do use safety stands (jackstands) for any undervehicle service. Jacks are for raising vehicles; jackstands are for making sure the vehicle stays raised until you want it to come down. Whenever the vehicle is raised, block the wheels remaining on the ground and set the parking brake.
• Do use adequate ventilation when working with any chemicals or hazardous materials. Like carbon monoxide, the asbestos dust resulting from some brake lining wear can be hazardous in sufficient quantities.
• Do disconnect the negative battery cable when working on the electrical system. The secondary ignition system contains EXTREMELY HIGH VOLTAGE. In some cases it can even exceed 50,000 volts.
• Do follow manufacturer's directions whenever working with potentially hazardous materials. Most chemicals and fluids are poisonous if taken internally.

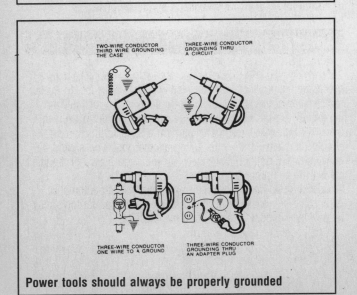

Power tools should always be properly grounded

1-8 GENERAL INFORMATION AND MAINTENANCE

- Do properly maintain your tools. Loose hammerheads, mushroomed punches and chisels, frayed or poorly grounded electrical cords, excessively worn screwdrivers, spread wrenches (open end), cracked sockets, slipping ratchets, or faulty droplight sockets can cause accidents.
- Likewise, keep your tools clean; a greasy wrench can slip off a bolt head, ruining the bolt and often harming your knuckles in the process.
- Do use the proper size and type of tool for the job at hand. Do select a wrench or socket that fits the nut or bolt. The wrench or socket should sit straight, not cocked.
- Do, when possible, pull on a wrench handle rather than push on it, and adjust your stance to prevent a fall.
- Do be sure that adjustable wrenches are tightly closed on the nut or bolt and pulled so that the force is on the side of the fixed jaw.
- Do strike squarely with a hammer; avoid glancing blows.
- Do set the parking brake and block the drive wheels if the work requires a running engine.

Don'ts

- Don't run the engine in a garage or anywhere else without proper ventilation—EVER! Carbon monoxide is poisonous; it takes a long time to leave the human body and you can build up a deadly supply of it in your system by simply breathing in a little every day. You may not realize you are slowly poisoning yourself. Always use power vents, windows, fans and/or open the garage door.
- Don't work around moving parts while wearing loose clothing. Short sleeves are much safer than long, loose sleeves. Hard-toed shoes with neoprene soles protect your toes and give a better grip on slippery surfaces. Jewelry such as watches, fancy belt buckles, beads or body adornment of any kind is not safe working around a vehicle. Long hair should be tied back under a hat or cap.
- Don't use pockets for toolboxes. A fall or bump can drive a screwdriver deep into your body. Even a rag hanging from your back pocket can wrap around a spinning shaft or fan.
- Don't smoke when working around gasoline, cleaning solvent or other flammable material.
- Don't smoke when working around the battery. When the battery is being charged, it gives off explosive hydrogen gas.
- Don't use gasoline to wash your hands; there are excellent soaps available. Gasoline contains dangerous additives which can enter the body through a cut or through your pores. Gasoline also removes all the natural oils from the skin so that bone dry hands will suck up oil and grease.
- Don't service the air conditioning system unless you are equipped with the necessary tools and training. When liquid or compressed gas refrigerant is released to atmospheric pressure it will absorb heat from whatever it contacts. This will chill or freeze anything it touches. Although refrigerant is normally non-toxic, R-12 becomes a deadly poisonous gas in the presence of an open flame. One good whiff of the vapors from burning refrigerant can be fatal.
- Don't use screwdrivers for anything other than driving screws! A screwdriver used as an prying tool can snap when you least expect it, causing injuries. At the very least, you'll ruin a good screwdriver.
- Don't use a bumper or emergency jack (that little ratchet, scissors, or pantograph jack supplied with the vehicle) for anything other than changing a flat! These jacks are only intended for emergency use out on the road; they are NOT designed as a maintenance tool. If you are serious about maintaining your vehicle yourself, invest in a hydraulic floor jack of at least a 1½ ton capacity, and at least two sturdy jackstands.

FASTENERS, MEASUREMENTS AND CONVERSIONS

Bolts, Nuts and Other Threaded Retainers

Although there are a great variety of fasteners found in the modern car or truck, the most commonly used retainer is the threaded fastener (nuts, bolts, screws, studs, etc). Most threaded retainers may be reused, provided that they are not damaged in use or during the repair. Some retainers (such as stretch bolts or torque prevailing nuts) are designed to deform when tightened or in use and should not be reinstalled.

Whenever possible, we will note any special retainers which should be replaced during a procedure. But you should always inspect the condition of a retainer when it is removed and replace any that show signs of damage. Check all threads for rust or corrosion which can increase the torque necessary to achieve the desired clamp load for which that fastener was originally selected. Additionally, be sure that the driver surface of the fastener has not been compromised by rounding or other damage. In some cases a driver surface may become only partially rounded, allowing the driver to catch in only one direction. In many of these occurrences, a fastener may be installed and tightened, but the driver would not be able to grip and loosen the fastener again. (This could lead to frustration down the line should that component ever need to be disassembled again).

If you must replace a fastener, whether due to design or damage, you must ALWAYS be sure to use the proper replacement. In all cases, a retainer of the same design, material and strength should be used. Markings on the heads of most bolts will help determine the proper strength of the fastener. The same material, thread and pitch must be selected to assure proper installation and safe operation of the vehicle afterwards.

Thread gauges are available to help measure a bolt or stud's thread. Most automotive and hardware stores keep gauges available to help you select the proper size. In a pinch, you can use another nut or bolt for a thread gauge. If the bolt you are replacing is not too badly damaged, you can select a match by finding another bolt which will thread in its place. If you find a nut which threads properly onto the damaged bolt, then use that nut to help select the replacement bolt. If however, the bolt you are replacing is so badly damaged (broken or drilled out) that its threads cannot be used as a gauge, you might start by looking for another bolt (from the same assembly or a similar location on your vehicle) which will thread into the damaged bolt's mounting. If so, the other bolt can be used to select a nut; the nut can then be used to select the replacement bolt.

GENERAL INFORMATION AND MAINTENANCE

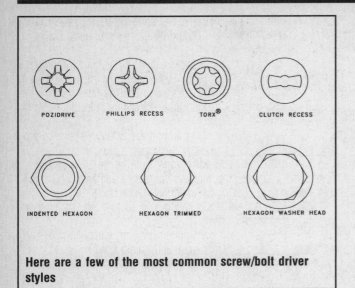

Here are a few of the most common screw/bolt driver styles

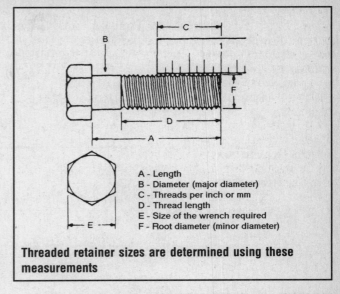

Threaded retainer sizes are determined using these measurements

- A - Length
- B - Diameter (major diameter)
- C - Threads per inch or mm
- D - Thread length
- E - Size of the wrench required
- F - Root diameter (minor diameter)

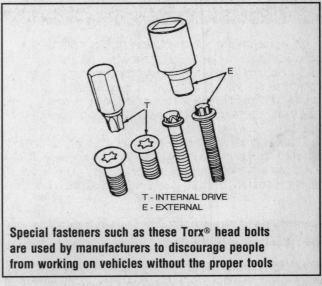

T - INTERNAL DRIVE
E - EXTERNAL

Special fasteners such as these Torx® head bolts are used by manufacturers to discourage people from working on vehicles without the proper tools

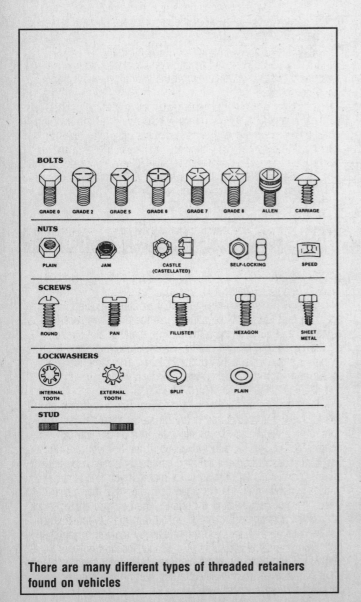

There are many different types of threaded retainers found on vehicles

In all cases, be absolutely sure you have selected the proper replacement. Don't be shy, you can always ask the store clerk for help.

✸✸ WARNING

Be aware that when you find a bolt with damaged threads, you may also find the nut or drilled hole it was threaded into has also been damaged. If this is the case, you may have to drill and tap the hole, replace the nut or otherwise repair the threads. NEVER try to force a replacement bolt to fit into the damaged threads.

Torque

Torque is defined as the measurement of resistance to turning or rotating. It tends to twist a body about an axis of rotation. A common example of this would be tightening a threaded retainer such as a nut, bolt or screw. Measuring torque is one of the most

GENERAL INFORMATION AND MAINTENANCE

Standard Torque Specifications and Fastener Markings

In the absence of specific torques, the following chart can be used as a guide to the maximum safe torque of a particular size/grade of fastener.
- There is no torque difference for fine or coarse threads.
- Torque values are based on clean, dry threads. Reduce the value by 10% if threads are oiled prior to assembly.
- The torque required for aluminum components or fasteners is considerably less.

U.S. Bolts

SAE Grade Number	1 or 2			5			6 or 7		
Number of lines always 2 less than the grade number.									
Bolt Size (Inches)—(Thread)	Maximum Torque			Maximum Torque			Maximum Torque		
	Ft./Lbs.	Kgm	Nm	Ft./Lbs.	Kgm	Nm	Ft./Lbs.	Kgm	Nm
¼—20	5	0.7	6.8	8	1.1	10.8	10	1.4	13.5
—28	6	0.8	8.1	10	1.4	13.6			
⁵⁄₁₆—18	11	1.5	14.9	17	2.3	23.0	19	2.6	25.8
—24	13	1.8	17.6	19	2.6	25.7			
⅜—16	18	2.5	24.4	31	4.3	42.0	34	4.7	46.0
—24	20	2.75	27.1	35	4.8	47.5			
⁷⁄₁₆—14	28	3.8	37.0	49	6.8	66.4	55	7.6	74.5
—20	30	4.2	40.7	55	7.6	74.5			
½—13	39	5.4	52.8	75	10.4	101.7	85	11.75	115.2
—20	41	5.7	55.6	85	11.7	115.2			
⁹⁄₁₆—12	51	7.0	69.2	110	15.2	149.1	120	16.6	162.7
—18	55	7.6	74.5	120	16.6	162.7			
⅝—11	83	11.5	112.5	150	20.7	203.3	167	23.0	226.5
—18	95	13.1	128.8	170	23.5	230.5			
¾—10	105	14.5	142.3	270	37.3	366.0	280	38.7	379.6
—16	115	15.9	155.9	295	40.8	400.0			
⅞—9	160	22.1	216.9	395	54.6	535.5	440	60.9	596.5
—14	175	24.2	237.2	435	60.1	589.7			
1—8	236	32.5	318.6	590	81.6	799.9	660	91.3	894.8
—14	250	34.6	338.9	660	91.3	849.8			

Metric Bolts

Relative Strength Marking	4.6, 4.8			8.8		
Bolt Markings						
Bolt Size Thread Size x Pitch (mm)	Maximum Torque			Maximum Torque		
	Ft./Lbs.	Kgm	Nm	Ft./Lbs.	Kgm	Nm
6 x 1.0	2–3	.2–.4	3–4	3–6	4–.8	5–8
8 x 1.25	6–8	.8–1	8–12	9–14	1.2–1.9	13–19
10 x 1.25	12–17	1.5–2.3	16–23	20–29	2.7–4.0	27–39
12 x 1.25	21–32	2.9–4.4	29–43	35–53	4.8–7.3	47–72
14 x 1.5	35–52	4.8–7.1	48–70	57–85	7.8–11.7	77–110
16 x 1.5	51–77	7.0–10.6	67–100	90–120	12.4–16.5	130–160
18 x 1.5	74–110	10.2–15.1	100–150	130–170	17.9–23.4	180–230
20 x 1.5	110–140	15.1–19.3	150–190	190–240	26.2–46.9	160–320
22 x 1.5	150–190	22.0–26.2	200–260	250–320	34.5–44.1	340–430
24 x 1.5	190–240	26.2–46.9	260–320	310–410	42.7–56.5	420–550

Standard and metric bolt torque specifications based on bolt strengths—WARNING: use only as a guide

GENERAL INFORMATION AND MAINTENANCE 1-11

common ways to help assure that a threaded retainer has been properly fastened.

When tightening a threaded fastener, torque is applied in three distinct areas, the head, the bearing surface and the clamp load. About 50 percent of the measured torque is used in overcoming bearing friction. This is the friction between the bearing surface of the bolt head, screw head or nut face and the base material or washer (the surface on which the fastener is rotating). Approximately 40 percent of the applied torque is used in overcoming thread friction. This leaves only about 10 percent of the applied torque to develop a useful clamp load (the force which holds a joint together). This means that friction can account for as much as 90 percent of the applied torque on a fastener.

TORQUE WRENCHES

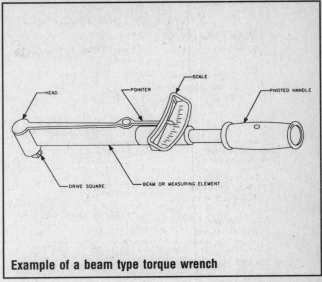

Example of a beam type torque wrench

In most applications, a torque wrench can be used to assure proper installation of a fastener. Torque wrenches come in various designs and most automotive supply stores will carry a variety to suit your needs. A torque wrench should be used any time we supply a specific torque value for a fastener. A torque wrench can also be used if you are following the general guidelines in the accompanying charts. Keep in mind that because there is no worldwide standardization of fasteners, the charts are a general guideline and should be used with caution. Again, the general rule of "if you are using the right tool for the job, you should not have to strain to tighten a fastener" applies here.

Beam Type

The beam type torque wrench is one of the most popular types. It consists of a pointer attached to the head that runs the length of the flexible beam (shaft) to a scale located near the handle. As the wrench is pulled, the beam bends and the pointer indicates the torque using the scale.

Click (Breakaway) Type

Another popular design of torque wrench is the click type. To use the click type wrench you pre-adjust it to a torque setting. Once the torque is reached, the wrench has a reflex signalling fea-

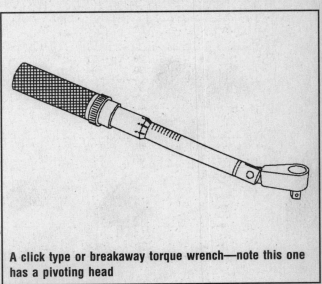

A click type or breakaway torque wrench—note this one has a pivoting head

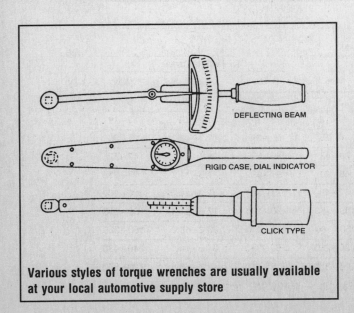

Various styles of torque wrenches are usually available at your local automotive supply store

ture that causes a momentary breakaway of the torque wrench body, sending an impulse to the operator's hand.

Pivot Head Type

Some torque wrenches (usually of the click type) may be equipped with a pivot head which can allow it to be used in areas of limited access. BUT, it must be used properly. To hold a pivot head wrench, grasp the handle lightly, and as you pull on the handle, it should be floated on the pivot point. If the handle comes in contact with the yoke extension during the process of pulling, there is a very good chance the torque readings will be inaccurate because this could alter the wrench loading point. The design of the handle is usually such as to make it inconvenient to deliberately misuse the wrench.

➡ It should be mentioned that the use of any U-joint, wobble or extension will have an effect on the torque readings, no matter what type of wrench you are using. For the most accurate readings, install the socket directly on the wrench driver. If necessary, straight extensions (which hold a

1-12 GENERAL INFORMATION AND MAINTENANCE

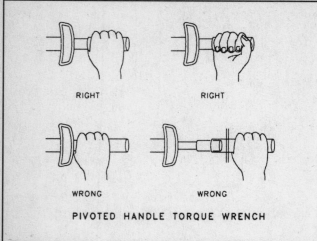

Torque wrenches with pivoting heads must be grasped and used properly to prevent an incorrect reading

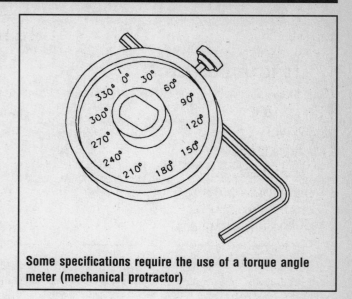

Some specifications require the use of a torque angle meter (mechanical protractor)

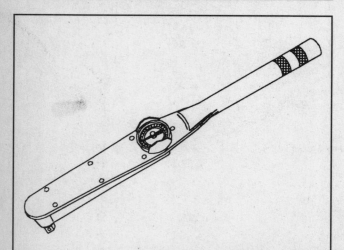

The rigid case (direct reading) torque wrench uses a dial indicator to show torque

socket directly under the wrench driver) will have the least effect on the torque reading. Avoid any extension that alters the length of the wrench from the handle to the head/driving point (such as a crow's foot). U-joint or Wobble extensions can greatly affect the readings; avoid their use at all times.

Rigid Case (Direct Reading)

A rigid case or direct reading torque wrench is equipped with a dial indicator to show torque values. One advantage of these wrenches is that they can be held at any position on the wrench without affecting accuracy. These wrenches are often preferred because they tend to be compact, easy to read and have a great degree of accuracy.

TORQUE ANGLE METERS

Because the frictional characteristics of each fastener or threaded hole will vary, clamp loads which are based strictly on torque will vary as well. In most applications, this variance is not significant enough to cause worry. But, in certain applications, a manufacturer's engineers may determine that more precise clamp loads are necessary (such is the case with many aluminum cylinder heads). In these cases, a torque angle method of installation would be specified. When installing fasteners which are torque angle tightened, a predetermined seating torque and standard torque wrench are usually used first to remove any compliance from the joint. The fastener is then tightened the specified additional portion of a turn measured in degrees. A torque angle gauge (mechanical protractor) is used for these applications.

Standard and Metric Measurements

Throughout this manual, specifications are given to help you determine the condition of various components on your vehicle, or to assist you in their installation. Some of the most common measurements include length (in. or cm/mm), torque (ft. lbs., inch lbs. or Nm) and pressure (psi, in. Hg, kPa or mm Hg). In most cases, we strive to provide the proper measurement as determined by the manufacturer's engineers.

Though, in some cases, that value may not be conveniently measured with what is available in your toolbox. Luckily, many of the measuring devices which are available today will have two scales so the Standard or Metric measurements may easily be taken. If any of the various measuring tools which are available to you do not contain the same scale as listed in the specifications, use the accompanying conversion factors to determine the proper value.

The conversion factor chart is used by taking the given specification and multiplying it by the necessary conversion factor. For instance, looking at the first line, if you have a measurement in inches such as "free-play should be 2 in." but your ruler reads only in millimeters, multiply 2 in. by the conversion factor of 25.4 to get the metric equivalent of 50.8mm. Likewise, if the specification was given only in a Metric measurement, for example in Newton Meters (Nm), then look at the center column first. If the measurement is 100 Nm, multiply it by the conversion factor of 0.738 to get 73.8 ft. lbs.

CONVERSION FACTORS

LENGTH–DISTANCE

Inches (in.)	x 25.4	= Millimeters (mm)	x .0394	= Inches
Feet (ft.)	x .305	= Meters (m)	x 3.281	= Feet
Miles	x 1.609	= Kilometers (km)	x .0621	= Miles

VOLUME

Cubic Inches (in3)	x 16.387	= Cubic Centimeters	x .061	= in3
IMP Pints (IMP pt.)	x .568	= Liters (L)	x 1.76	= IMP pt.
IMP Quarts (IMP qt.)	x 1.137	= Liters (L)	x .88	= IMP qt.
IMP Gallons (IMP gal.)	x 4.546	= Liters (L)	x .22	= IMP gal.
IMP Quarts (IMP qt.)	x 1.201	= US Quarts (US qt.)	x .833	= IMP qt.
IMP Gallons (IMP gal.)	x 1.201	= US Gallons (US gal.)	x .833	= IMP gal.
Fl. Ounces	x 29.573	= Milliliters	x .034	= Ounces
US Pints (US pt.)	x .473	= Liters (L)	x 2.113	= Pints
US Quarts (US qt.)	x .946	= Liters (L)	x 1.057	= Quarts
US Gallons (US gal.)	x 3.785	= Liters (L)	x .264	= Gallons

MASS–WEIGHT

Ounces (oz.)	x 28.35	= Grams (g)	x .035	= Ounces
Pounds (lb.)	x .454	= Kilograms (kg)	x 2.205	= Pounds

PRESSURE

Pounds Per Sq. In. (psi)	x 6.895	= Kilopascals (kPa)	x .145	= psi
Inches of Mercury (Hg)	x .4912	= psi	x 2.036	= Hg
Inches of Mercury (Hg)	x 3.377	= Kilopascals (kPa)	x .2961	= Hg
Inches of Water (H_2O)	x .07355	= Inches of Mercury	x 13.783	= H_2O
Inches of Water (H_2O)	x .03613	= psi	x 27.684	= H_2O
Inches of Water (H_2O)	x .248	= Kilopascals (kPa)	x 4.026	= H_2O

TORQUE

Pounds–Force Inches (in–lb)	x .113	= Newton Meters (N·m)	x 8.85	= in–lb
Pounds–Force Feet (ft–lb)	x 1.356	= Newton Meters (N·m)	x .738	= ft–lb

VELOCITY

Miles Per Hour (MPH)	x 1.609	= Kilometers Per Hour (KPH)	x .621	= MPH

POWER

Horsepower (Hp)	x .745	= Kilowatts	x 1.34	= Horsepower

FUEL CONSUMPTION*

Miles Per Gallon IMP (MPG)	x .354	= Kilometers Per Liter (Km/L)
Kilometers Per Liter (Km/L)	x 2.352	= IMP MPG
Miles Per Gallon US (MPG)	x .425	= Kilometers Per Liter (Km/L)
Kilometers Per Liter (Km/L)	x 2.352	= US MPG

*It is common to covert from miles per gallon (mpg) to liters/100 kilometers (1/100 km), where mpg (IMP) x 1/100 km = 282 and mpg (US) x 1/100 km = 235.

TEMPERATURE

Degree Fahrenheit (°F) = (°C x 1.8) + 32

Degree Celsius (°C) = (°F − 32) x .56

Standard and metric conversion factors chart

1-14 GENERAL INFORMATION AND MAINTENANCE

HISTORY

In 1932, Ferdinand Porsche produced prototypes for the NSU company of Germany which eventually led to the design of the Volkswagen. The prototypes had a rear mounted, air-cooled engine, torsion bar suspension, and the spare tire mounted at an angle in the front luggage compartment. In 1936, Porsche produced three Volkswagen prototypes, one of which was a 995 cc, horizontally opposed, four cylinder automobile. Passenger car development was sidetracked during World War II, when all attention was on military vehicles. In 1945, Volkswagen production began and 1,785 Beetles were built. The Volkswagen convertible was introduced in 1949, the same year that only two Volkswagens were sold in the United States. 1950 marked the beginning of the sunroof models and the transporter series. The Karmann Ghia was introduced in 1956, and remained in the same basic styling format until its demise in 1974. The 1500 Squareback was introduced in the United States in 1966 to start the Type 3 series. The Type 4 was imported into the U.S.A. beginning with the 1971 model. 1977 marked the last year for the Beetle. The Beetle convertible was available through 1980 and the new VW bus, the Vanagon, was introduced in 1980.

MODEL IDENTIFICATION

Type numbers are the way Volkswagen designates its various groups of models. The type 1 group contains the Beetle, Super Beetle, and the Karmann Ghia. Type 2 vehicles are the Delivery Van, the Micro Bus, The Vanagon, the Kombi and the Campmobile. The Type 3 designation is for the Fastback and Squareback sedans. The Type 4 is for the 411 and 412 sedans and wagon. These type numbers will be used throughout the book when it is necessary to refer to models.

An explanation of the terms suitcase engine and upright fan engine is, perhaps, necessary. The upright fan engine refers to the engine used in the Type 1 and 2 (1970–71) vehicles. This engine has the engine cooling fan mounted on the top of the engine and is driven by the generator. The fan is mounted vertically in contrast to a horizontally mounted fan as found on the Chevrolet Corvair engine. The suitcase engine is a comparatively compact unit to fit in the Type 3, 4 and 1972 and later Type 2 engine compartments. On this engine, the cooling fan is mounted on the crankshaft giving the engine a rectangular shape similar to that of a suitcase.

SERIAL NUMBER IDENTIFICATION

Vehicle (Chassis)

The chassis number consists of ten digits. The first two numbers indicate the model type, and the third number gives the model year. For example, a 2 as the third digit means that the car was produced during the 1972 model year run.

The chassis number is stamped on a metal plate. On Type 1, 3, and 4 models, the plate is located in the luggage compartment, on the frame tunnel under the back seat, and on the driver's side of the instrument panel (visible through the windshield). On Type 2 models, the plate is located behind the front passenger's seat, on the left-hand engine cover plate, and on top of the driver's side of the instrument panel.

Vehicle Certification Label

♦ See Figures 1 and 2

The vehicle certification label is a decal affixed to the left door jamb. It indicates that the vehicle meets all U.S. federal safety standards as of the date of manufacture. The label also gives the chassis number of the car. Beginning with the 1973 model year, the label lists the gross vehicle weight rating and the gross axle weight rating. The gross vehicle weight rating is useful in determining the load carrying capacity of your car. Merely subtract the curb weight from the posted gross weight and what is left over is about how much you can haul. The gross axle weight rating is a good guide to the weight distribution of your car.

The vehicle certification label is constructed of special material

The Vehicle Identification Number (VIN) is located throughout the vehicle, such as on the upper left-hand corner of the dashboard

to guard against its alteration. If it is tampered with or removed, it will be destroyed or the word "VOID" will appear.

Engine

♦ See Figure 3

The engine can be identified by a letter or pair of letters preceding the serial number. Engine specifications are listed according to the letter code and model year.

GENERAL INFORMATION AND MAINTENANCE   1-15

The VIN can also be found on the body plate, located near the front hood latch

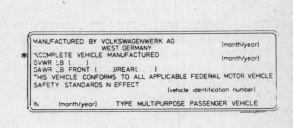

Fig. 2 A typical vehicle certification label used on all 1973–81 models (Type 2 1980–81 Vanagon shown)

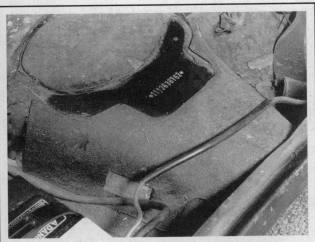

The VIN is also stamped into the frame tube of the vehicle under the rear bench seat

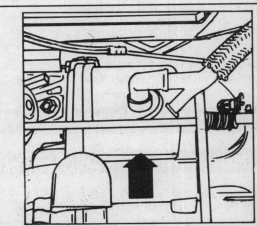

Fig. 3 On Type 2, 3 and 4 models utilizing the "suitcase" engine, the engine ID number is stamped into the engine block along the crankcase joint, near the oil breather

Fig. 1 An example of a vehicle certification label used on 1970–72 models (Type 1 shown)

On all Type 1 models, and on 1970–71 Type 2/1600 models using the upright fan engine, the engine number is stamped into the crankcase flange for the generator support. The number can readily be seen by looking through the center of the fan belt.

On all Type 3 and 4 models, and on 1972–79 Type 2/1700, Type 2/1800, and Type 2/2000 models using the "suitcase" engine, the engine number is stamped into the crankcase along the crankcase joint near the oil breather. On the 1980–81 Type 2 (Vanagon) the engine number is located on the right side of the engine compartment, directly in front of the fan housing.

Transaxle

Transaxle identification marks are stamped either into the bell housing or on the final drive housing.

Chassis Number Chart

Model Year	Vehicle	Model No.	From			To	
					Chassis Number		
1970	Beetle	113	110	2000	001	110 3096	945
	Karmann Ghia	14	140	2000	001	140 3100	000
	Beetle Convertible	15	150	2000	001	150 3100	000
	Van	21	210	2000	001	210 2300	000
	Bus	22	220	2000	001	220 2300	000
	Camper, Kombi	23	230	2000	001	230 2300	000
	Type 3 Fastback	31	310	2000	001	310 2500	000
	Type 3 Squareback	36	360	2000	001	360 2500	000
1971	Beetle/Super Beetle	111/113	111	2000	001	111 3143	118
	Karmann Ghia	14	141	2000	001	141 3200	000
	Beetle Convertible	15	151	2000	001	151 3200	000
	Van	21	211	2000	001	211 2300	000
	Bus	22	221	2000	001	221 2300	000
	Camper, Kombi	23	231	2000	001	231 2300	000
	Type 3 Fastback	31	311	2000	001	311 2500	000
	Type 3 Squareback	36	361	2000	001	361 2500	000
	411 2 Door	41	411	2000	001	411 2100	000
	411 4 Door	42	421	2000	001	421 2100	000
	411 Wagon	46	461	2000	001	461 2100	000
1972	Beetle/Super Beetle	111/113	112	2000	001	112 2961	362
	Karmann Ghia	14	142	2000	001	142 3200	000
	Beetle Convertible	15	152	2000	001	152 3200	000
	Van	21	212	2000	001	212 2300	000
	Bus	22	222	2000	001	222 2300	000
	Camper, Kombi	23	232	2000	001	232 2300	000

Chassis Number Chart (cont.)

Model Year	Vehicle	Model No.	Chassis Number From			Chassis Number To		
1972	Type 3 Fastback	31	312	2000	001	312	2500	000
	Type 3 Squareback	36	362	2000	001	362	2500	000
	411 2 Door	41	412	2000	001	412	2100	000
	411 4 Door	42	422	2000	001	422	2100	000
	411 Wagon	46	462	2000	001	462	2100	000
1973	Beetle	111	113	2000	001	113	3021	954
	Super Beetle	113	133	2000	001	133	3021	860
	Karmann Ghia	14	143	2000	001	143	3200	000
	Beetle Convertible	15	153	2000	001	153	3200	000
	Van	21	213	2000	001	213	2300	000
	Bus	22	223	2000	001	223	2300	000
	Camper, Kombi	23	233	2000	001	233	2300	000
	Type 3 Fastback	31	313	2000	001	313	2500	000
	Type 3 Squareback	36	363	2000	001	363	2500	000
	412 2 Door	41	413	2000	001	413	2100	000
	412 4 Door	42	423	2000	001	423	2100	000
	412 Wagon	46	463	2000	001	463	2100	000
1974	Beetle	111	114	2000	001	114	2818	456
	Super Beetle	113	134	2000	001	134	2798	165
	Karmann Ghia	14	144	2000	001	144	3200	000
	Beetle Convertible	15	154	2000	001	154	3200	000
	Van	21	214	2000	001	214	2300	000
	Bus	22	224	2000	001	224	2300	000
	Camper, Kombi	23	234	2000	001	234	2300	000
	412 2 Door	41	414	2000	001	414	2100	000

1-18 GENERAL INFORMATION AND MAINTENANCE

Chassis Number Chart (cont.)

Model Year	Vehicle	Model No.	Chassis Number From			Chassis Number To		
1974	412 4 Door	42	424	2000	001	424	2100	000
	412 Wagon	46	464	2000	001	464	2100	000
1975	Beetle	111	115	2000	001	115	3200	000
	Super Beetle (La Grande Bug)	113	135	2000	001	135	3200	000
	Beetle Convertible	15	155	2000	001	155	3200	000
	Van	21	215	2000	001	215	2300	000
	Bus	22	225	2000	001	225	2300	000
	Camper, Kombi	23	235	2000	001	235	2300	000
1976	Beetle	111	116	2000	001	116	3200	000
	Beetle Convertible	15	156	2000	001	156	2000	001
	Bus	22	226	2000	001	226	2300	000
	Camper, Kombi	23	236	2000	001	236	2300	001
1977	Beetle	111	117	2000	001	—		
	Beetle Convertible	15	157	2000	001	—		
	Bus	22	227	2000	001	—		
	Camper, Kombi	23	237	2000	001	—		
1978	Beetle Convertible	15	158	2000	001	—		
	Bus	22	228	2000	001	—		
	Camper	23	238	2000	001	—		
1979–80	Beetle Convertible	15	159	2000	001	—		
	Bus	22	229	2000	001	—		
	Camper	23	239	2000	001	—		
1980–81	Vanagon	24	24A	0000	001	—		

GENERAL INFORMATION AND MAINTENANCE 1-19

Engine Identification Chart

Engine Code Letter	Type Vehicle	First Production Year	Last Production Year	Engine Type	Common Designation
B	1, 2	1967	1970	Upright Fan	1600
AE	1, 2	1971	1972	Upright Fan	1600
AH (Calif)	1	1972	1974	Upright Fan	1600
AK	1	1973	1974	Upright Fan	1600
AJ	1	1975	1979	Upright Fan	1600
CB	2	1972	1973	Suitcase	1700
CD	2	1973	1973	Suitcase	1700
AW	2	1974	1974	Suitcase	1800
ED	2	1975	1975	Suitcase	1800
GD, GE, CV	2	1976	1981	Suitcase	2000
U	3	1968	1973	Suitcase	1600
X	3	1972	1973	Suitcase	1600
W	4	1971	1971	Suitcase	1700
EA	4	1972	1974	Suitcase	1700
EB (Calif)	4	1973	1973	Suitcase	1700
EC	4	1974	1974	Suitcase	1800

1-20 GENERAL INFORMATION AND MAINTENANCE

ROUTINE MAINTENANCE

GENERAL MAINTENANCE ITEMS – UPRIGHT 1600cc ENGINE SHOWN

1. Spark plugs
2. Hot air heater hose
3. Distributor cap and rotor (inside cap)
4. Fuel line
5. Spark plug wires
6. Inline fuel filter
7. Breather hose
8. Air cleaner assembly
9. Pre-heated intake air hose
10. Oil filler cap
11. Automatic stick shift dipstick and filler cap
12. Generator/alternator drive belt
13. Top dead center (TDC) indicator on crankcase
14. Timing marks on crankshaft pulley
15. Engine oil dipstick

GENERAL INFORMATION AND MAINTENANCE 1-21

GENERAL MAINTENANCE ITEMS (CONTINUED) - UPRIGHT 1600cc ENGINE SHOWN

1. Automatic stick shift fluid cover
2. Engine oil drain plug
3. Engine oil strainer plate
4. Inner CU boots
5. Outer CU boots
6. Charcoal (EVAP) canister
7. Automatic stick shift fluid reservoir

1-22 GENERAL INFORMATION AND MAINTENANCE

GENERAL MAINTENANCE ITEMS (CONTINUED) - SUPER BEETLE SHOWN
1. Spare tire (under floor mat)
2. Windshield washer fluid reservoir
3. Brake master cylinder reservoir
4. Windshield wiper blades
5. Fuel filler door and cap

Air Cleaner

REMOVAL & INSTALLATION

Oil Bath Type

This type cleaner should be cleaned at 6,000 mile intervals, or when the oil is changed.

TYPE 1 AND 2 (1970–73)

1. To clean the air cleaner, remove the hoses attached to the air cleaner.

※※ CAUTION

Be careful to note the places where the hoses are attached. Interchanging the hoses will affect the operation of the engine.

2. Next, loosen the air cleaner support bracket screw and the air cleaner clamp screw.
3. On 1970 models, disconnect the warm air flap cable. Lift the air cleaner off the engine. Keep the carburetor hole down to prevent spilling the oil out of the air cleaner.
4. Loosen the spring clips which secure the top of the air

GENERAL INFORMATION AND MAINTENANCE

To remove the air cleaner lid, unclasp the 4 lid clamps and lift the lid off of the air cleaner body—oil bath type

. . . loosen the air cleaner-to-carburetor clamp . . .

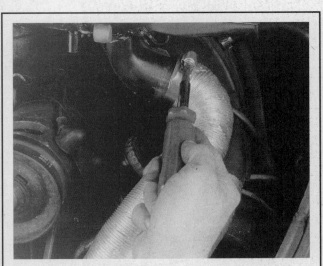
To remove the entire air cleaner assembly, disconnect the hot air hose from the air cleaner snorkel . . .

. . . and lift the air cleaner assembly off of the carburetor—oil bath type

. . . detach the breather hose from the air cleaner body

cleaner to the bottom and then separate the halves. Do not invert the upper half.

5. Put the upper half of the air cleaner down with the filter element facing downward. Thoroughly clean the bottom half.

6. Fill the air cleaner with 0.9 pints of SAE 30 (SAE 10W in sub-freezing climates) oil or, if present, to the oil level mark stamped into the side of the air cleaner.

7. Reassemble the air cleaner and install it on the engine.

ALL TYPE 3, TYPE 4 (1971–72)

▶ See Figure 4

1. Disconnect the activated charcoal filter hose, the rubber elbow, and the crankcase ventilation hose. Remove the wing nut in the center of the air cleaner and lift the air cleaner assembly off of the engine.

2. Release the spring clips which keep the air cleaner halves together and take the cleaner apart. Do not invert the upper half.

3. Clean the lower half and refill it with 0.085 pints of SAE 30 oil (SAE 10W in subfreezing climates) to the level mark. When re-

1-24 GENERAL INFORMATION AND MAINTENANCE

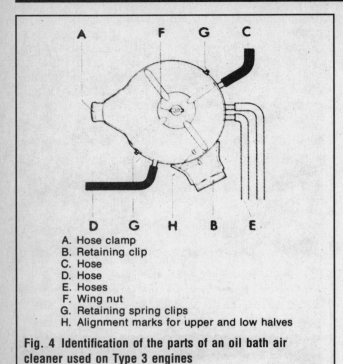

A. Hose clamp
B. Retaining clip
C. Hose
D. Hose
E. Hoses
F. Wing nut
G. Retaining spring clips
H. Alignment marks for upper and low halves

Fig. 4 Identification of the parts of an oil bath air cleaner used on Type 3 engines

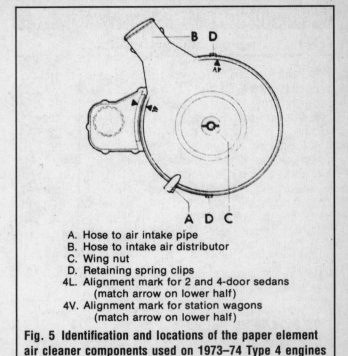

A. Hose to air intake pipe
B. Hose to intake air distributor
C. Wing nut
D. Retaining spring clips
4L. Alignment mark for 2 and 4-door sedans (match arrow on lower half)
4V. Alignment mark for station wagons (match arrow on lower half)

Fig. 5 Identification and locations of the paper element air cleaner components used on 1973–74 Type 4 engines

assembling the air cleaner, align the marks for the upper and lower halves.

4. Reinstall the air cleaner on the engine. Make sure it is properly seated.

Paper Element Type

TYPE 1 (1973–74), AND TYPE 4 (1973–74)

◆ See Figures 5 and 6

1. Label and disconnect the hoses from the air cleaner.

※※ CAUTION

Do not interchange the position of the hoses.

2. Loosen the air cleaner clamp and remove the air cleaner from the engine. Release spring clamps which keep the halves of the cleaner together and separate the halves.
3. Clean the inside of the air cleaner housing.
4. The paper element should be replaced every 18,000 miles under normal service. It should be replaced more often under severe operating conditions. A paper element may be cleaned by blowing through the element from the inside with compressed air. Never use a liquid solvent to clean a paper element.
5. Install the air cleaner element in the air cleaner housing and install the spring clips, making sure the halves are properly aligned. Install the cleaner on the engine.

TYPE 1 (1975–80)

1. Release the four clips at the top, side and bottom of the air cleaner housing and pull the front cover off the housing just enough to slide off the cardboard vent pipe at the bottom of the housing.

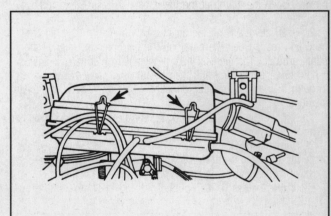

Fig. 6 The air cleaner lid on 1973–74 Type 1 Beetle and Super Beetle air cleaners is held in place with wire clamps (arrows)

2. Take the filter out and clean it by striking it against a hard surface or blowing through it with compressed air. The filter should be replaced every 18,000 miles under normal conditions and more frequently under severe conditions.
3. When installing filter, make sure it is seated properly and that the bottom hose is connected.

TYPE 2/1700 ENGINE (1972–73), TYPE 2/1800 ENGINE (1974)

The air filter is removed through the hatch in the interior of the bus above the engine on some models. Remove the rear mat to gain access.

1. Label and disconnect the hoses from the air cleaner.

GENERAL INFORMATION AND MAINTENANCE

⚠️ CAUTION
Do not interchange the position of the hoses.

2. Release the two clamps which secure the air cleaner to the carburetors. Release the clips which secure the air ducts to each carburetor.
3. Remove the air ducts separately. Remove the air cleaner housing.
4. Release the four spring clips which secure the cleaner halves together and then separate the halves.
5. Clean the inside of the housing. The paper element should be replaced every 18,000 miles under normal service. It should be replaced more often under severe operating conditions. A paper element may be cleaned by blowing through the element from the inside with compressed air. Never use a liquid solvent.
6. Assemble the air cleaner halves, making sure that they are properly aligned. Install the air cleaner by reversing the above. Make sure that the rubber sleeves on the air ducts and the rubber seals on the carburetors are seated properly.

TYPE 2 (1975–79)
▶ See Figure 7

1. Disconnect the upper part of hose A from the heater air blower. Open the clamp at the lower part of hose A and remove the hose.
2. Open clamps B on the air cleaner housing (2 at both the front and rear). Open the cover on the left side and remove the air filter cartridge. The element may be cleaned by striking it against a hard surface face first and then by blowing compressed air through it from the opposite direction of air flow then the filter is installed.

To remove the right section of the air cleaner housing to either clean it or remove the battery, proceed as follows:

3. Open clamp C.
4. Lift the right section of the air cleaner housing up and pull it out of the engine compartment.

Clean the insides of the housing with a cloth. Reverse the procedure to install both the housing and filter. Observe the following:

When inserting the right section of the housing, insert the locating projection into the side grommet first, then pull the housing toward the rear as you insert the lower locating projection in its seat. The paper element should be replaced every 18,000 miles under normal operating conditions, more often if the vehicle is used in severe climates or conditions.

TYPE 2 (1980–81)

1. Open the hatch inside the luggage compartment at the rear of the vehicle.
2. The air cleaner is located off to the side of the engine. Unfasten the five clips holding the top cover to the air cleaner housing and remove the top cover.
3. Pull the filter out and shake it to remove dust, or replace it as necessary.
4. When replacing, install the filter in the lower housing and install the top cover, securing it first with the top clamps.
5. Secure the remaining clamps. The air filter must be replaced every 18,000 miles, more often if the vehicle is used in severe conditions.

Fuel Filter

SERVICING

On carbureted models, the fuel filter is located in the mechanical fuel pump. There are three types of fuel pumps. Two types have a single screw holding a cover on the top of the pump. To remove the filter screen, undo the screw and carefully lift the cover off the pump. Remove the cover gasket and filter screen taking careful note of the position of the screen. Blow the screen out with air and replace the screen and cover using a new gasket if necessary.

The third type of fuel pump has four screws securing the top cover to the pump. This type of pump has a large plug with a hexagonal head. Remove this plug and washer (gasket) to gain ac-

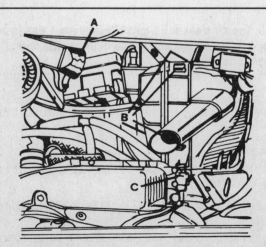

Fig. 7 Identification of air cleaner related components used on fuel injected Type 2 engines

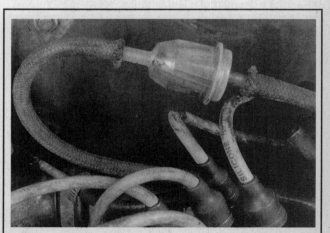

The inline fuel filter is installed in the fuel hose running from the fuel pump to the carburetor—note that the fuel filter hose clamps have already been removed

cess to the cylindrical filter screen located beneath the plug. Blow the screen out with air and replace it in its bore with the open end facing into the pump. Install the washer and plug. Do not overtighten the plug.

Fuel injected engines have an electric fuel pump located near the front axle on all models except the Type 2, in which the fuel pump is located near the fuel tank. This type of engine has an in-line fuel filter located atop the fuel pump in the suction line for the fuel pump. The suction line is the line running from the gas tank to the "S" connection at the fuel pump. To change the fuel filter, clamp the lines shut, then release the retaining pin and bracket, disconnect the gas lines from either end of the filter and insert a new filter. Filter should be installed with arrow (if any) pointing in the direction of fuel flow. This type of filter cannot be cleaned. VW recommends replacement at 12,000 mile intervals. A small speck of dirt entering a fuel injector may completely block the flow of fuel, necessitating disassembly of the injection system.

Crankcase Ventilation

SERVICING

Type 1/1600, 2/1600, 2/1700, 2/1800 (Carbureted)

The crankcase is vented by a hose running from the crankcase breather to the air cleaner. In some cases the hose is attached to the air inlet for the air cleaner. No PCV valve is used. No regular service is required.

Type 2/1800, 3/1600, 4/1700, 4/1800 (Fuel Injected)

Air is drawn in from the air cleaner to the cylinder head covers and pushrod tubes where it passes into the crankcase. From there, blow-by fumes then pass into the crankcase breather where they are drawn into the intake air distributor. On some models a PCV valve is used to control the flow of crankcase fumes. These systems usually need no maintenance other than keeping the hoses clear and all connections tight.

➡ Many VW engines do not use a conventional PCV valve because engine design limits crankcase pressure pulsing and allows almost no oil vapor to go into the PCV system.

Fuel Evaporation Control System

SERVICING

♦ See Figure 8

This system consists of an expansion chamber, an activated charcoal filter, and a hose which connects the parts into a closed system.

When fuel in the gas tank expands due to heat, the fuel travels to the expansion chamber. Any fumes generated either in the gas tank or the expansion chamber are trapped in the activated charcoal filter found in a line connecting the tank and chamber. The fumes are purged from the filter when the engine is started. Air from the engine cooling fan is forced through the filter when the engine is started. From the filter, this air/fuel vapor mixture is routed to the inside of the air cleaner where it is sent to the engine to be burned.

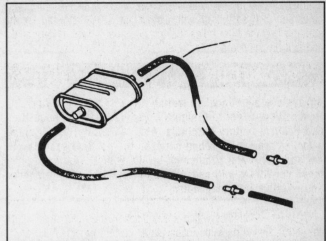

Fig. 8 The evaporative control canister is mounted inline between the fuel tank and air cleaner

1976 and later Type 2 models have an Evaporative Emission Control (EEC) cutoff valve which prevents fuel fumes from entering the air cleaner when the engine is stopped or idling. The cutoff is located in the air cleaner body. To test the valve, turn off the engine and disconnect the charcoal filter to air cleaner hose from the charcoal filter (this hose is usually transparent). Blow into the hose. The valve should be closed and no air should be going into the air cleaner. If the cutoff valve is open and air is getting through, the valve must be replaced.

The only maintenance required on the system is checking the tightness of all hose connections, and replacement of the charcoal filter element at 48,000 miles or 2 year intervals (whichever occurs first).

The filter canister is located under the right rear fender on Beetles and Super Beetles, at the lower right-hand side of the engine compartment on Karmann Ghias, at the upper right-hand side of the engine compartment on Type 3 models, beneath the floor near the forward end of the transaxle on Type 4 models and in the engine compartment on Type 2 models.

Battery

GENERAL MAINTENANCE

All batteries, regardless of type, should be carefully secured by a battery hold-down device. If this is not done, the battery terminals or casing may crack from stress applied to the battery during vehicle operation. A battery which is not secured may allow acid to leak out, making it discharge faster; such leaking corrosive acid can also eat away components under the hood. A battery that is not sealed must be checked periodically for electrolyte level. You cannot add water to a sealed maintenance-free battery (though not all maintenance-free batteries are sealed), but a sealed battery must also be checked for proper electrolyte level as indicated by the color of the built-in hydrometer "eye."

Keep the top of the battery clean, as a film of dirt can help completely discharge a battery that is not used for long periods. A

GENERAL INFORMATION AND MAINTENANCE

solution of baking soda and water may be used for cleaning, but be careful to flush this off with clear water. DO NOT let any of the solution into the filler holes. Baking soda neutralizes battery acid and will de-activate a battery cell.

** CAUTION

Always use caution when working on or near the battery. Never allow a tool to bridge the gap between the negative and positive battery terminals. Also, be careful not to allow a tool to provide a ground between the positive cable/terminal and any metal component on the vehicle. Either of these conditions will cause a short circuit leading to sparks and possible personal injury.

Batteries in vehicles which are not operated on a regular basis can fall victim to parasitic loads (small current drains which are constantly drawing current from the battery). Normal parasitic loads may drain a battery on a vehicle that is in storage and not used for 6–8 weeks. Vehicles that have additional accessories such as a cellular phone, an alarm system or other devices that increase parasitic load may discharge a battery sooner. If the vehicle is to be stored for 6–8 weeks in a secure area and the alarm system, if present, is not necessary, the negative battery cable should be disconnected at the onset of storage to protect the battery charge.

Remember that constantly discharging and recharging will shorten battery life. Take care not to allow a battery to be needlessly discharged.

BATTERY FLUID

** CAUTION

Battery electrolyte contains sulfuric acid. If you should splash any on your skin or in your eyes, flush the affected area with plenty of clear water. If it lands in your eyes, get medical help immediately.

The fluid (sulfuric acid solution) contained in the battery cells will tell you many things about the condition of the battery. Because the cell plates must be kept submerged below the fluid level in order to operate, maintaining the fluid level is extremely important. And, because the specific gravity of the acid is an indication of electrical charge, testing the fluid can be an aid in determining if the battery must be replaced. A battery in a vehicle with a properly operating charging system should require little maintenance, but careful, periodic inspection should reveal problems before they leave you stranded.

Fluid Level

Check the battery electrolyte level at least once a month, or more often in hot weather or during periods of extended vehicle operation. On non-sealed batteries, the level can be checked either through the case on translucent batteries or by removing the cell caps on opaque-cased types. The electrolyte level in each cell should be kept filled to the split ring inside each cell, or the line marked on the outside of the case.

If the level is low, add only distilled water through the opening until the level is correct. Each cell is separate from the others, so

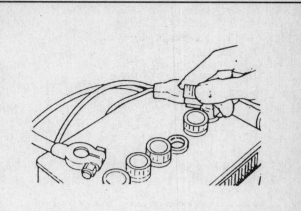

On non-maintenance free batteries, the level can be checked through the case on translucent batteries; the cell caps must be removed on other models

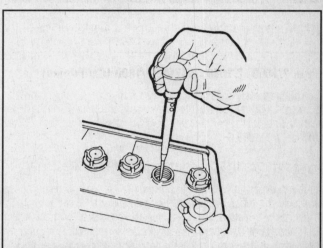

Check the specific gravity of the battery's electrolyte with a hydrometer

each must be checked and filled individually. Distilled water should be used, because the chemicals and minerals found in most drinking water are harmful to the battery and could significantly shorten its life.

If water is added in freezing weather, the vehicle should be driven several miles to allow the water to mix with the electrolyte. Otherwise, the battery could freeze.

Although some maintenance-free batteries have removable cell caps for access to the electrolyte, the electrolyte condition and level on all sealed maintenance-free batteries must be checked using the built-in hydrometer "eye." The exact type of eye varies between battery manufacturers, but most apply a sticker to the battery itself explaining the possible readings. When in doubt, refer to the battery manufacturer's instructions to interpret battery condition using the built-in hydrometer.

➥**Although the readings from built-in hydrometers found in sealed batteries may vary, a green eye usually indicates a properly charged battery with sufficient fluid level. A dark eye is normally an indicator of a battery with sufficient fluid, but one which may be low in charge. And a light or**

1-28 GENERAL INFORMATION AND MAINTENANCE

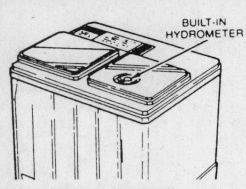

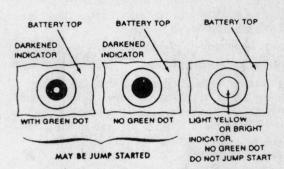

Check the appearance of the charge indicator on top of the battery before attempting a jump start; if it's not green or dark, do not jump start the car

A typical sealed (maintenance-free) battery with a built-in hydrometer—NOTE that the hydrometer eye may vary between battery manufacturers; always refer to the battery's label

yellow eye is usually an indication that electrolyte supply has dropped below the necessary level for battery (and hydrometer) operation. In this last case, sealed batteries with an insufficient electrolyte level must usually be discarded.

Specific Gravity

As stated earlier, the specific gravity of a battery's electrolyte level can be used as an indication of battery charge. At least once a year, check the specific gravity of the battery. It should be between 1.20 and 1.26 on the gravity scale. Most auto supply stores carry a variety of inexpensive battery testing hydrometers. These can be used on any non-sealed battery to test the specific gravity in each cell.

The battery testing hydrometer has a squeeze bulb at one end and a nozzle at the other. Battery electrolyte is sucked into the hydrometer until the float is lifted from its seat. The specific gravity is then read by noting the position of the float. If gravity is low in one or more cells, the battery should be slowly charged and checked again to see if the gravity has come up. Generally, if after charging, the specific gravity between any two cells varies more than 50 points (0.50), the battery should be replaced as it can no longer produce sufficient voltage to guarantee proper operation.

On sealed batteries, the built-in hydrometer is the only way of checking specific gravity. Again, check with your battery's manufacturer for proper interpretation of its built-in hydrometer readings.

CABLES

Once a year (or as necessary), the battery terminals and the cable clamps should be cleaned. Loosen the clamps and remove the cables, negative cable first. On batteries with posts on top, the use of a puller specially made for this purpose is recommended. These are inexpensive and available in most auto parts stores. Side terminal battery cables are secured with a small bolt.

Clean the cable clamps and the battery terminal with a wire brush, until all corrosion, grease, etc., is removed and the metal is shiny. It is especially important to clean the inside of the clamp (an old knife is useful here) thoroughly, since a small deposit of foreign material or oxidation there will prevent a sound electrical connection and inhibit either starting or charging. Special tools are available for cleaning these parts, one type for conventional top post batteries and another type for side terminal batteries.

Before installing the cables, loosen the battery hold-down clamp or strap, remove the battery and check the battery tray. Clear it of any debris, and check it for soundness (the battery tray can be cleaned with a baking soda and water solution). Rust should be wire brushed away, and the metal given a couple coats of anti-rust paint. Install the battery and tighten the hold-down clamp or strap securely. Do not overtighten, as this can crack the battery case.

After the clamps and terminals are clean, reinstall the cables, negative cable last; DO NOT hammer the clamps onto post batteries. Tighten the clamps securely, but do not distort them. Give the clamps and terminals a thin external coating of grease after installation, to retard corrosion.

Check the cables at the same time that the terminals are cleaned. If the cable insulation is cracked or broken, or if the ends are frayed, the cable should be replaced with a new cable of the same length and gauge.

Maintenance is performed with household items and with special tools like this post cleaner

GENERAL INFORMATION AND MAINTENANCE   1-29

CHARGING

> ✲✲ **CAUTION**
>
> The chemical reaction which takes place in all batteries generates explosive hydrogen gas. A spark can cause the battery to explode and splash acid. To avoid serious personal injury, be sure there is proper ventilation and take appropriate fire safety precautions when connecting, disconnecting, or charging a battery and when using jumper cables.

A battery should be charged at a slow rate to keep the plates inside from getting too hot. However, if some maintenance-free batteries are allowed to discharge until they are almost "dead," they may have to be charged at a high rate to bring them back to "life." Always follow the charger manufacturer's instructions on charging the battery.

The underside of this special battery tool has a wire brush to clean post terminals

Place the tool over the terminals and twist to clean the post

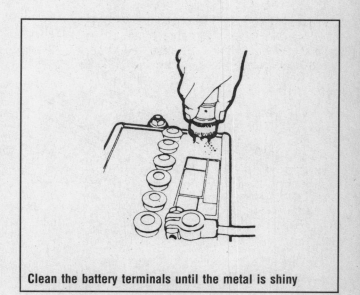
Clean the battery terminals until the metal is shiny

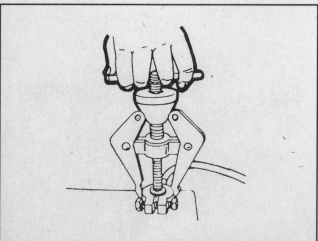
A special tool is available to pull the clamp from the post

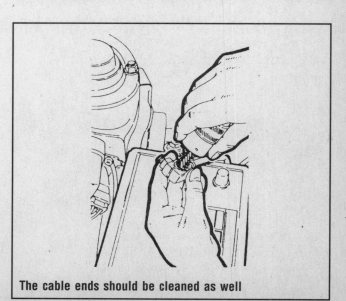
The cable ends should be cleaned as well

1-30 GENERAL INFORMATION AND MAINTENANCE

REPLACEMENT

When it becomes necessary to replace the battery, select one with a rating equal to or greater than the battery originally installed. Deterioration and just plain aging of the battery cables, starter motor, and associated wires makes the battery's job harder in successive years. The slow increase in electrical resistance over time makes it prudent to install a new battery with a greater capacity than the old.

Belts

INSPECTION

Inspect the belts for signs of glazing or cracking. A glazed belt will be perfectly smooth from slippage, while a good belt will

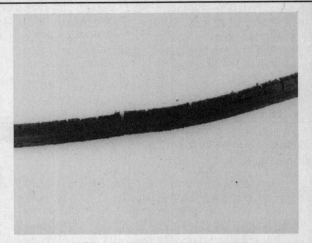

Deep cracks in this belt will cause flex, building up heat that will eventually lead to belt failure

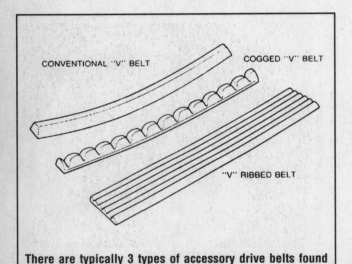

There are typically 3 types of accessory drive belts found on vehicles today

The cover of this belt is worn, exposing the critical reinforcing cords to excessive wear

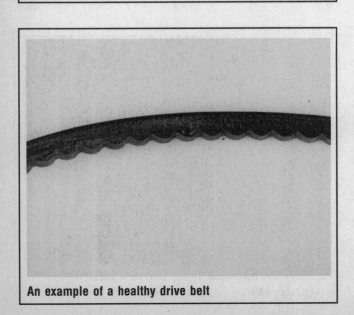

An example of a healthy drive belt

Installing too wide a belt can result in serious belt wear and/or breakage

GENERAL INFORMATION AND MAINTENANCE

have a slight texture of fabric visible. Cracks will usually start at the inner edge of the belt and run outward. All worn or damaged drive belts should be replaced immediately. It is best to replace all drive belts at one time, as a preventive maintenance measure, during this service operation.

ADJUSTMENT

Generator/Alternator Drive Belt
♦ See Figure 9

Improper fan belt adjustment can lead to either overheating of the engine or to loss in generating power, or both. In the Type 1, Type 2 or Type 4 a loose fan belt can cause both, while the slipping of the generator or alternator belt of the Type 3 engine will cause loss of generator efficiency only. In any case, it is important that the fan belt adjustment be checked and, if necessary, corrected at periodic intervals. When adjusted properly, the belt of any Volkswagen engine should deflect approximately ½ in. when pressed firmly in the center with the thumb. Check the tension at 6,000 mile intervals.

TYPE 1/1600, TYPE 2/1600

Adjustment of the Type 1/1600 and Type 2/1600 fan belt is made as follows: loosen the fan pulley by unscrewing the nut while at the same time holding the pulley from rotating by using a screwdriver inserted into the slot cut into the inner half of the generator pulley and supported against the upper generator bolt to cause a counter-torque. Remove the nut from the generator shaft pulley and remove the outer half of the pulley. The spacer washers must then be arranged so as to make the fan belt tension either greater or less. The greater the number of washers between the pulley halves, the smaller the effective diameter of the pulley, and the less the fan belt tension will be. Conversely, the subtraction of washers from between the pulley halves will lead to a larger effective diameter and to a greater degree of fan belt tension. If it is impossible to achieve proper adjustment with all the washers removed, then the fan belt is excessively stretched, and must be replaced. If it is impossible to adjust a new belt properly by using some combination of the available washers, the belt is

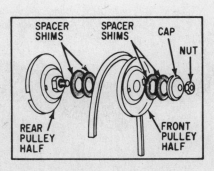

Fig. 9 The proper tension of the generator/alternator drive belt is set and maintained by the installation or removal of shims between the rear pulley and front pulley halves

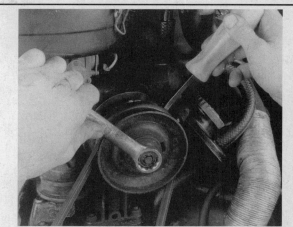

To remove the generator pulley nut, use a small prybar to hold the pulley stable while loosening the nut with a wrench . . .

. . . then remove the nut, spacer and washers from the pulley

Remove the outer half of the generator pulley and generator belt

1-32 GENERAL INFORMATION AND MAINTENANCE

The tension of the drive belt is controlled by the number of washers used between the two pulley halves (the fewer the washers, the tighter the drive belt)—upright 1600cc engine shown

the wrong size and must not be used. After the correct number of washers has been applied between the pulley halves, install the outer pulley half and place all surplus washers between the outer pulley half and the pulley nut so that they will be available if needed in a subsequent adjustment. Tighten the pulley nut and recheck the adjustment. If incorrect, add or subtract washers between the pulley halves until the proper amount of deflection is achieved. If the belt is too tight, the generator bearings will be subjected to undue stress and to very early failure. On the other hand, if the belt is too loose, it will slip and cause overheating. Cracked or frayed belts should be replaced. There is no comparison between the cost of a fan belt and that of repairing a badly overheated engine. If it is necessary to replace the belt, remove the three sheet metal screws and the crankshaft pulley cover plate to gain access to the pulley.

TYPE 3/1600

Adjustment of the fan belt on the Type 3 engine is much the same as that of the smaller Volkswagen engines. On the Type 3 engine, the fan belt is subject to a great deal less stress because it has no fan to turn. Therefore, a loose adjustment is not quite so critical as on the Beetle models. However, the ½ in. deflection should nevertheless be maintained, because a loose fan belt could possibly climb over the pulley and foul the fan. In addition, loose fan belts have a shorter service life. In adjusting the Type 3 fan belt, the first step is to remove the cover of the air intake housing. Next, hold the generator pulley with a suitable wrench, and unscrew the retaining nut. (Note: a 27 mm and a 21 mm wrench will come in handy here. Also, be careful that no adjusting washers fall off the shaft into the air intake housing, for they could be quite difficult to remove.) Loosen the generator strap and push the generator slightly forward. Remove the outer pulley half, sleeve and washers included. Arrange the spacer washers as was described in the Type 1/1600, 2/1600 belt adjustment; i.e., more washers between halves mean a looser belt, and fewer washers mean a tighter belt. Install outer half of pulley. Install unused washers on outside of outer pulley half so that the total number of washers on the shaft will remain the same. Fit the nut into place and tighten down the generator strap after pulling the generator back to the rear. Tighten the retaining nut and make sure that the generator belt is parallel to the air intake housing and at least 4 mm away from it at all points. Install housing cover.

TYPE 2/1700, 2/1800, 2/2000 AND TYPE 4

To adjust the alternator/cooling belt tension on these models, first remove the plastic insert in the cover plate over the alternator. Then, loosen the 12 point allen head adjusting bolt and the hex-head mounting bolt. Adjust the tension so that light thumb pressure applied midway in the belts longest run causes a deflection of approximately ½ in. Tighten the bolts.

When installing a new belt, move the alternator fully to the left and slip off the old belt. Install a new belt and tighten by moving the alternator to the right.

Air Injection Air Pump Drive Belt

1973–74 TYPE 2

To provide proper air pump output for the emission control system on 1973–74 Type 2 models, the belt tension must be checked at 6,000 mile intervals. Deflection is correct when light thumb pressure applied midway in the longest run of the belt deflects about ¼ in. To adjust, loosen the adjusting and mounting bolts (black arrows). Hold the air pump in position while tightening the bolts.

Air Conditioning

➡ **Be sure to consult the laws in your area before servicing the air conditioning system. In most areas, it is illegal to perform repairs involving refrigerant unless the work is done by a certified technician. Also, it is quite likely that you will not be able to purchase refrigerant without proof of certification.**

SAFETY PRECAUTIONS

There are two major hazards associated with air conditioning systems and they both relate to the refrigerant gas. First, the refrigerant gas (R-12) is an extremely cold substance. When exposed to air, it will instantly freeze any surface it comes in contact with, including your eyes. The other hazard relates to fire. Although normally non-toxic, the R-12 gas becomes highly poisonous in the presence of an open flame. One good whiff of the vapor formed by burning R-12 can be fatal. Keep all forms of fire (including cigarettes) well clear of the air conditioning system.

Because of the inherent dangers involved with working on air conditioning systems and R-12 refrigerant, these safety precautions must be strictly followed.

• Avoid contact with a charged refrigeration system, even when working on another part of the air conditioning system or vehicle. If a heavy tool comes into contact with a section of tubing or a heat exchanger, it can easily cause the relatively soft material to rupture.

• When it is necessary to apply force to a fitting which contains refrigerant, as when checking that all system couplings are securely tightened, use a wrench on both parts of the fitting involved, if possible. This will avoid putting torque on refrigerant tubing. (It is also advisable to use tube or line wrenches when tightening these flare nut fittings.)

GENERAL INFORMATION AND MAINTENANCE 1-33

➡ **R-12 refrigerant is a chlorofluorocarbon which, when released into the atmosphere, can contribute to the depletion of the ozone layer in the upper atmosphere. Ozone filters out harmful radiation from the sun.**

• Do not attempt to discharge the system without the proper tools. Precise control is possible only when using the service gauges and a proper A/C refrigerant recovery station. Wear protective gloves when connecting or disconnecting service gauge hoses.

• Discharge the system only in a well ventilated area, as high concentrations of the gas which might accidentally escape can exclude oxygen and act as an anesthetic. When leak testing or soldering, this is particularly important, as toxic gas is formed when R-12 contacts any flame.

• Never start a system without first verifying that both service valves are properly installed, and that all fittings throughout the system are snugly connected.

• Avoid applying heat to any refrigerant line or storage vessel. Charging may be aided by using water heated to less than 125°F (50°C) to warm the refrigerant container. Never allow a refrigerant storage container to sit out in the sun, or near any other source of heat, such as a radiator or heater.

• Always wear goggles to protect your eyes when working on a system. If refrigerant contacts the eyes, it is advisable in all cases to consult a physician immediately.

• Frostbite from liquid refrigerant should be treated by first gradually warming the area with cool water, and then gently applying petroleum jelly. A physician should be consulted.

• Always keep refrigerant drum fittings capped when not in use. If the container is equipped with a safety cap to protect the valve, make sure the cap is in place when the can is not being used. Avoid sudden shock to the drum, which might occur from dropping it, or from banging a heavy tool against it. Never carry a drum in the passenger compartment of a vehicle.

• Always completely discharge the system into a suitable recovery unit before painting the vehicle (if the paint is to be baked on), or before welding anywhere near refrigerant lines.

• When servicing the system, minimize the time that any refrigerant line or fitting is open to the air in order to prevent moisture or dirt from entering the system. Contaminants such as moisture or dirt can damage internal system components. Always replace O-rings on lines or fittings which are disconnected. Prior to installation coat, but do not soak, replacement O-rings with suitable compressor oil.

GENERAL SERVICING PROCEDURES

➡ **It is recommended, and possibly required by law, that a qualified technician perform the following services.**

The most important aspect of air conditioning service is the maintenance of a pure and adequate charge of refrigerant in the system. A refrigeration system cannot function properly if a significant percentage of the charge is lost. Leaks are common because the severe vibration encountered underhood in an automobile can easily cause a sufficient cracking or loosening of the air conditioning fittings; allowing, the extreme operating pressures of the system to force refrigerant out.

The problem can be understood by considering what happens to the system as it is operated with a continuous leak. Because the expansion valve regulates the flow of refrigerant to the evaporator, the level of refrigerant there is fairly constant. The receiver/drier stores any excess refrigerant, and so a loss will first appear there as a reduction in the level of liquid. As this level nears the bottom of the vessel, some refrigerant vapor bubbles will begin to appear in the stream of liquid supplied to the expansion valve. This vapor decreases the capacity of the expansion valve very little as the valve opens to compensate for its presence. As the quantity of liquid in the condenser decreases, the operating pressure will drop there and throughout the high side of the system. As the R-12 continues to be expelled, the pressure available to force the liquid through the expansion valve will continue to decrease, and, eventually, the valve's orifice will prove to be too much of a restriction for adequate flow even with the needle fully withdrawn.

At this point, low side pressure will start to drop, and a severe reduction in cooling capacity, marked by freeze-up of the evaporator coil, will result. Eventually, the operating pressure of the evaporator will be lower than the pressure of the atmosphere surrounding it, and air will be drawn into the system wherever there are leaks in the low side.

Because all atmospheric air contains at least some moisture, water will enter the system and mix with the R-12 and the oil. Trace amounts of moisture will cause sludging of the oil, and corrosion of the system. Saturation and clogging of the filter/drier, and freezing of the expansion valve orifice will eventually result. As air fills the system to a greater and greater extent, it will interfere more and more with the normal flows of refrigerant and heat.

From this description, it should be obvious that much of the repairman's focus in on detecting leaks, repairing them, and then restoring the purity and quantity of the refrigerant charge. A list of general rules should be followed in addition to all safety precautions:

• Keep all tools as clean and dry as possible.

• Thoroughly purge the service gauges/hoses of air and moisture before connecting them to the system. Keep them capped when not in use.

• Thoroughly clean any refrigerant fitting before disconnecting it, in order to minimize the entrance of dirt into the system.

• Plan any operation that requires opening the system beforehand, in order to minimize the length of time it will be exposed to open air. Cap or seal the open ends to minimize the entrance of foreign material.

• When adding oil, pour it through an extremely clean and dry tube or funnel. Keep the oil capped whenever possible. Do not use oil that has not been kept tightly sealed.

• Use only R-12 refrigerant. Purchase refrigerant intended for use only in automatic air conditioning systems.

• Completely evacuate any system that has been opened for service, or that has leaked sufficiently to draw in moisture and air. This requires evacuating air and moisture with a good vacuum pump for at least one hour. If a system has been open for a considerable length of time it may be advisable to evacuate the system for up to 12 hours (overnight).

• Use a wrench on both halves of a fitting that is to be disconnected, so as to avoid placing torque on any of the refrigerant lines.

• When overhauling a compressor, pour some of the oil into a clean glass and inspect it. If there is evidence of dirt, metal particles, or both, flush all refrigerant components with clean refrigerant before evacuating and recharging the system. In addition, if metal particles are present, the compressor should be replaced.

• Schrader valves may leak only when under full operating pressure. Therefore, if leakage is suspected but cannot be located, operate the system with a full charge of refrigerant and look for leaks from all Schrader valves. Replace any faulty valves.

Additional Preventive Maintenance

USING THE SYSTEM

The easiest and most important preventive maintenance for your A/C system is to be sure that it is used on a regular basis. Running the system for five minutes each month (no matter what the season) will help assure that the seals and all internal components remain lubricated.

ANTIFREEZE

In order to prevent heater core freeze-up during A/C operation, it is necessary to maintain a proper antifreeze protection. Use a hand-held antifreeze tester (hydrometer) to periodically check the condition of the antifreeze in your engine's cooling system.

➡ **Antifreeze should not be used longer than the manufacturer specifies.**

RADIATOR CAP

For efficient operation of an air conditioned vehicle's cooling system, the radiator cap should have a holding pressure which meets manufacturer's specifications. A cap which fails to hold these pressures should be replaced.

CONDENSER

Any obstruction of or damage to the condenser configuration will restrict the air flow which is essential to its efficient operation. It is therefore a good rule to keep this unit clean and in proper physical shape.

➡ **Bug screens which are mounted in front of the condenser, (unless they are original equipment), are regarded as obstructions.**

CONDENSATION DRAIN TUBE

This single molded drain tube expels the condensation, which accumulates on the bottom of the evaporator housing, into the engine compartment. If this tube is obstructed, the air conditioning performance can be restricted and condensation buildup can spill over onto the vehicle's floor.

SYSTEM INSPECTION

➡ **R-12 refrigerant is a chlorofluorocarbon which, when released into the atmosphere, can contribute to the depletion of the ozone layer in the upper atmosphere. Ozone filters out harmful radiation from the sun.**

The easiest and often most important check for the air conditioning system consists of a visual inspection of the system components. Visually inspect the air conditioning system for refrigerant leaks, damaged compressor clutch, compressor drive belt tension and condition, plugged evaporator drain tube, blocked condenser fins, disconnected or broken wires, blown fuses, corroded connections and poor insulation.

A refrigerant leak will usually appear as an oily residue at the

An antifreeze tester can be used to determine the freezing and boiling levels of the coolant

leakage point in the system. The oily residue soon picks up dust or dirt particles from the surrounding air and appears greasy. Through time, this will build up and appear to be a heavy dirt impregnated grease. Most leaks are caused by damaged or missing O-ring seals at the component connections, damaged charging valve cores or missing service gauge port caps.

For a thorough visual and operational inspection, check the following:

1. Check the surface of the radiator and condenser for dirt, leaves or other material which might block air flow.
2. Check for kinks in hoses and lines. Check the system for leaks.
3. Make sure the drive belt is under the proper tension. When the air conditioning is operating, make sure the drive belt is free of noise or slippage.
4. Make sure the blower motor operates at all appropriate positions, then check for distribution of the air from all outlets with the blower on **HIGH**.

➡ **Keep in mind that under conditions of high humidity, air discharged from the A/C vents may not feel as cold as expected, even if the system is working properly. This is because the vaporized moisture in humid air retains heat more effectively than does dry air, making the humid air more difficult to cool.**

Make sure the air passage selection lever is operating correctly. Start the engine and warm it to normal operating temperature, then make sure the hot/cold selection lever is operating correctly.

DISCHARGING, EVACUATING AND CHARGING

Discharging, evacuating and charging the air conditioning system must be performed by a properly trained and certified mechanic in a facility equipped with refrigerant recovery/recycling equipment that meets SAE standards for the type of system to be serviced.

If you don't have access to the necessary equipment, we recommend that you take your vehicle to a reputable service station to have the work done. If you still wish to perform repairs on the ve-

GENERAL INFORMATION AND MAINTENANCE   1-35

hicle, have them discharge the system, then take your vehicle home and perform the necessary work. When you are finished, return the vehicle to the station for evacuation and charging. Just be sure to cap ALL A/C system fittings immediately after opening them and keep them protected until the system is recharged.

Windshield Wipers

ELEMENT (REFILL) CARE AND REPLACEMENT

For maximum effectiveness and longest element life, the windshield and wiper blades should be kept clean. Dirt, tree sap, road tar and so on will cause streaking, smearing and blade deterioration if left on the glass. It is advisable to wash the windshield carefully with a commercial glass cleaner at least once a month. Wipe off the rubber blades with the wet rag afterwards. Do not attempt to move wipers across the windshield by hand; damage to the motor and drive mechanism will result.

To inspect and/or replace the wiper blade elements, place the wiper switch in the **LOW** speed position and the ignition switch in the **ACC** position. When the wiper blades are approximately vertical on the windshield, turn the ignition switch to **OFF**.

Examine the wiper blade elements. If they are found to be cracked, broken or torn, they should be replaced immediately. Replacement intervals will vary with usage, although ozone deterioration usually limits element life to about one year. If the wiper pattern is smeared or streaked, or if the blade chatters across the glass, the elements should be replaced. It is easiest and most sensible to replace the elements in pairs.

If your vehicle is equipped with aftermarket blades, there are several different types of refills and your vehicle might have any kind. Aftermarket blades and arms rarely use the exact same type blade or refill as the original equipment. Here are some typical aftermarket blades; not all may be available for your vehicle:

The Anco® type uses a release button that is pushed down to allow the refill to slide out of the yoke jaws. The new refill slides back into the frame and locks in place.

Some Trico® refills are removed by locating where the metal

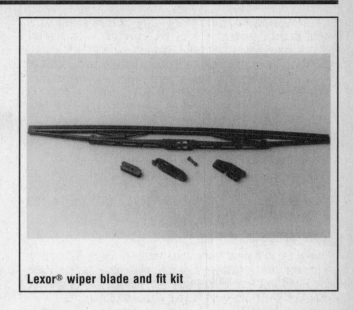
Lexor® wiper blade and fit kit

Pylon® wiper blade and adaptor

Bosch® wiper blade and fit kit

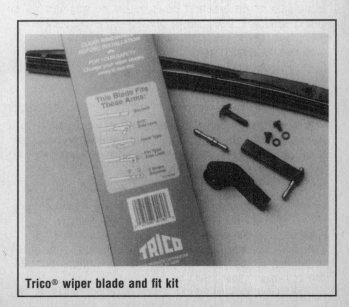
Trico® wiper blade and fit kit

1-36 GENERAL INFORMATION AND MAINTENANCE

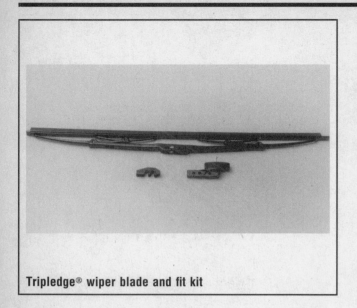

Tripledge® wiper blade and fit kit

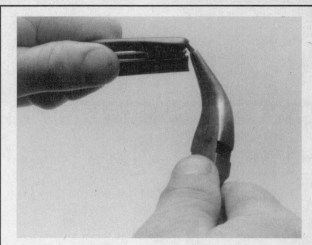

On Trico® wiper blades, the tab at the end of the blade must be turned up . . .

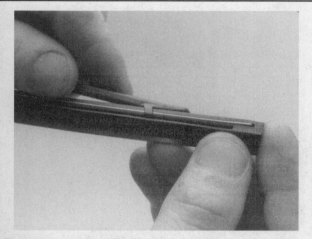

To remove and install a Lexor® wiper blade refill, slip out the old insert and slide in a new one

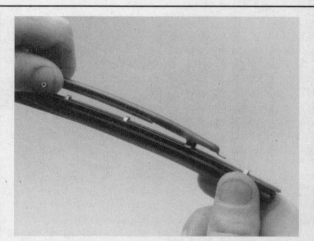

. . . then the insert can be removed. After installing the replacement insert, bend the tab back

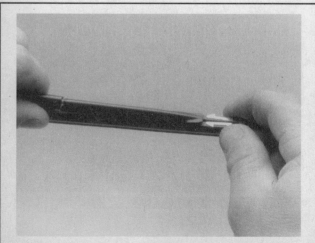

On Pylon® inserts, the clip at the end has to be removed prior to sliding the insert off

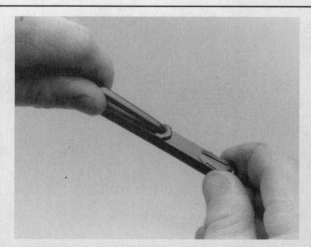

The Tripledge® wiper blade insert is removed and installed using a securing clip

GENERAL INFORMATION AND MAINTENANCE   1-37

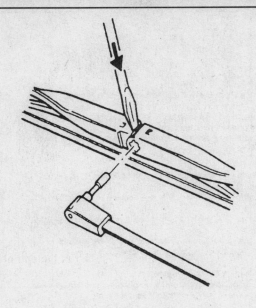

BLADE REPLACEMENT

1. CYCLE ARM AND BLADE ASSEMBLY TO UP POSITION ON THE WINDSHIELD WHERE REMOVAL OF BLADE ASSEMBLY CAN BE PERFORMED WITHOUT DIFFICULTY. TURN IGNITION KEY OFF AT DESIRED POSITION.

2. TO REMOVE BLADE ASSEMBLY, INSERT SCREWDRIVER IN SLOT, PUSH DOWN ON SPRING LOCK AND PULL BLADE ASSEMBLY FROM PIN (VIEW A)

3. TO INSTALL, PUSH THE BLADE ASSEMBLY ON THE PIN SO THAT THE SPRING LOCK ENGAGES THE PIN (VIEW A). BE SURE THE BLADE ASSEMBLY IS SECURELY ATTACHED TO PIN

VIEW A

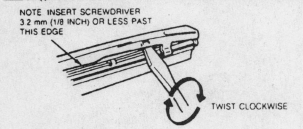

ELEMENT REPLACEMENT

1. INSERT SCREWDRIVER BETWEEN THE EDGE OF THE SUPER STRUCTURE AND THE BLADE BACKING DRIP (VIEW B) TWIST SCREWDRIVER SLOWLY UNTIL ELEMENT CLEARS ONE SIDE OF THE SUPER STRUCTURE CLAW

2. SLIDE THE ELEMENT INTO THE SUPER STRUCTURE CLAWS

VIEW B

4. INSERT ELEMENT INTO ONE SIDE OF THE END CLAWS (VIEW D) AND WITH A ROCKING MOTION PUSH ELEMENT UPWARD UNTIL IT SNAPS IN (VIEW E)

VIEW D

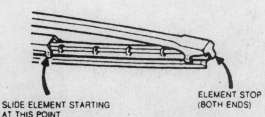

3. SLIDE THE ELEMENT INTO THE SUPER STRUCTURE CLAWS, STARTING WITH SECOND SET FROM EITHER END (VIEW C) AND CONTINUE TO SLIDE THE BLADE ELEMENT INTO ALL THE SUPER STRUCTURE CLAWS TO THE ELEMENT STOP (VIEW C)

VIEW C

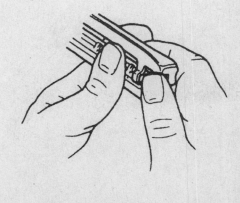

VIEW E

Trico® wiper blade insert (element) replacement

1-38 GENERAL INFORMATION AND MAINTENANCE

BLADE REPLACEMENT

1. Cycle arm and blade assembly to a position on the windshield where removal of blade assembly can be performed without difficulty. Turn ignition key off at desired position.
2. To remove blade assembly from wiper arm, pull up on spring lock and pull blade assembly from pin (View A). Be sure spring lock is not pulled excessively or it will become distorted.
3. To install, push the blade assembly onto the pin so that the spring lock engages the pin (View A). Be sure the blade assembly is securely attached to pin.

ELEMENT REPLACEMENT

1. In the plastic backing strip which is part of the rubber blade assembly, there is an 11.11mm (7/16 inch) long notch located approximately one inch from either end. Locate either notch.
2. Place the frame of the wiper blade assembly on a firm surface with either notched end of the backing strip visible.
3. Grasp the frame portion of the wiper blade assembly and push down until the blade assembly is tightly bowed.
4. With the blade assembly in the bowed position, grasp the tip of the backing strip firmly, pulling up and twisting C.C.W. at the same time. The backing strip will then snap out of the retaining tab on the end of the frame.
5. Lift the wiper blade assembly from the surface and slide the backing strip down the frame until the notch lines up with the next retaining tab, twist slightly, and the backing strip will snap out. Continue this operation with the remaining tabs until the blade element is completely detached from the frame.
6. To install blade element, reverse the above procedure, making sure all six (6) tabs are locked to the backing strip before installing blade to wiper arm.

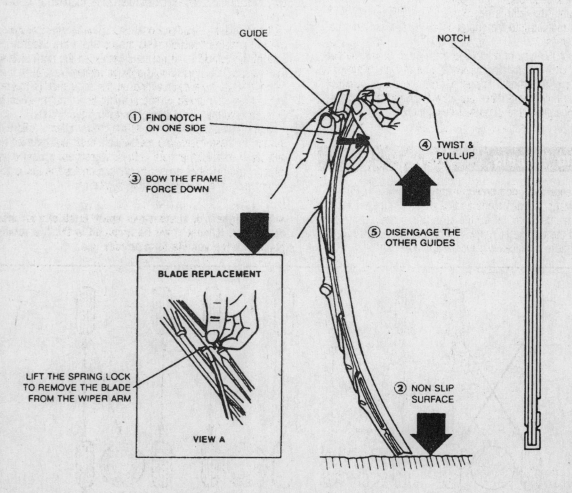

Tridon® wiper blade insert (element) replacement

GENERAL INFORMATION AND MAINTENANCE

backing strip or the refill is wider. Insert a small screwdriver blade between the frame and metal backing strip. Press down to release the refill from the retaining tab.

Other types of Trico® refills have two metal tabs which are unlocked by squeezing them together. The rubber filler can then be withdrawn from the frame jaws. A new refill is installed by inserting the refill into the front frame jaws and sliding it rearward to engage the remaining frame jaws. There are usually four jaws; be certain when installing that the refill is engaged in all of them. At the end of its travel, the tabs will lock into place on the front jaws of the wiper blade frame.

Another type of refill is made from polycarbonate. The refill has a simple locking device at one end which flexes downward out of the groove into which the jaws of the holder fit, allowing easy release. By sliding the new refill through all the jaws and pushing through the slight resistance when it reaches the end of its travel, the refill will lock into position.

To replace the Tridon® refill, it is necessary to remove the wiper blade. This refill has a plastic backing strip with a notch about 1 in. (25mm) from the end. Hold the blade (frame) on a hard surface so that the frame is tightly bowed. Grip the tip of the backing strip and pull up while twisting counterclockwise. The backing strip will snap out of the retaining tab. Do this for the remaining tabs until the refill is free of the blade. The length of these refills is molded into the end and they should be replaced with identical types.

Regardless of the type of refill used, be sure to follow the part manufacturer's instructions closely. Make sure that all of the frame jaws are engaged as the refill is pushed into place and locked. If the metal blade holder and frame are allowed to touch the glass during wiper operation, the glass will be scratched.

Tires and Wheels

Common sense and good driving habits will afford maximum tire life. Fast starts, sudden stops and hard cornering are hard on tires and will shorten their useful life span. Make sure that you don't overload the vehicle or run with incorrect pressure in the tires. Both of these practices will increase tread wear.

➡ **For optimum tire life, keep the tires properly inflated, rotate them often and have the wheel alignment checked periodically.**

Inspect your tires frequently. Be especially careful to watch for bubbles in the tread or sidewall, deep cuts or underinflation. Replace any tires with bubbles in the sidewall. If cuts are so deep that they penetrate to the cords, discard the tire. Any cut in the sidewall of a radial tire renders it unsafe. Also look for uneven tread wear patterns that may indicate the front end is out of alignment or that the tires are out of balance.

TIRE ROTATION

Tires must be rotated periodically to equalize wear patterns that vary with a tire's position on the vehicle. Tires will also wear in an uneven way as the front steering/suspension system wears to the point where the alignment should be reset.

Rotating the tires will ensure maximum life for the tires as a set, so you will not have to discard a tire early due to wear on only part of the tread. Regular rotation is required to equalize wear.

When rotating "unidirectional tires," make sure that they always roll in the same direction. This means that a tire used on the left side of the vehicle must not be switched to the right side and vice-versa. Such tires should only be rotated front-to-rear or rear-to-front, while always remaining on the same side of the vehicle. These tires are marked on the sidewall as to the direction of rotation; observe the marks when reinstalling the tire(s).

Some styled or "mag" wheels may have different offsets front to rear. In these cases, the rear wheels must not be used up front and vice-versa. Furthermore, if these wheels are equipped with unidirectional tires, they cannot be rotated unless the tire is re-mounted for the proper direction of rotation.

➡ **The compact or space-saver spare is strictly for emergency use. It must never be included in the tire rotation or placed on the vehicle for everyday use.**

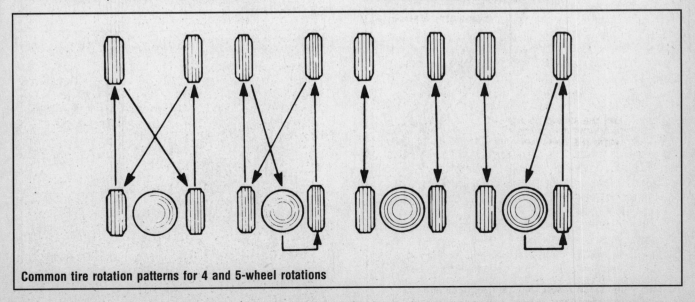

Common tire rotation patterns for 4 and 5-wheel rotations

1-40 GENERAL INFORMATION AND MAINTENANCE

Unidirectional tires are identifiable by sidewall arrows and/or the word "rotation"

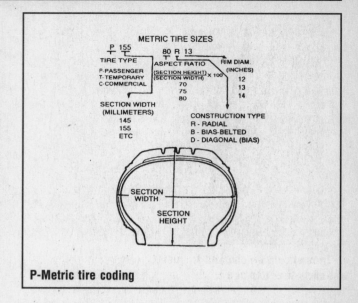

P-Metric tire coding

TIRE DESIGN

For maximum satisfaction, tires should be used in sets of four. Mixing of different types (radial, bias-belted, fiberglass belted) must be avoided. In most cases, the vehicle manufacturer has designated a type of tire on which the vehicle will perform best. Your first choice when replacing tires should be to use the same type of tire that the manufacturer recommends.

When radial tires are used, tire sizes and wheel diameters should be selected to maintain ground clearance and tire load capacity equivalent to the original specified tire. Radial tires should always be used in sets of four.

✱✱ CAUTION

Radial tires should never be used on only the front axle.

When selecting tires, pay attention to the original size as marked on the tire. Most tires are described using an industry size code sometimes referred to as P-Metric. This allows the exact identification of the tire specifications, regardless of the manufacturer. If selecting a different tire size or brand, remember to check the installed tire for any sign of interference with the body or suspension while the vehicle is stopping, turning sharply or heavily loaded.

Snow Tires

Good radial tires can produce a big advantage in slippery weather, but in snow, a street radial tire does not have sufficient tread to provide traction and control. The small grooves of a street tire quickly pack with snow and the tire behaves like a billiard ball on a marble floor. The more open, chunky tread of a snow tire will self-clean as the tire turns, providing much better grip on snowy surfaces.

To satisfy municipalities requiring snow tires during weather emergencies, most snow tires carry either an M + S designation after the tire size stamped on the sidewall, or the designation "all-season." In general, no change in tire size is necessary when buying snow tires.

Most manufacturers strongly recommend the use of 4 snow tires on their vehicles for reasons of stability. If snow tires are fitted only to the drive wheels, the opposite end of the vehicle may become very unstable when braking or turning on slippery surfaces. This instability can lead to unpleasant endings if the driver can't counteract the slide in time.

Note that snow tires, whether 2 or 4, will affect vehicle handling in all non-snow situations. The stiffer, heavier snow tires will noticeably change the turning and braking characteristics of the vehicle. Once the snow tires are installed, you must re-learn the behavior of the vehicle and drive accordingly.

➡ **Consider buying extra wheels on which to mount the snow tires. Once done, the "snow wheels" can be installed and removed as needed. This eliminates the potential damage to tires or wheels from seasonal removal and installation. Even if your vehicle has styled wheels, see if inexpensive steel wheels are available. Although the look of the vehicle will change, the expensive wheels will be protected from salt, curb hits and pothole damage.**

TIRE STORAGE

If they are mounted on wheels, store the tires at proper inflation pressure. All tires should be kept in a cool, dry place. If they are stored in the garage or basement, do not let them stand on a concrete floor; set them on strips of wood, a mat or a large stack of newspaper. Keeping them away from direct moisture is of paramount importance. Tires should not be stored upright, but in a flat position.

INFLATION & INSPECTION

The importance of proper tire inflation cannot be overemphasized. A tire employs air as part of its structure. It is designed around the supporting strength of the air at a specified pressure. For this reason, improper inflation drastically reduces the tires's ability to perform as intended. A tire will lose some air in day-to-day use; having to add a few pounds of air periodically is not necessarily a sign of a leaking tire.

GENERAL INFORMATION AND MAINTENANCE 1-41

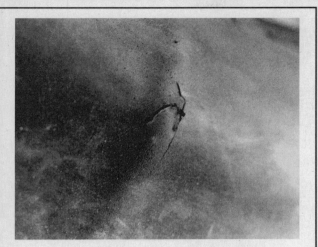

Tires should be checked frequently for any sign of puncture or damage

Tires with deep cuts, or cuts which show bulging should be replaced immediately

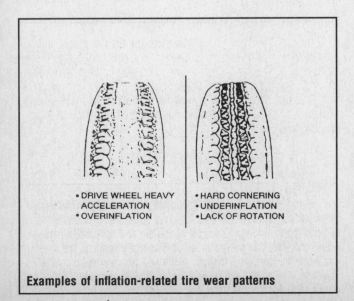

Examples of inflation-related tire wear patterns

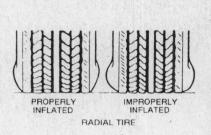

Radial tires have a characteristic sidewall bulge; don't try to measure pressure by looking at the tire. Use a quality air pressure gauge

Two items should be a permanent fixture in every glove compartment: an accurate tire pressure gauge and a tread depth gauge. Check the tire pressure (including the spare) regularly with a pocket type gauge. Too often, the gauge on the end of the air hose at your corner garage is not accurate because it suffers too much abuse. Always check tire pressure when the tires are cold, as pressure increases with temperature. If you must move the vehicle to check the tire inflation, do not drive more than a mile before checking. A cold tire is generally one that has not been driven for more than three hours.

A plate or sticker is normally provided somewhere in the vehicle (door post, hood, tailgate or trunk lid) which shows the proper pressure for the tires. Never counteract excessive pressure build-up by bleeding off air pressure (letting some air out). This will cause the tire to run hotter and wear quicker.

✲✲ CAUTION

Never exceed the maximum tire pressure embossed on the tire! This is the pressure to be used when the tire is at maximum loading, but it is rarely the correct pressure for everyday driving. Consult the owner's manual or the tire pressure sticker for the correct tire pressure.

Once you've maintained the correct tire pressures for several weeks, you'll be familiar with the vehicle's braking and handling personality. Slight adjustments in tire pressures can fine-tune these characteristics, but never change the cold pressure specification by more than 2 psi. A slightly softer tire pressure will give a softer ride but also yield lower fuel mileage. A slightly harder tire will give crisper dry road handling but can cause skidding on wet surfaces. Unless you're fully attuned to the vehicle, stick to the recommended inflation pressures.

All tires made since 1968 have built-in tread wear indicator bars that show up as ½ in. (13mm) wide smooth bands across the tire when 1/16 in. (1.5mm) of tread remains. The appearance of tread wear indicators means that the tires should be replaced. In fact, many states have laws prohibiting the use of tires with less than this amount of tread.

You can check your own tread depth with an inexpensive gauge or by using a Lincoln head penny. Slip the Lincoln penny

1-42 GENERAL INFORMATION AND MAINTENANCE

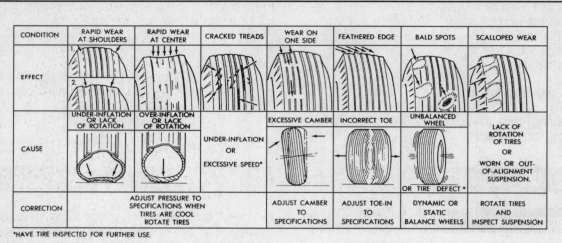

Common tire wear patterns and causes

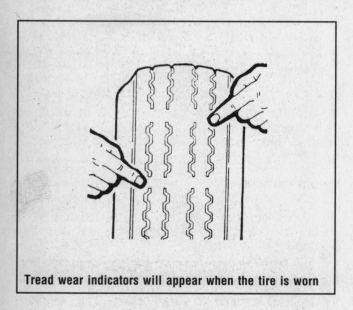

Tread wear indicators will appear when the tire is worn

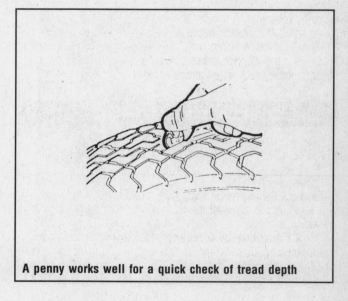

A penny works well for a quick check of tread depth

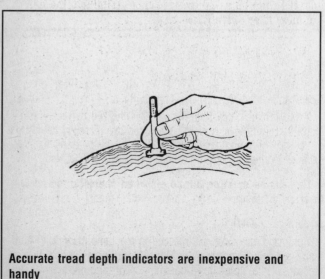

Accurate tread depth indicators are inexpensive and handy

(with Lincoln's head upside-down) into several tread grooves. If you can see the top of Lincoln's head in 2 adjacent grooves, the tire has less than 1/16 in. (1.5mm) tread left and should be replaced. You can measure snow tires in the same manner by using the "tails" side of the Lincoln penny. If you can see the top of the Lincoln memorial, it's time to replace the snow tire(s).

CARE OF SPECIAL WHEELS

If you have invested money in magnesium, aluminum alloy or sport wheels, special precautions should be taken to make sure your investment is not wasted and that your special wheels look good for the life of the vehicle.

Special wheels are easily damaged and/or scratched. Occasionally check the rims for cracking, impact damage or air leaks. If any of these are found, replace the wheel. But in order to prevent this type of damage and the costly replacement of a special wheel, observe the following precautions:

GENERAL INFORMATION AND MAINTENANCE

- Use extra care not to damage the wheels during removal, installation, balancing, etc. After removal of the wheels from the vehicle, place them on a mat or other protective surface. If they are to be stored for any length of time, support them on strips of wood. Never store tires and wheels upright; the tread may develop flat spots.
- When driving, watch for hazards; it doesn't take much to crack a wheel.
- When washing, use a mild soap or non-abrasive dish detergent (keeping in mind that detergent tends to remove wax). Avoid cleansers with abrasives or the use of hard brushes. There are many cleaners and polishes for special wheels.
- If possible, remove the wheels during the winter. Salt and sand used for snow removal can severely damage the finish of a wheel.
- Make certain the recommended lug nut torque is never exceeded or the wheel may crack. Never use snow chains on special wheels; severe scratching will occur.

FLUIDS AND LUBRICANTS

Fluid Disposal

Used fluids such as engine oil, transmission fluid, antifreeze and brake fluid are hazardous wastes and must be disposed of properly. Before draining any fluids, consult with the local authorities; in many areas, waste oil, etc. is being accepted as a part of recycling programs. A number of service stations and auto parts stores are also accepting waste fluids for recycling.

Be sure of the recycling center's policies before draining any fluids, as many will not accept different fluids that have been mixed together, such as oil and antifreeze.

Fluid Level Checks

ENGINE OIL

To check the engine oil level, park the car on level ground and wait 5 minutes to allow all the oil in the engine to drain into the crankcase.

Check the oil level by withdrawing the dipstick and wiping it clean. Insert the dipstick into its hole and note the position of the oil level on the bottom of the stick. The level should be between the two marks on the bottom of the stick, preferably closer to the top mark. The distance between the two marks represents one quart of oil.

On upright fan engines, the dipstick is located directly beneath the generator or alternator; oil is added through the capped opening beside the generator/alternator support post. On Type 2 suitcase engines through 1979, the dipstick is located next to the alternator with the oil filler right beside it. On the 1980–81 Type 2 the dipstick is accessible by pulling down the hinged license plate holder: the filler cap is below it. On the Type 3 the dipstick and filler are located in the lower door jamb of the rear compartment lid. On Type 4 two door and four door models, the dipstick is located at the center of the engine next to the oil filler cap: on wagon models, it is under the rear door jamb.

TRANSAXLE

Manual Transaxle

The oil level is checked by removing the 17 mm socket head plug located on the driver's side of the transaxle. The oil level should be even with the hole when the vehicle is level. Check it with your finger.

✽✽ CAUTION

Do not fill the transaxle too quickly because it may overflow from the filler hole and give the impression that the unit has been filled when it has not.

Top up as necessary with SAE 90 gear oil.

Automatic Stick Shift Transaxle

TYPE 1

The automatic Stick Shift transaxle is checked by means of a dipstick. The oil level should be between two marks at the bottom of the stick. The engine should be warm when the transmission oil level is checked. Top up as necessary with DEXRON®.

➡ The engine must be turned off when checking the transmission oil level.

TYPES 2, 3 AND 4

Automatic transaxles are checked in the same manner as Automatic Stick Shift transaxles, except that the engine should be running at an idle, transaxle in Neutral, and parking brake firmly applied. Top up as necessary with DEXRON® through the transaxle

The oil dipstick is located near the right-hand side of the crankshaft pulley—upright 1600cc engine shown

1-44 GENERAL INFORMATION AND MAINTENANCE

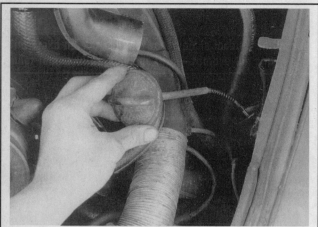

On automatic stick shift equipped models, the transaxle fluid dipstick and fill tube are located on the right inner fender in the engine bay

... and fill the reservoir to the proper level with clean brake fluid

dipstick tube located above the distributor (Type 2) or above the air manifold pipes (Types 3 and 4). The difference between the two marks on the dipstick is less than one pint. On all Type 2 and 4 models and on Squareback Type 3 models, the dipstick is accessible through the hatch in the luggage compartment. On Fastback Type 3 models, the dipstick is reached through the rear engine lid.

BRAKE FLUID

The brake fluid reservoir is located above the clutch pedal on 1970 Type 2 models, behind the drivers seat on 1971–72 Type 2 models, under the driver's seat on 1973–79 Type 2's and under the instrument cluster on 1980–81 Vanagons. It is located in the front luggage compartment on all other models. The fluid level in all vehicles should be above the upper seam of the reservoir. On 1973–79 Type 2 models, the fluid level is visible through a cutout beneath the seat. Fill the reservoir only with the new, clean heavy-duty brake fluid. If the vehicle is equipped with disc brakes make sure the fluid is marked for use with disc brakes. All fluid used should meet DOT 3, DOT 4, or SAE J1703 specifications.

DIFFERENTIAL

Automatic Transaxle Only—All Types

The filler hole is in basically the same position as on the manual transaxle. Make sure the oil level is even with the bottom of the hole when the vehicle is level. Fill the differential housing with 90W hypoid oil.

➡ The differential gears are lubricated by the transaxle oil on manual transaxle models.

STEERING GEAR (EXCEPT RACK AND PINION)

➡ The rack and pinion steering systems used on the 1975 and later Type 1 Super Beetle and Convertible and the 1980–81 Type 2 Vanagon are sealed systems which do not need regular maintenance.

Types 1 and 3 except the 1971–74 Super Beetle and 1971–74 Beetle Convertible, are filled with 5.4 ozs of gear oil which is added to a plug at the top of the gearbox. The 1971–74 Super Beetle and 1971–74 Convertible hold 5.9 ozs of steering gear oil. The Type 2 holds 9.4 ozs of gear oil in the steering gear box. Type 4 holds 9 ozs.

Unless the steering gear box has been rebuilt or is leaking severely, there is no reason to add or change gear box oil.

Engine Oil and Fuel Recommendations

Only oils which are high detergent and are graded MS or SE should be used in the engine. Oils should be selected for the SAE viscosity which will perform satisfactorily in the temperature expected before the next oil change.

Factory recommendations for fuel are regular gasoline with an octane rating of 91 RON or higher for Types 1, 2, 3, 1972–73

The brake master cylinder reservoir is located on the left inner fender in the luggage area—remove the cap . . .

GENERAL INFORMATION AND MAINTENANCE

Oil Viscosity Selection Chart

	Anticipated Temperature Range	SAE Viscosity
Multigrade	Above 32° F	10W–40 10W–50 20W–40 20W–50 10W–30
	May be used as low as –10° F	10W–30 10W–40
	Consistently below 10° F	5W–20 5W–30
Single-grade	Above 32° F	30
	Between 0°–32° F	20
	Temperature below 0° F	10W

After raising and safely supporting the vehicle, loosen and remove the oil drain plug . . .

Type 4 and 1974 Type 4 equipped with automatic transmission. The 1971 Type 4 and 1974 Type 4 equipped with manual transmission require premium gasoline with an octane rating of 98 RON or better.

All 1975 and later models sold in California and all 1977 and later models sold in the other 49 states use catalytic converters to lower exhaust emissions, requiring the use of only lead-free regular (91 RON) fuel. Failure to do so will render the converter ineffective.

If the vehicle is used for towing, it may be necessary to buy a higher grade of gasoline. The extra load caused by the trailer may be sufficient to cause elevated engine knock or ping. This condition, when allowed to continue over a period of time, will cause extreme damage to the engine. Furthermore, ping or knock has several causes besides low octane. An engine in need of a tune-up will ping. If there is an excessive carbon build-up and lead deposits in the combustion chamber, ping will also result.

Fluid Changes

ENGINE

Oil and Filter

The engine oil should be changed only after the engine has been warmed up to operating temperatures. In this way, the oil holds in suspension many of the contaminants that would otherwise remain in the engine. As the oil drains, it carries dirt and sludge from the engine. After the initial oil change at 600 miles the oil should be changed regularly at a period not to exceed 3,000 miles or three months. If the car is being operated mainly

. . . then allow the oil to empty into an aptly-sized catch pan until all the oil is completely drained

To clean the oil strainer, remove the 6 capscrews . . .

1-46 GENERAL INFORMATION AND MAINTENANCE

. . . then remove the strainer plate from the crankcase

. . . pull the oil strainer out of its hole . . .

Make certain to thoroughly clean all components of all dirt and grime, as can be seen on this oil strainer plate

. . . then remove the upper oil strainer gasket from the crankcase—1600cc engine shown

Remove the lower oil strainer gasket . . .

The oil dipstick is located near the right-hand side of the crankshaft pulley—upright 1600cc engine shown

GENERAL INFORMATION AND MAINTENANCE

After installing the oil strainer, gaskets, strainer plate and drain plug, fill the engine with the proper type and amount of clean engine oil

for short low speed trips, it may be advisable to change the oil at 2,000 mile intervals.

TYPES 1, 2/1600, 3

When changing the oil in Type 1, 2/1600, and 3 vehicles, first unscrew the drain plug in the crankcase (1970–72 models) or strainer cover cap nuts (1973–80 models); and allow dirty oil to drain. The oil strainer should also be cleaned. This wire mesh strainer is held in place by six cap nuts, and should be cleaned thoroughly with solvent. The strainer plate should also be cleaned. The lower part of the crankcase collects a great deal of sludge in the course of 3,000 miles. Replace the assembly using a new paper gasket and copper washers. Refill the crankcase with 2.5 quarts of oil. Tighten the cap nuts to no more than 5 ft lbs. The drain plug (1970–72 models) is tightened to 9 ft. lbs.

TYPES 2/1700, 2/1800, 2/2000 AND TYPE 4

Types 2/1700, 2/1800, 2/2000 and 4 require a different oil changing procedure. The crankcase drain plug is to one side of the oil strainer. The oil should be drained before removing the strainer. The strainer is located in the center of the crankcase and is held in position by a single plug. Remove the plug and remove the strainer assembly from the crankcase. Clean the strainer in solvent and reinstall it in the crankcase with a new paper gasket and copper washers. Tighten the drain plug and strainer nut to 9 ft. lbs.

The Type 2/1700, 2/1800, 2/2000 and Type 4 also have a spin-on oil filter located near the engine cooling fan. This filter is removed by unscrewing it from its fitting using a special adaptor extension. When reinstalling the new filter, lubricate the filter gasket with oil. Refill the crankcase and start the engine. Run the engine until it picks up oil pressure and then stop the engine. Recheck the engine oil level. The filter should be changed every 6,000 miles (carbureted engine) or 15,000 miles (fuel injected engines).

TRANSAXLE

Automatic Stick Shift

TYPE 1

The Automatic Stick Shift transaxle uses hypoid gear oil in the final drive section of the transaxle, see below for draining procedures. The front section of the transaxle uses ATF, however it does not have to be changed. The level should be checked every 6,000 miles.

➡ The engine should be off when the level is checked.

Use Type A or Dexron®.

Automatic Transaxle

TYPES 2, 3 AND 4
◆ See Figure 10

The Automatic Transmission Fluid (ATF) should be changed every 30,000 miles, or every 18,000 miles under heavy duty operating conditions. Heavy operating conditions include: continued stop and go driving; extended mountain driving; extremely high outside temperatures.

Before installing a new oil filter, lightly coat the rubber gasket with clean oil

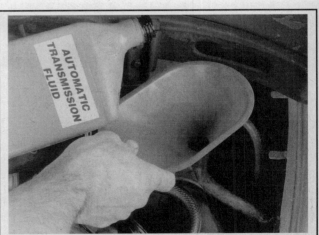

Using a funnel and a short piece of hose helps avoid messy spills when filling an automatic stick shift transaxle

1-48 GENERAL INFORMATION AND MAINTENANCE

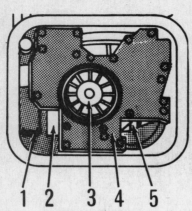

1. Manual valve
2. Kickdown solenoid
3. ATF strainer
4. Transfer plate
5. Valve body

Fig. 10 Automatic stick shift transaxle strainer location and related component identification—Types 2, 3 and 4

Drain the ATF by removing the drain plug (E) from the pan. Remove the pan and clean the ATF strainer. Install the pan using a new gasket and tighten the pan screws in a cross-cross pattern to 7 ft. lbs. Retighten the screws two or three times at five minute intervals to compensate for settling of the gasket. Refill the transaxle with the proper type transmission fluid using a funnel with a 20 in. long neck. Use Type A or Dexron®.

Manual Transaxle

The manual transaxle is drained by removing the 17 mm plug in the bottom of the transaxle case. It is refilled through another 17 mm plug in the side of the transaxle case. The plug in the side of the case also functions as the fluid level hole.

➡ When refilling the transaxle, do not fill it too fast because it may overflow from the filler hole and give the impression that the unit has been filled when it has not.

After the transaxle oil is changed at 600 miles, it is generally not necessary to change the oil. However, the factory does recommend the fluid level be checked every 6,000 miles and topped up if necessary. Refill the transaxle with SAE 90 hypoid gear oil.

FINAL DRIVE HOUSING

Automatic and Automatic Stick Shift Transaxles

The automatic transaxle unit has a separate housing for final drive gears which uses 90 weight hypoid gear oil. The unit is drained in much the same manner as the manual transaxle.

Chassis Greasing

There are four grease fittings on Type 1 and 3 models. They are located at the end of each front torsion bar housing. There is a fifth fitting at the center of the front torsion bar housing for the steering linkage on Type 2 vehicles through 1979. Wipe off each fitting before greasing. Super Beetles, 1971–80 Super Beetle Convertibles, and Type 4 models require no greasing. The vehicle should be greased every 6,000 miles (1970–72 Types 1 and 3 and 1970 Type 2) or 18,000 miles (1973–80 Type 1, 1973 Type 3, 1971–79 Type 2). The 1980–81 Type 2 does not need chassis lubrication.

Wheel Bearings

REMOVAL, PACKING & INSTALLATION

For procedures dealing with the removal, packing and installation of wheel bearings, please refer to Section 9 in this manual.

TRAILER TOWING

General Recommendations

Your vehicle was primarily designed to carry passengers and cargo. It is important to remember that towing a trailer will place additional loads on your vehicles engine, drivetrain, steering, braking and other systems. However, if you decide to tow a trailer, using the prior equipment is a must.

Local laws may require specific equipment such as trailer brakes or fender mounted mirrors. Check your local laws.

Trailer Weight

The weight of the trailer is the most important factor. A good weight-to-horsepower ratio is about 35:1, 35 lbs. of Gross Combined Weight (GCW) for every horsepower your engine develops. Multiply the engine's rated horsepower by 35 and subtract the weight of the vehicle passengers and luggage. The number remaining is the approximate ideal maximum weight you should tow, although a numerically higher axle ratio can help compensate for heavier weight.

Hitch (Tongue) Weight

Calculate the hitch weight in order to select a proper hitch. The weight of the hitch is usually 9–11% of the trailer gross weight and should be measured with the trailer loaded. Hitches fall into various categories: those that mount on the frame and rear bumper, the bolt-on type, or the weld-on distribution type used

GENERAL INFORMATION AND MAINTENANCE

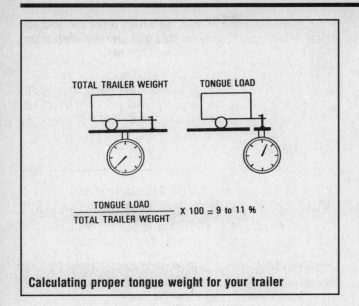

Calculating proper tongue weight for your trailer

for larger trailers. Axle mounted or clamp-on bumper hitches should never be used.

Check the gross weight rating of your trailer. Tongue weight is usually figured as 10% of gross trailer weight. Therefore, a trailer with a maximum gross weight of 2000 lbs. will have a maximum tongue weight of 200 lbs. Class I trailers fall into this category. Class II trailers are those with a gross weight rating of 2000–3000 lbs., while Class III trailers fall into the 3500–6000 lbs. category. Class IV trailers are those over 6000 lbs. and are for use with fifth wheel trucks, only.

When you've determined the hitch that you'll need, follow the manufacturer's installation instructions, exactly, especially when it comes to fastener torques. The hitch will subjected to a lot of stress and good hitches come with hardened bolts. Never substitute an inferior bolt for a hardened bolt.

Cooling

ENGINE

Overflow Tank

One of the most common, if not THE most common, problems associated with trailer towing is engine overheating. If you have a cooling system without an expansion tank, you'll definitely need to get an aftermarket expansion tank kit, preferably one with at least a 2 quart capacity. These kits are easily installed on the radiator's overflow hose, and come with a pressure cap designed for expansion tanks.

Flex Fan

Another helpful accessory for vehicles using a belt-driven radiator fan is a flex fan. These fans are large diameter units designed to provide more airflow at low speeds, by using fan blades that have deeply cupped surfaces. The blades then flex, or flatten out, at high speed, when less cooling air is needed. These fans are far lighter in weight than stock fans, requiring less horsepower to drive them. Also, they are far quieter than stock fans. If you do decide to replace your stock fan with a flex fan, note that if your vehicle has a fan clutch, a spacer will be needed between the flex fan and water pump hub.

Oil Cooler

Aftermarket engine oil coolers are helpful for prolonging engine oil life and reducing overall engine temperatures. Both of these factors increase engine life. While not absolutely necessary in towing Class I and some Class II trailers, they are recommended for heavier Class II and all Class III towing. Engine oil cooler systems usually consist of an adapter, screwed on in place of the oil filter, a remote filter mounting and a multi-tube, finned heat exchanger, which is mounted in front of the radiator or air conditioning condenser.

TRANSMISSION

An automatic transmission is usually recommended for trailer towing. Modern automatics have proven reliable and, of course, easy to operate, in trailer towing. The increased load of a trailer, however, causes an increase in the temperature of the automatic transmission fluid. Heat is the worst enemy of an automatic transmission. As the temperature of the fluid increases, the life of the fluid decreases.

It is essential, therefore, that you install an automatic transmission cooler. The cooler, which consists of a multi-tube, finned heat exchanger, is usually installed in front of the radiator or air conditioning compressor, and hooked in-line with the transmission cooler tank inlet line. Follow the cooler manufacturer's installation instructions.

Select a cooler of at least adequate capacity, based upon the combined gross weights of the vehicle and trailer.

Cooler manufacturers recommend that you use an aftermarket cooler in addition to, and not instead of, the present cooling tank in your radiator. If you do want to use it in place of the radiator cooling tank, get a cooler at least two sizes larger than normally necessary.

➡ **A transmission cooler can, sometimes, cause slow or harsh shifting in the transmission during cold weather, until the fluid has a chance to come up to normal operating temperature. Some coolers can be purchased with or retrofitted with a temperature bypass valve which will allow fluid flow through the cooler only when the fluid has reached above a certain operating temperature.**

Handling A Trailer

Towing a trailer with ease and safety requires a certain amount of experience. It's a good idea to learn the feel of a trailer by practicing turning, stopping and backing in an open area such as an empty parking lot.

1-50 GENERAL INFORMATION AND MAINTENANCE

TOWING THE VEHICLE

A vehicle equipped with an automatic or Automatic Stick Shift cannot be pushed or tow started. To push start a vehicle with a manual transmission, switch on the ignition, select the highest forward gear, and keep the clutch pedal depressed until suitable speed has been provided by pushing the vehicle. When the vehicle is going about 15 mph, slowly release the clutch to start the engine.

There are two towing eyes on all models except the 1973–74 Type 4. The front eye is located on the lower right front and the rear eye is located under the right rear bumper bracket. Tow the vehicle with the transmission in Neutral and the brakes off. Tow the 1973–74 Type 4 by its bumper bracket.

When towing an automatic transmission vehicle, it is always wise to tow with the drive wheels in a towing cradle or off the ground to avoid damage to the transmission.

JACKING AND HOISTING

♦ See Figures 11, 12, 13 and 14

Jacking points are provided at the sides of all models for the standard equipment jack. The jack supplied with the car should never be used for any service operation other than tire changing.

NEVER get under the car while it is supported by just a jack. If the jack should slip or tip over, as jacks often do, it would be exceedingly difficult to raise the car again while pinned underneath. Always block the wheels when changing tires.

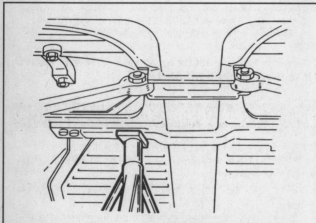

Fig. 11 Support Type 1 Super Beetle and Type 4 models at the front end with jackstands under the crossmember as shown

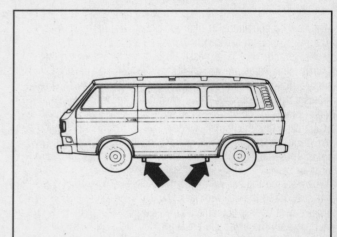

Fig. 13 When jacking a Type 2 models to change a tire, use the points indicated (arrows)

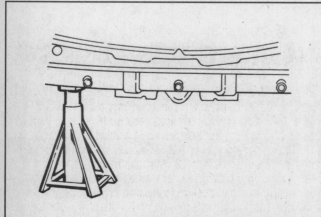

Fig. 12 On Type 1 Karmann Ghia, Beetle, 1970–79 Type 2 and all Type 3 models, the front end should be supported by jackstands beneath the front axle

Fig. 14 Jack a Beetle or Super Beetle at the point indicated to change a tire

GENERAL INFORMATION AND MAINTENANCE

Make certain to position the floor jack under the jacking point when lifting the rear of the vehicle . . .

. . . or under the center frame when lifting the front of the vehicle

The service operations in this book often require that one end or the other, or both, of the car be raised and supported safely. The best arrangement is a grease pit or a vehicle hoist. It is realized that these items are not often found in the home garage, but there are reasonable and safe substitutes. Small hydraulic, screw, or scissors jacks are satisfactory for raising the car. Heavy wooden blocks or adjustable jackstands should be used to support the car while it is being worked on.

Drive-on trestles, or ramps, are a handy and safe way to raise the car. These can be bought or constructed from suitable heavy boards or steel.

When raising the car with a floor, screw or scissors jack, or when supporting the car with jack stands, care should be taken in the placement of this equipment. The front of the car may be supported beneath the axle tube on Type 1 Beetles, 1970 Beetle convertibles, all Karmann Ghias, Type 2 models through 1979, and all Type 3 models. The front of all Super Beetles, 1971–77 Beetle convertibles, and all Type 4 models may be supported at the reinforced member to the rear of the lower control arms. 1980–81 Type 2 models should be supported at the frame crossmembers.

In any case, it is always best to spend a little extra time to make sure that the car is lifted and supported safely.

➡ Concrete blocks are not recommended. They may break if the load is not evenly distributed.

Jacking Precautions

The following safety points cannot be overemphasized:
- Always block the opposite wheel or wheels to keep the vehicle from rolling off the jack.
- When raising the front of the vehicle, firmly apply the parking brake.
- When the drive wheels are to remain on the ground, leave the vehicle in gear to help prevent it from rolling.
- Always use jackstands to support the vehicle when you are working underneath. Place the stands beneath the vehicle's jacking brackets. Before climbing underneath, rock the vehicle a bit to make sure it is firmly supported.

JUMP STARTING A DEAD BATTERY

Whenever a vehicle is jump started, precautions must be followed in order to prevent the possibility of personal injury. Remember that batteries contain a small amount of explosive hydrogen gas which is a by-product of battery charging. Sparks should always be avoided when working around batteries, especially when attaching jumper cables. To minimize the possibility of accidental sparks, follow the procedure carefully.

❋ CAUTION

NEVER hook the batteries up in a series circuit or the entire electrical system will go up in smoke, including the starter!

Vehicles equipped with a diesel engine may utilize two 12 volt batteries. If so, the batteries are connected in a parallel circuit (positive terminal to positive terminal, negative terminal to negative terminal). Hooking the batteries up in parallel circuit increases battery cranking power without increasing total battery voltage output. Output remains at 12 volts. On the other hand, hooking two 12 volt batteries up in a series circuit (positive terminal to negative terminal, positive terminal to negative terminal) increases total battery output to 24 volts (12 volts plus 12 volts).

1-52 GENERAL INFORMATION AND MAINTENANCE

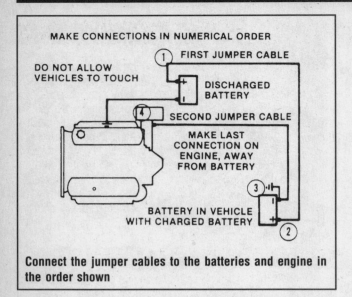

Connect the jumper cables to the batteries and engine in the order shown

Jump Starting Precautions

- Be sure that both batteries are of the same voltage. Vehicles covered by this manual and most vehicles on the road today utilize a 12 volt charging system.
- Be sure that both batteries are of the same polarity (have the same terminal, in most cases NEGATIVE grounded).
- Be sure that the vehicles are not touching or a short could occur.
- On serviceable batteries, be sure the vent cap holes are not obstructed.
- Do not smoke or allow sparks anywhere near the batteries.
- In cold weather, make sure the battery electrolyte is not frozen. This can occur more readily in a battery that has been in a state of discharge.
- Do not allow electrolyte to contact your skin or clothing.

Jump Starting Procedure

1. Make sure that the voltages of the 2 batteries are the same. Most batteries and charging systems are of the 12 volt variety.
2. Pull the jumping vehicle (with the good battery) into a position so the jumper cables can reach the dead battery and that vehicle's engine. Make sure that the vehicles do NOT touch.
3. Place the transmissions/transaxles of both vehicles in **Neutral** (MT) or **P** (AT), as applicable, then firmly set their parking brakes.

➡If necessary for safety reasons, the hazard lights on both vehicles may be operated throughout the entire procedure without significantly increasing the difficulty of jumping the dead battery.

4. Turn all lights and accessories OFF on both vehicles. Make sure the ignition switches on both vehicles are turned to the **OFF** position.
5. Cover the battery cell caps with a rag, but do not cover the terminals.
6. Make sure the terminals on both batteries are clean and free of corrosion or proper electrical connection will be impeded. If necessary, clean the battery terminals before proceeding.
7. Identify the positive (+) and negative (−) terminals on both batteries.
8. Connect the first jumper cable to the positive (+) terminal of the dead battery, then connect the other end of that cable to the positive (+) terminal of the booster (good) battery.
9. Connect one end of the other jumper cable to the negative (−) terminal on the booster battery and the final cable clamp to an engine bolt head, alternator bracket or other solid, metallic point on the engine with the dead battery. Try to pick a ground on the engine that is positioned away from the battery in order to minimize the possibility of the 2 clamps touching should one loosen during the procedure. DO NOT connect this clamp to the negative (−) terminal of the bad battery.

✱✱ CAUTION

Be very careful to keep the jumper cables away from moving parts (cooling fan, belts, etc.) on both engines.

10. Check to make sure that the cables are routed away from any moving parts, then start the donor vehicle's engine. Run the engine at moderate speed for several minutes to allow the dead battery a chance to receive some initial charge.
11. With the donor vehicle's engine still running slightly above idle, try to start the vehicle with the dead battery. Crank the engine for no more than 10 seconds at a time and let the starter cool for at least 20 seconds between tries. If the vehicle does not start in 3 tries, it is likely that something else is also wrong or that the battery needs additional time to charge.
12. Once the vehicle is started, allow it to run at idle for a few seconds to make sure that it is operating properly.
13. Turn ON the headlights, heater blower and, if equipped, the rear defroster of both vehicles in order to reduce the severity of voltage spikes and subsequent risk of damage to the vehicles' electrical systems when the cables are disconnected. This step is especially important to any vehicle equipped with computer control modules.
14. Carefully disconnect the cables in the reverse order of connection. Start with the negative cable that is attached to the engine ground, then the negative cable on the donor battery. Disconnect the positive cable from the donor battery and finally, disconnect the positive cable from the formerly dead battery. Be careful when disconnecting the cables from the positive terminals not to allow the alligator clips to touch any metal on either vehicle or a short and sparks will occur.

GENERAL INFORMATION AND MAINTENANCE

HOW TO BUY A USED VEHICLE

Many people believe that a two or three year old used car or truck is a better buy than a new vehicle. This may be true as most new vehicles suffer the heaviest depreciation in the first two years and, at three years old, a vehicle is usually not old enough to present a lot of costly repair problems. But keep in mind, when buying a non-warranted automobile, there are no guarantees. Whatever the age of the used vehicle you might want to purchase, this section and a little patience should increase your chances of selecting one that is safe and dependable.

Tips

1. First decide what model you want, and how much you want to spend.
2. Check the used car lots and your local newspaper ads. Privately owned vehicles are usually less expensive, however, you may not get a warranty that, in many cases, comes with a used vehicle purchased from a lot. Of course, some aftermarket warranties may not be worth the extra money, so this is a point you will have to debate and consider based on your priorities.
3. Never shop at night. The glare of the lights make it easy to miss faults on the body caused by accident or rust repair.
4. Try to get the name and phone number of the previous owner. Contact him/her and ask about the vehicle. If the owner of a lot refuses this information, look for a vehicle somewhere else.

A private seller can tell you about the vehicle and maintenance. But remember, there's no law requiring honesty from private citizens selling used vehicles. There is a law that forbids tampering with or turning back the odometer mileage. This includes both the private citizen and the lot owner. The law also requires that the seller or anyone transferring ownership of the vehicle must provide the buyer with a signed statement indicating the mileage on the odometer at the time of transfer.

5. You may wish to contact the National Highway Traffic Safety Administration (NHTSA) to find out if the vehicle has ever been included in a manufacturer's recall. Write down the year, model and serial number before you buy the vehicle, then contact NHTSA (there should be a 1-800 number that your phone company's information line can supply). If the vehicle was listed for a recall, make sure the needed repairs were made.
6. Refer to the Used Vehicle Checklist in this section and check all the items on the vehicle you are considering. Some items are more important than others. Only you know how much money you can afford for repairs, and depending on the price of the vehicle, may consider performing any needed work yourself. Beware, however, of trouble in areas that will affect operation, safety or emission. Problems in the Used Vehicle Checklist break down as follows:

• Numbers 1–8: Two or more problems in these areas indicate a lack of maintenance. You should beware.
• Numbers 9–13: Problems here tend to indicate a lack of proper care, however, these can usually be corrected with a tune-up or relatively simple parts replacement.
• Numbers 14–17: Problems in the engine or transmission can be very expensive. Unless you are looking for a project, walk away from any vehicle with problems in 2 or more of these areas.

7. If you are satisfied with the apparent condition of the vehicle, take it to an independent diagnostic center or mechanic for a complete check. If you have a state inspection program, have it inspected immediately before purchase, or specify on the bill of sale that the sale is conditional on passing state inspection.
8. Road test the vehicle—refer to the Road Test Checklist in this section. If your original evaluation and the road test agree—the rest is up to you.

USED VEHICLE CHECKLIST

➡ The numbers on the illustrations refer to the numbers on this checklist.

1. Mileage: Average mileage is about 12,000–15,000 miles per year. More than average mileage may indicate hard usage or could indicate many highway miles (which could be less detrimental than half as many tough around town miles).
2. Paint: Check around the tailpipe, molding and windows for overspray indicating that the vehicle has been repainted.

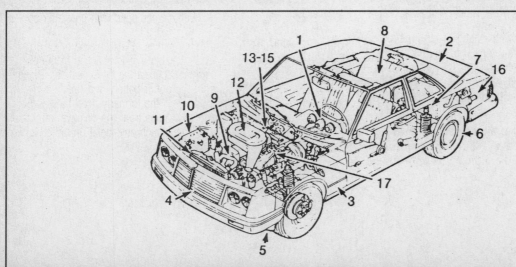
Each of the numbered items should be checked when purchasing a used vehicle

GENERAL INFORMATION AND MAINTENANCE

3. Rust: Check fenders, doors, rocker panels, window moldings, wheelwells, floorboards, under floormats, and in the trunk for signs of rust. Any rust at all will be a problem. There is no way to permanently stop the spread of rust, except to replace the part or panel.

➡ If rust repair is suspected, try using a magnet to check for body filler. A magnet should stick to the sheet metal parts of the body, but will not adhere to areas with large amounts of filler.

4. Body appearance: Check the moldings, bumpers, grille, vinyl roof, glass, doors, trunk lid and body panels for general overall condition. Check for misalignment, loose hold-down clips, ripples, scratches in glass, welding in the trunk, severe misalignment of body panels or ripples, any of which may indicate crash work.

5. Leaks: Get down and look under the vehicle. There are no normal leaks, other than water from the air conditioner evaporator.

6. Tires: Check the tire air pressure. One old trick is to pump the tire pressure up to make the vehicle roll easier. Check the tread wear, then open the trunk and check the spare too. Uneven wear is a clue that the front end may need an alignment.

7. Shock absorbers: Check the shock absorbers by forcing downward sharply on each corner of the vehicle. Good shocks will not allow the vehicle to bounce more than once after you let go.

8. Interior: Check the entire interior. You're looking for an interior condition that agrees with the overall condition of the vehicle. Reasonable wear is expected, but be suspicious of new seat covers on sagging seats, new pedal pads, and worn armrests. These indicate an attempt to cover up hard use. Pull back the carpets and look for evidence of water leaks or flooding. Look for missing hardware, door handles, control knobs, etc. Check lights and signal operations. Make sure all accessories (air conditioner, heater, radio, etc.) work. Check windshield wiper operation.

9. Belts and Hoses: Open the hood, then check all belts and hoses for wear, cracks or weak spots.

10. Battery: Low electrolyte level, corroded terminals and/or cracked case indicate a lack of maintenance.

11. Radiator: Look for corrosion or rust in the coolant indicating a lack of maintenance.

12. Air filter: A severely dirty air filter would indicate a lack of maintenance.

13. Ignition wires: Check the ignition wires for cracks, burned spots, or wear. Worn wires will have to be replaced.

14. Oil level: If the oil level is low, chances are the engine uses oil or leaks. Beware of water in the oil (there is probably a cracked block or bad head gasket), excessively thick oil (which is often used to quiet a noisy engine), or thin, dirty oil with a distinct gasoline smell (this may indicate internal engine problems).

15. Automatic Transmission: Pull the transmission dipstick out when the engine is running. The level should read FULL, and the fluid should be clear or bright red. Dark brown or black fluid that has distinct burnt odor, indicates a transmission in need of repair or overhaul.

16. Exhaust: Check the color of the exhaust smoke. Blue smoke indicates, among other problems, worn rings. Black smoke can indicate burnt valves or carburetor problems. Check the exhaust system for leaks; it can be expensive to replace.

17. Spark Plugs: Remove one or all of the spark plugs (the most accessible will do, though all are preferable). An engine in good condition will show plugs with a light tan or gray deposit on the firing tip.

ROAD TEST CHECKLIST

1. Engine Performance: The vehicle should be peppy whether cold or warm, with adequate power and good pickup. It should respond smoothly through the gears.

2. Brakes: They should provide quick, firm stops with no noise, pulling or brake fade.

3. Steering: Sure control with no binding harshness, or looseness and no shimmy in the wheel should be expected. Noise or vibration from the steering wheel when turning the vehicle means trouble.

4. Clutch (Manual Transmission/Transaxle): Clutch action should give quick, smooth response with easy shifting. The clutch pedal should have free-play before it disengages the clutch. Start the engine, set the parking brake, put the transmission in first gear and slowly release the clutch pedal. The engine should begin to stall when the pedal is ½–¾ of the way up.

5. Automatic Transmission/Transaxle: The transmission should shift rapidly and smoothly, with no noise, hesitation, or slipping.

6. Differential: No noise or thumps should be present. Differentials have no normal leaks.

7. Driveshaft/Universal Joints: Vibration and noise could mean driveshaft problems. Clicking at low speed or coast conditions means worn U-joints.

8. Suspension: Try hitting bumps at different speeds. A vehicle that bounces excessively has weak shock absorbers or struts. Clunks mean worn bushings or ball joints.

9. Frame/Body: Wet the tires and drive in a straight line. Tracks should show two straight lines, not four. Four tire tracks indicate a frame/body bent by collision damage. If the tires can't be wet for this purpose, have a friend drive along behind you and see if the vehicle appears to be traveling in a straight line.

Capacities Chart

Year	Type and Model	Engine Displacement (cc)	Engine Crankcase (qts) With Filter	Engine Crankcase (qts) Without	Transaxle (pts) Manual	Transaxle (pts) Automatic Conv	Transaxle (pts) Automatic Final Drive	Gasoline Tank (gals)
1970-79	1, 111, 114	1600	—	2.5	6.3	7.6	6.3 ①	10.6
1970-80	1, 113, 15	1600	—	2.5	6.3	7.6	6.3 ①	11.1
1970-71	2, All	1600	—	2.5	7.4	12.6 ②	3.0	15.8
1972-81	2, All	1700, 1800, 2000	3.7	3.2	7.4	12.6 ②	3.0	15.8 ③
1970-73	3, All	1600	—	2.5	6.3	12.6 ②	2.1	10.6
1971-74	4, All	1700, 1800	3.7	3.2	5.3	12.6 ②	2.1	13.2

Conv—Torque Converter
① 5.3 when changed
② 6.3 when changed
③ 1980-81—15.9 gals

ENGLISH TO METRIC CONVERSION: MASS (WEIGHT)

Current mass measurement is expressed in pounds and ounces (lbs. & ozs.). The metric unit of mass (or weight) is the kilogram (kg). Even although this table does not show conversion of masses (weights) larger than 15 lbs, it is easy to calculate larger units by following the data immediately below.

To convert ounces (oz.) to grams (g): multiply th number of ozs. by 28
To convert grams (g) to ounces (oz.): multiply the number of grams by .035

To convert pounds (lbs.) to kilograms (kg): multiply the number of lbs. by .45
To convert kilograms (kg) to pounds (lbs.): multiply the number of kilograms by 2.2

lbs	kg	lbs	kg	oz	kg	oz	kg
0.1	0.04	0.9	0.41	0.1	0.003	0.9	0.024
0.2	0.09	1	0.4	0.2	0.005	1	0.03
0.3	0.14	2	0.9	0.3	0.008	2	0.06
0.4	0.18	3	1.4	0.4	0.011	3	0.08
0.5	0.23	4	1.8	0.5	0.014	4	0.11
0.6	0.27	5	2.3	0.6	0.017	5	0.14
0.7	0.32	10	4.5	0.7	0.020	10	0.28
0.8	0.36	15	6.8	0.8	0.023	15	0.42

ENGLISH TO METRIC CONVERSION: TEMPERATURE

To convert Fahrenheit (°F) to Celsius (°C): take number of °F and subtract 32; multiply result by 5; divide result by 9
To convert Celsius (°C) to Fahrenheit (°F): take number of °C and multiply by 9; divide result by 5; add 32 to total

Fahrenheit (F)		Celsius (C)		Fahrenheit (F)		Celsius (C)		Fahrenheit (F)		Celsius (C)	
°F	°C	°C	°F	°F	°C	°C	°F	°F	°C	°C	°F
−40	−40	−38	−36.4	80	26.7	18	64.4	215	101.7	80	176
−35	−37.2	−36	−32.8	85	29.4	20	68	220	104.4	85	185
−30	−34.4	−34	−29.2	90	32.2	22	71.6	225	107.2	90	194
−25	−31.7	−32	−25.6	95	35.0	24	75.2	230	110.0	95	202
−20	−28.9	−30	−22	100	37.8	26	78.8	235	112.8	100	212
−15	−26.1	−28	−18.4	105	40.6	28	82.4	240	115.6	105	221
−10	−23.3	−26	−14.8	110	43.3	30	86	245	118.3	110	230
−5	−20.6	−24	−11.2	115	46.1	32	89.6	250	121.1	115	239
0	−17.8	−22	−7.6	120	48.9	34	93.2	255	123.9	120	248
1	−17.2	−20	−4	125	51.7	36	96.8	260	126.6	125	257
2	−16.7	−18	−0.4	130	54.4	38	100.4	265	129.4	130	266
3	−16.1	−16	3.2	135	57.2	40	104	270	132.2	135	275
4	−15.6	−14	6.8	140	60.0	42	107.6	275	135.0	140	284
5	−15.0	−12	10.4	145	62.8	44	112.2	280	137.8	145	293
10	−12.2	−10	14	150	65.6	46	114.8	285	140.6	150	302
15	−9.4	−8	17.6	155	68.3	48	118.4	290	143.3	155	311
20	−6.7	−6	21.2	160	71.1	50	122	295	146.1	160	320
25	−3.9	−4	24.8	165	73.9	52	125.6	300	148.9	165	329
30	−1.1	−2	28.4	170	76.7	54	129.2	305	151.7	170	338
35	1.7	0	32	175	79.4	56	132.8	310	154.4	175	347
40	4.4	2	35.6	180	82.2	58	136.4	315	157.2	180	356
45	7.2	4	39.2	185	85.0	60	140	320	160.0	185	365
50	10.0	6	42.8	190	87.8	62	143.6	325	162.8	190	374
55	12.8	8	46.4	195	90.6	64	147.2	330	165.6	195	383
60	15.6	10	50	200	93.3	66	150.8	335	168.3	200	392
65	18.3	12	53.6	205	96.1	68	154.4	340	171.1	205	401
70	21.1	14	57.2	210	98.9	70	158	345	173.9	210	410
75	23.9	16	60.8	212	100.0	75	167	350	176.7	215	414

ENGLISH TO METRIC CONVERSION: LENGTH

To convert inches (ins.) to millimeters (mm): multiply number of inches by 25.4

To convert millimeters (mm) to inches (ins.): multiply number of millimeters by .04

Inches		Decimals	Milli-meters	Inches to millimeters inches	mm	Inches		Decimals	Milli-meters	Inches to millimeters inches	mm
	1/64	0.051625	0.3969	0.0001	0.00254		33/64	0.515625	13.0969	0.6	15.24
1/32		0.03125	0.7937	0.0002	0.00508	17/32		0.53125	13.4937	0.7	17.78
	3/64	0.046875	1.1906	0.0003	0.00762		35/64	0.546875	13.8906	0.8	20.32
1/16		0.0625	1.5875	0.0004	0.01016	9/16		0.5625	14.2875	0.9	22.86
	5/64	0.078125	1.9844	0.0005	0.01270		37/64	0.578125	14.6844	1	25.4
3/32		0.09375	2.3812	0.0006	0.01524	19/32		0.59375	15.0812	2	50.8
	7/64	0.109375	2.7781	0.0007	0.01778		39/64	0.609375	15.4781	3	76.2
1/8		0.125	3.1750	0.0008	0.02032	5/8		0.625	15.8750	4	101.6
	9/64	0.140625	3.5719	0.0009	0.02286		41/64	0.640625	16.2719	5	127.0
5/32		0.15625	3.9687	0.001	0.0254	21/32		0.65625	16.6687	6	152.4
	11/64	0.171875	4.3656	0.002	0.0508		43/64	0.671875	17.0656	7	177.8
3/16		0.1875	4.7625	0.003	0.0762	11/16		0.6875	17.4625	8	203.2
	13/64	0.203125	5.1594	0.004	0.1016		45/64	0.703125	17.8594	9	228.6
7/32		0.21875	5.5562	0.005	0.1270	23/32		0.71875	18.2562	10	254.0
	15/64	0.234375	5.9531	0.006	0.1524		47/64	0.734375	18.6531	11	279.4
1/4		0.25	6.3500	0.007	0.1778	3/4		0.75	19.0500	12	304.8
	17/64	0.265625	6.7469	0.008	0.2032		49/64	0.765625	19.4469	13	330.2
9/32		0.28125	7.1437	0.009	0.2286	25/32		0.78125	19.8437	14	355.6
	19/64	0.296875	7.5406	0.01	0.254		51/64	0.796875	20.2406	15	381.0
5/16		0.3125	7.9375	0.02	0.508	13/16		0.8125	20.6375	16	406.4
	21/64	0.328125	8.3344	0.03	0.762		53/64	0.828125	21.0344	17	431.8
11/32		0.34375	8.7312	0.04	1.016	27/32		0.84375	21.4312	18	457.2
	23/64	0.359375	9.1281	0.05	1.270		55/64	0.859375	21.8281	19	482.6
3/8		0.375	9.5250	0.06	1.524	7/8		0.875	22.2250	20	508.0
	25/64	0.390625	9.9219	0.07	1.778		57/64	0.890625	22.6219	21	533.4
13/32		0.40625	10.3187	0.08	2.032	29/32		0.90625	23.0187	22	558.8
	27/64	0.421875	10.7156	0.09	2.286		59/64	0.921875	23.4156	23	584.2
7/16		0.4375	11.1125	0.1	2.54	15/16		0.9375	23.8125	24	609.6
	29/64	0.453125	11.5094	0.2	5.08		61/64	0.953125	24.2094	25	635.0
15/32		0.46875	11.9062	0.3	7.62	31/32		0.96875	24.6062	26	660.4
	31/64	0.484375	12.3031	0.4	10.16		63/64	0.984375	25.0031	27	690.6
1/2		0.5	12.7000	0.5	12.70						

ENGLISH TO METRIC CONVERSION: TORQUE

To convert foot-pounds (ft. lbs.) to Newton-meters: multiply the number of ft. lbs. by 1.3

To convert inch-pounds (in. lbs.) to Newton-meters: multiply the number of in. lbs. by .11

in lbs	N-m	in lbs	N-m	in lbs	N-m	in lbs	N-m	in lbs	N-m
0.1	0.01	1	0.11	10	1.13	19	2.15	28	3.16
0.2	0.02	2	0.23	11	1.24	20	2.26	29	3.28
0.3	0.03	3	0.34	12	1.36	21	2.37	30	3.39
0.4	0.04	4	0.45	13	1.47	22	2.49	31	3.50
0.5	0.06	5	0.56	14	1.58	23	2.60	32	3.62
0.6	0.07	6	0.68	15	1.70	24	2.71	33	3.73
0.7	0.08	7	0.78	16	1.81	25	2.82	34	3.84
0.8	0.09	8	0.90	17	1.92	26	2.94	35	3.95
0.9	0.10	9	1.02	18	2.03	27	3.05	36	4.0

GENERAL INFORMATION AND MAINTENANCE

ENGLISH TO METRIC CONVERSION: TORQUE

Torque is now expressed as either foot-pounds (ft./lbs.) or inch-pounds (in./lbs.). The metric measurement unit for torque is the Newton-meter (Nm). This unit—the Nm—will be used for all SI metric torque references, both the present ft./lbs. and in./lbs.

ft lbs	N-m	ft lbs	N-m	ft lbs	N-m	ft lbs	N-m
0.1	0.1	33	44.7	74	100.3	115	155.9
0.2	0.3	34	46.1	75	101.7	116	157.3
0.3	0.4	35	47.4	76	103.0	117	158.6
0.4	0.5	36	48.8	77	104.4	118	160.0
0.5	0.7	37	50.7	78	105.8	119	161.3
0.6	0.8	38	51.5	79	107.1	120	162.7
0.7	1.0	39	52.9	80	108.5	121	164.0
0.8	1.1	40	54.2	81	109.8	122	165.4
0.9	1.2	41	55.6	82	111.2	123	166.8
1	1.3	42	56.9	83	112.5	124	168.1
2	2.7	43	58.3	84	113.9	125	169.5
3	4.1	44	59.7	85	115.2	126	170.8
4	5.4	45	61.0	86	116.6	127	172.2
5	6.8	46	62.4	87	118.0	128	173.5
6	8.1	47	63.7	88	119.3	129	174.9
7	9.5	48	65.1	89	120.7	130	176.2
8	10.8	49	66.4	90	122.0	131	177.6
9	12.2	50	67.8	91	123.4	132	179.0
10	13.6	51	69.2	92	124.7	133	180.3
11	14.9	52	70.5	93	126.1	134	181.7
12	16.3	53	71.9	94	127.4	135	183.0
13	17.6	54	73.2	95	128.8	136	184.4
14	18.9	55	74.6	96	130.2	137	185.7
15	20.3	56	75.9	97	131.5	138	187.1
16	21.7	57	77.3	98	132.9	139	188.5
17	23.0	58	78.6	99	134.2	140	189.8
18	24.4	59	80.0	100	135.6	141	191.2
19	25.8	60	81.4	101	136.9	142	192.5
20	27.1	61	82.7	102	138.3	143	193.9
21	28.5	62	84.1	103	139.6	144	195.2
22	29.8	63	85.4	104	141.0	145	196.6
23	31.2	64	86.8	105	142.4	146	198.0
24	32.5	65	88.1	106	143.7	147	199.3
25	33.9	66	89.5	107	145.1	148	200.7
26	35.2	67	90.8	108	146.4	149	202.0
27	36.6	68	92.2	109	147.8	150	203.4
28	38.0	69	93.6	110	149.1	151	204.7
29	39.3	70	94.9	111	150.5	152	206.1
30	40.7	71	96.3	112	151.8	153	207.4
31	42.0	72	97.6	113	153.2	154	208.8
32	43.4	73	99.0	114	154.6	155	210.2

TUNE-UP PROCEDURES 2-3
SPARK PLUGS 2-4
    SPARK PLUG HEAT RANGE 2-4
    REMOVAL & INSTALLATION 2-4
    RETHREADING SPARK PLUG
      HOLE 2-7
    CHECKING & REPLACING SPARK
      PLUG CABLES 2-7
FIRING ORDERS 2-8
POINT TYPE IGNITION 2-8
BREAKER POINTS AND
  CONDENSER 2-8
    REMOVAL & INSTALLATION 2-8
    POINT GAP ADJUSTMENT 2-8
    DWELL ANGLE ADJUSTMENT 2-8
ELECTRONIC IGNITION 2-9
PRECAUTIONS 2-9
MAINTENANCE 2-10
    TROUBLESHOOTING 2-10
IGNITION TIMING 2-10
VALVE LASH 2-12
ADJUSTMENT 2-12
**IDLE SPEED AND MIXTURE
  ADJUSTMENTS 2-13**
CARBURETED VEHICLES 2-13
    SOLEX 30 PICT-3 (1970 TYPE 1
      AND TYPE 2 MODELS) 2-14
    SOLEX PICT-3 (1971–74 TYPE 1
      MODELS, 1971 TYPE 2 MODELS)
      AND SOLEX 34 PICT-4 (1973–74
      TYPE 1 CALIFORNIA
      MODELS) 2-14
    SOLEX 34 PDSIT-2/3 TWIN
      CARBURETORS (1972–74) TYPE 2
      MODELS 2-14
FUEL INJECTED VEHICLES 2-16
    ALL TYPE 3 AND TYPE 4; 1975 AND
    LATER TYPE 1 AND TYPE 2 2-16
SPECIFICATION CHART
  TUNE-UP SPECIFICATIONS 2-2

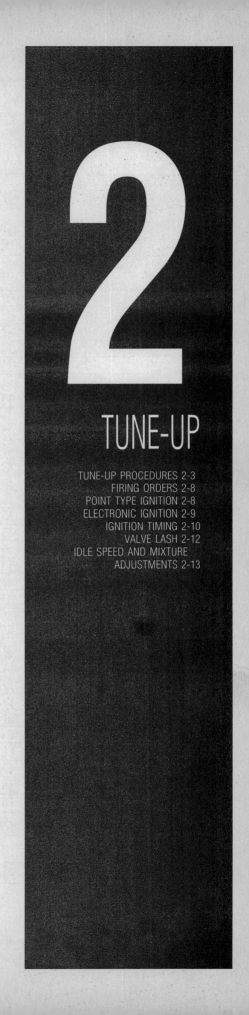

2
TUNE-UP

TUNE-UP PROCEDURES 2-3
FIRING ORDERS 2-8
POINT TYPE IGNITION 2-8
ELECTRONIC IGNITION 2-9
IGNITION TIMING 2-10
VALVE LASH 2-12
IDLE SPEED AND MIXTURE
ADJUSTMENTS 2-13

2-2 TUNE-UP

Tune-Up Specifications

Year	Engine Code	Type	Common Designation	Factory Recommended Spark Plugs Type	Gap (in.)	Distributor Point Dwell (deg)	Point Gap (in.)	Ignition Timing (deg) MT	Ignition Timing (deg) AT	Fuel Pump Pressure (psi) @ 4000 rpm	Compression Pressure (psi)	Idle Speed (rpm) MT	Idle Speed (rpm) AT	Valve Clearance (in.) Cold In	Valve Clearance (in.) Cold Ex
1970	B	1, 2	1600	Bosch W145T1 Champion L88A	.024	44–50	.016	TDC②	TDC②	3.5	114–142	800–900	900–1000	.006	.006
	U	3	1600	Bosch W145T1 Champion L88A	.024	44–50	.016	TDC②	TDC②	28	114–142	800–900	900–1000	.006	.006
1971	AE	1, 2	1600	Bosch W145T1 Champion L88A	.024	44–50	.016	5ATDC①	5ATDC①	3.5	114–142	800–900	900–1000	.006	.006
	U	3	1600	Bosch W145T1 Champion L88A	.024	44–50	.016	TDC②	TDC②	28	114–142	800–900	900–1000	.006	.006
	W	4	1700	Bosch W175T2	.024	44–50	.016	27BTDC③	27BTDC③	28	128–156	800–900	900–1000	.006	.006
1972	AE	1	1600	Bosch W145T1 Champion L88A	.024	44–50	.016	5ATDC①	5ATDC①	3.5	107–135	800–900	900–1000	.006	.006
	AH (Calif only)	1	1600	Bosch W145T1 Champion L88A	.024	44–50	.016	5ATDC①	5ATDC①	3.5	107–135	800–900	900–1000	.006	.006
	CB	2	1700	Bosch W145T2 Champion N88	.024	44–50	.016	5ATDC①	—	5.0	100–135	800–900	—	.006	.006
	U, X	3	1600	Bosch W145T1 Champion L88A	.024	44–50	.016	5BTDC②	5BTDC②	28	107–135	800–900	900–1000	.006	.006
	EA	4	1700	Bosch W175T2	.024	44–50	.016	27BTDC③	27BTDC③	28	128–156	800–900	900–1000	.006	.006
1973	AK	1	1600	Bosch W145T1 Champion L88A	.024	44–50	.016	5ATDC①④	5ATDC①④	3.5	107–135	800–900	900–1000	.006	.006
	AH (Calif only)	1	1600	Bosch W145T1 Champion L88A	.024	44–50	.016	5ATDC①	5ATDC①	3.5	107–135	800–900	900–1000	.006	.006
	CB	2	1700	Bosch W145T2 Champion N88	.024	44–50	.016	10ATDC①	—	5.0	100–135	800–900	—	.006	.008
	CD	2	1700	Bosch W145T2 Champion N88	.024	44–50	.016	—	5ATDC①	5.0	100–135	—	900–1000	.006	.008
	U, X	3	1600	Bosch W145T1 Champion L88A	.024	44–50	.016	5BTDC②	5BTDC②	28	107–135	800–900	900–1000	.006	.006
	EA	4	1700	Bosch W175T2 Champion N88	.024	44–50	.016	27BTDC③	27BTDC③	28	128–156	800–900	900–1000	.006	.006
	EB (Calif only)	4	1700	Bosch W175T2 Champion N88	.024	44–50	.016	27BTDC③	27BTDC③	28	107–135	800–900	900–1000	.006	.006
1974	AK	1	1600	Bosch W145T1 Champion L88A	.024	44–50	.016	7½BTDC②	7½BTDC②	3.5	107–135	800–900	900–1000	.006	.006
	AH (Calif only)	1	1600	Bosch W145T1 Champion L88A	.024	44–50	.016	5ATDC①	5ATDC①	3.5	107–135	800–900	900–1000	.006	.006
	AW	2	1800	Bosch W175T2 Champion N88	.024	44–50	.016	10ATDC①	5ATDC①	5.0	85–135	800–900	900–1000	.006	.008
	EA	4	1700	Bosch W175T2 Champion N88	.024	44–50	.016	27BTDC③	—	28	128–156	800–900	—	.006	.006
	EC	4	1800	Bosch W175T2 Champion N88	.024	44–50	.016	—	7½BTDC②	28	85–135	—	900–1000	.006	.006

TUNE-UP 2-3

Tune-Up Specifications (cont.)

Year	Engine Code	Type	Common Designation	Factory Recommended Spark Plugs Type	Gap (in.)	Distributor Point Dwell (deg)	Distributor Point Gap (in.)	Ignition Timing (deg) MT	Ignition Timing (deg) AT	Fuel Pump Pressure (psi) @ 4000 rpm	Compression Pressure (psi)	Idle Speed (rpm) MT	Idle Speed (rpm) AT	Valve Clearance (in.) Cold In	Valve Clearance (in.) Cold Ex
1975	AJ	1	1600	Bosch W145M1 Champion L288	.024	44–50	.016	5ATDC ⑤	TDC ⑤	28	85–135	875	875	.006	.006
	ED	2	1800	Bosch W145M2 Champion N288	.024	44–50	.016	5ATDC ⑤	5ATDC ⑤	28	85–135	900	900	.006	.006
1976–77	AJ	1	1600	Bosch W145M1 Champion L288	.024 ⑥	44–50	.016	5ATDC ⑤	TDC ⑤	28	85–135	875	925	.006	.006
	GD	2	2000	Bosch W145M2 Champion N288	.028	44–50	.016	7½BTDC ⑤	7½BTDC ⑤	28	85–135	900	950	.006	.006
1978	AJ	1	1600	Bosch W145M1 Champion L288	.028 .028	44–50	.016	5ATDC	5ATDC	28	85–135	800–950	800–950	.006	.006
	GE	2	2000	Bosch W145M2 Champion N288	.028 .028	44–50	.016	7½BTDC	7½BTDC	28	85–135	800–950	900–1000	Hydraulic	Hydraulic
1979–80	AJ	1	1600	Bosch W145M1 Champion L288	.028 .028	44–50	.016	5ATDC	5ATDC	28	85–135	800–950	800–950	.006	.006
1979–81	GE, CV	2 (49 states)	2000	Bosch W145M2	.028	44–50	.016	7½BTDC ⑧	7½BTDC ⑧	28	85–135	850–950	850–1000	Hydraulic	Hydraulic
	GE, CV	2 (California)	2000	Bosch W145M2	.028	Electronic		5ATDC ⑨ ⑩	5ATDC ⑨ ⑩	28	85–135	850–950	850–950	Hydraulic	Hydraulic

① At idle, throttle valve closed (Types 1 & 2), vacuum hose(s) on
② At idle, throttle valve closed (Types 1 & 2), vacuum hose(s) off
③ At 3,500 rpm, vacuum hose(s) off
④ From March 1973, vehicles with single diaphragm distributor (one vacuum hose); adjust timing to 7½° BTDC with hose disconnected and plugged. The starting serial numbers for those type 1 vehicles using the single diaphragm distributors are # 113 2674 897 (manual trans) and 113 2690 032 (auto stick shift)
⑤ Carbon canister hose at air cleaner disconnected; at idle; vacuum hose(s) on
MT Manual Transmission
AT Automatic Transmission
BTDC Before Top Dead Center

ATDC After Top Dead Center
⑥ 1977—.028
⑦ 5ATDC—California
⑧ Vacuum hose(s) off
⑨ Vacuum hose(s) on
⑩ Idle Stabilizer bypassed
Part numbers in this chart are not recommendations by Chilton for any product by brand name. They are references that can be used with interchange manuals and aftermarket supplier catalogs to locate each brand supplier's discrete part number.

TUNE-UP PROCEDURES

The tune-up is a routine maintenance operation which is essential for the efficient and economical operation, as well as the long life of your car's engine. The interval between tune-ups is a variable factor which depends upon the way you drive your car, the conditions under which you drive it (weather, road type, etc.), and the type of engine installed in your car. It is generally correct to say that these cars should not be driven more than 12,000 miles between tune-ups. If you plan to drive your car extremely hard or under severe weather conditions, the tune-ups should be performed at closer intervals. High-performance engines require more frequent tuning than other engines, regardless of weather or driving conditions.

The replaceable parts involved in a tune-up include the spark plugs, breaker points, condenser, distributor cap, rotor, spark plug wires and the ignition coil high-tension (secondary) wire. In addition to these parts and the adjustments involved in properly adapting them to your engine, there are several adjustments of other parts involved in completing the job. These include carburetor idle speed and air/fuel mixture, ignition timing, and dwell angle.

This section gives specific procedures on how to tune-up your Volkswagen and is intended to be as complete and basic as possible.

✲✲ CAUTION

When working with a running engine, make sure that there is proper ventilation. Also make sure that the transmission is in Neutral (unless otherwise specified) and the parking brake is fully applied. Always keep hands, long hair, clothing, neckties and tools well clear of the engine. On a warm engine, keep clear of the hot exhaust manifold(s) and exhaust pipe. When the ignition is turned on and the engine running, do not grasp the ignition wires, distributor cap, or coil wire, as a shock in excess of 20,000 volts may result. Whenever working around the distributor, even if the engine is not running, make sure that the ignition is switched off.

2-4 TUNE-UP

Spark Plugs

Before attempting any work on the cylinder head, it is very important to note that the cylinder head is cast aluminum alloy. This means that it is extremely easy to damage threads in the cylinder head. Care must be taken not to cross-thread the spark plugs or any bolts or studs. Never overtighten the spark plugs, bolts, or studs.

✳✳ CAUTION

To prevent seizure, always lubricate the spark plug threads with liquid silicon or Never-Seez®.

To avoid cross-threading the spark plugs, always start the plugs in their threads with your fingers. Never force the plugs into the cylinder head. Do not use a wrench until you are certain that the plug is correctly threaded.

VW spark plugs should be cleaned and regapped every 6,000 miles and replaced every 12,000 miles.

SPARK PLUG HEAT RANGE

While spark plug heat range has always seemed to be somewhat of a mystical subject for many people, in reality the entire subject is quite simple. Basically, it boils down to this; the amount of heat the plug absorbs is determined by the length of the lower insulator. The longer the insulator (or the farther it extends into the engine), the hotter the plug will operate; the shorter the insulator the cooler it will operate. A plug that absorbs little heat and remains too cool will quickly accumulate deposits of oil and carbon since it is not hot enough to burn them off. This leads to plug fouling and consequently to misfiring. A plug that absorbs too much heat will have no deposits, but, due to the excessive heat, the electrodes will burn away quickly and in some instances, preignition may result. Preignition takes place when plug tips get so hot that they glow sufficiently to ignite the fuel/air mixture before the actual spark occurs. This early ignition will usually cause a pinging during low speeds and heavy loads. In severe cases, the heat may become high enough to start the fuel/air

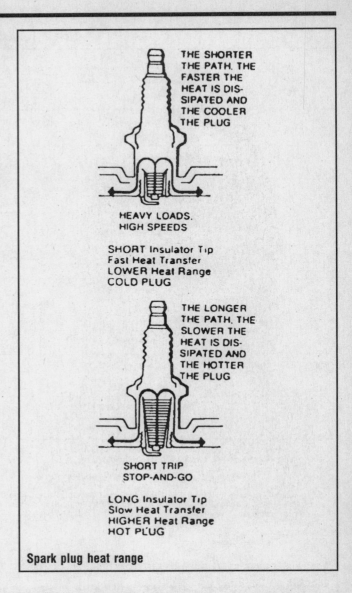

Spark plug heat range

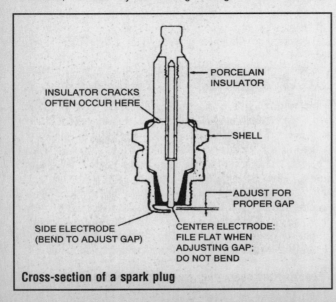

Cross-section of a spark plug

mixture burning throughout the combustion chamber rather than just to the front of the plug as in normal operation. At this time, the piston is rising in the cylinder making its compression stroke. The burning mass is compressed and an explosion results, forcing the piston back down in the cylinder while it is still trying to go up. Obviously, something must go, and it does—pistons are often damaged.

The general rule of thumb for choosing the correct heat range when picking a spark plug is: if most of your driving is long distance, high speed travel, use a colder plug; if most of your driving is stop and go, use a hotter plug. Factory-installed plugs are, of course, compromise plugs, since the factory has no way of knowing what sort of driving you do. It should be noted that most people never have occasion to change their plugs from the factory-recommended heat range.

REMOVAL & INSTALLATION

To remove the spark plugs, remove the spark plug wire from the plug. Grasp the plug connector and twist/pull the spark plug cable off of the spark plug; while removing, do not pull on the

TUNE-UP 2-5

A normally worn spark plug should have light tan or gray deposits on the firing tip

A carbon fouled plug, identified by soft, sooty, black deposits, may indicate an improperly tuned vehicle. Check the air cleaner, ignition components and engine control system

A variety of tools and gauges are needed for spark plug service

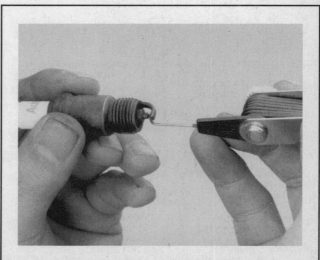

Checking the spark plug gap with a feeler gauge

2-6 TUNE-UP

A physically damaged spark plug may be evidence of severe detonation in that cylinder. Watch that cylinder carefully between services, as a continued detonation will not only damage the plug, but could also damage the engine

An oil fouled spark plug indicates an engine with worn piston rings and/or bad valve seals allowing excessive oil to enter the chamber

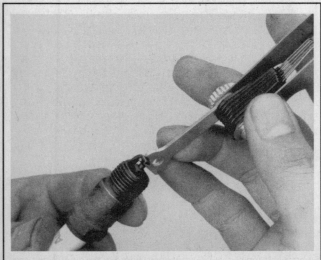

Adjusting the spark plug gap

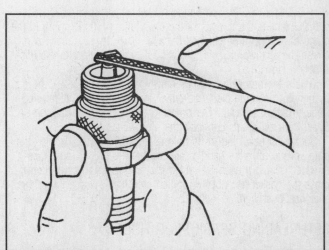

If the standard plug is in good condition, the electrode may be filed flat—CAUTION: do not file platinum plugs

TUNE-UP 2-7

This spark plug has been left in the engine too long, as evidenced by the extreme gap—Plugs with such an extreme gap can cause misfiring and stumbling accompanied by a noticeable lack of power

A bridged or almost bridged spark plug, identified by a build-up between the electrodes caused by excessive carbon or oil build-up on the plug

wire. Using a 13/16 in. spark plug socket, remove the old spark plugs. Examine the threads of the old plugs; if one or more of the plugs have aluminum clogged threads, it will be necessary to rethread the spark plug hole. See the following section for the necessary information.

Obtain the proper heat range and type of new plug. Set the gap by bending the side electrode only. Do not bend the center electrode to adjust the gap. The proper gap is listed in the "Tune-Up Specifications" chart. Lubricate the plug threads.

Start each new plug in its hole using your fingers. Tighten the plug several turns by hand to assure that the plug is not crossthreaded. Using a wrench, tighten the plug just enough to compress the gasket. Do not overtighten the plug. Consult the torque specifications chart.

RETHREADING SPARK PLUG HOLE

It is possible to repair light damage to spark plug hole threads by using a spark plug hole tap of the proper diameter and thread.

Plenty of grease should be used on the tap to catch any metal chips. Exercise caution when using the tap as it is possible to cut a second set of threads instead of straightening the old ones.

If the old threads are beyond repair, then the hole must be drilled and tapped to accept a steel bushing or Heli-Coil®. It is not always necessary to remove the cylinder head to rethread the spark plug holes. Bushing kits, Heli-Coil® kits, and spark plug hole taps are available at most auto parts stores. Heli-Coil® information is contained in the "Engine Rebuilding" section of this book.

CHECKING & REPLACING SPARK PLUG CABLES

Visually inspect the spark plug cables for burns, cuts, or breaks in the insulation. Check the spark plug boots and the nipples on the distributor cap and coil. Replace any damaged wiring. If no physical damage is obvious, the wires can be checked with an ohmmeter for excessive resistance. Remove the distributor cap and leave the wires connected to the cap. Connect one lead of the

2-8 TUNE-UP

ohmmeter to the corresponding electrode inside the cap and the other lead to the spark plug terminal (remove it from the spark plug for the test). Remove the static suppressor boot from the end of the cable by un-screwing it before testing cable. Replace any wire which shows over 50,000 ohms. Generally speaking, however, resistance should not run over 35,000 ohms and 50,000 ohms should be considered the outer limits of acceptability. Test the coil wire by connecting the ohmmeter between the center contact in the cap and either of the primary terminals at the coil. If the total resistance of the coil and cable is more than 25,000 ohms, remove the cable from the coil and check the resistance of the cable alone. If the resistance is higher than 15,000 ohms, replace the cable. It should be remembered that wire resistance is a function of length, and that the longer the cable, the greater the resistance. Thus, if the cables on your car are longer than the factory originals, resistance will be higher and quite possibly outside of these limits. Test the static suppressor boots separately—resistance should not exceed 5,000–10,000 ohms.

When installing a new set of spark plug cables, replace the cables one at a time so there will be no mixup. Start by replacing the longest cable first. Install the boot firmly over the spark plug. Route the wire exactly the same as the original. Insert the nipple firmly into the tower on the distributor cap. Repeat the process for each cable.

FIRING ORDERS

♦ See Figure 1

➡ To avoid confusion, remove and tag the spark plug wires one at a time, for replacement.

If a distributor is not keyed for installation with only one orientation, it could have been removed previously and rewired. The resultant wiring would hold the correct firing order, but could change the relative placement of the plug towers in relation to the engine. For this reason it is imperative that you label all wires before disconnecting any of them. Also, before removal, compare the current wiring with the accompanying illustrations. If the current wiring does not match, make notes in your book to reflect how your engine is wired.

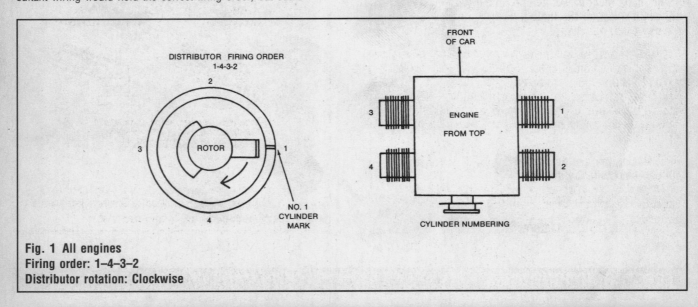

Fig. 1 All engines
Firing order: 1–4–3–2
Distributor rotation: Clockwise

POINT TYPE IGNITION

Breaker Points and Condenser

♦ See Figure 2

REMOVAL & INSTALLATION

1. Release the spring clips which secure the distributor cap and lift the cap from the distributor. Pull the rotor from the distributor shaft.
2. Disconnect the points wire from the condenser snap connection inside the distributor.
3. Remove the locking screw from the stationary breaker point.
4. To remove the condenser, which is located on the outside of the distributor, remove the screw which secures the condenser bracket and condenser connection to the distributor.
5. Disconnect the condenser wire from the coil.
6. With a clean rag, wipe the excess oil from the breaker plate.

➡ Make sure that the new point contacts are clean and oil free.

7. Installation of the point set and condenser is the reverse of the above; however, it will be necessary to adjust the point gap, also known as the dwell angle, and check the timing. Lubricate the point cam with a small amount of lithium or white grease. Set the dwell angle, or gap, before the ignition timing.

TUNE-UP 2-9

POINT GAP ADJUSTMENT

♦ See Figure 3

1. Remove the distributor cap and rotor.
2. Turn the engine by hand until the fiber rubbing block on the movable breaker point rests on a high point of the cam lobe. The point gap is the maximum distance between the points and must be set at the top of a cam lobe.
3. Using a screwdriver, slightly loosen the locking screw. Make sure that the feeler gauge is clean. After tightening the screw, recheck the gap.
4. Move the stationary point plate so that the gap is set as specified and then tighten the screw. Make sure that the feeler gauge is clean. After tightening the screw, recheck the gap.
5. It is important to set the point gap before setting the timing.

DWELL ANGLE ADJUSTMENT

Setting the dwell angle with a dwell meter achieves the same effect as setting the point gap but offers better accuracy.

➡ The dwell must be set before setting the timing. Setting the dwell will alter the timing, but when the timing is set, the dwell will not change.

Attach the positive lead of the dwell meter to the coil terminal which has a wire leading to the distributor. The negative lead should be attached to a good ground.

Remove the distributor cap and rotor. Turn the ignition ON and turn the engine over using the starter or a remote starter switch. Read the dwell from the meter and open or close the points to adjust the dwell.

➡ Increasing the gap decreases the dwell and decreasing the gap increases the dwell.

Dwell specifications are listed in the "Tune-Up Specifications" chart.
Reinstall the cap and rotor and start the engine.

Fig. 2 Exploded view of typical breaker points and condenser mounting used on all 1970-81 Volkswagen models

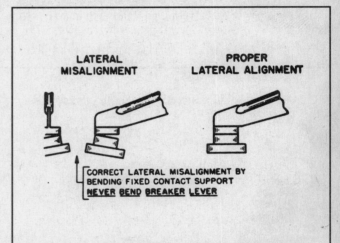

Fig. 3 It is vital that the surfaces of the breaker points are properly aligned to avoid premature wear

ELECTRONIC IGNITION

1979–81 Type 2's destined for California are equipped with a Hall Effect electronic ignition system which does away with the conventional breaker points and condenser used on all other models in this book.

Located in the distributor, in addition to the normal rotor cap, is a round, four bladed rotor unit. This unit is attached to the distributor shaft below the rotor cap and in about the same position as the breaker points cam on conventional ignition systems. Mounted on the distributor base, in about the same position as the breaker points on conventional ignition systems, is a pick-up coil unit.

The basic operating principle of the Hall Effect ignition is this: the rotor unit revolves with the top rotor cap and, as each of its four blades passes between the stationary pick-up coils, a magnetic shift occurs which signals the control unit to break primary ignition current in the coil and thus generate the spark. The control unit is used in conjunction with an idle stabilizer unit which controls idle speed according to engine load by adjusting the ignition timing. The idle stabilizer must be by-passed to adjust the idle speed and timing.

Precautions

To avoid damaging the electronic ignition system, always observe the following:
• Do not disconnect and connect ignition system wires with the ignition switch in the ON position (whether the engine is running or not).
• Do not install any ignition coil except the factory recommended one (part No. 211 905 115 C).
• To disable the engine, to prevent it from starting while performing any of the several tests in this book (compression test, etc.), disconnect the high tension wire from the center of the distributor cap and connect it to ground.

2-10 TUNE-UP

- Never attempt to connect a condenser to terminal 1 (−) of the ignition coil.
- Do not connect a quick charger to the battery for longer than one minute when attempting to boost start the engine.
- Do not disconnect the battery with the engine running.
- Do not wash the engine while it is running.
- Do not electric weld the vehicle with the battery connected.
- Disconnect the plug at the ignition control unit when towing a car with a damaged ignition system.
- Do not connect test instruments with a 12V supply on terminal 15 of ignition coil.

Maintenance

Periodic maintenance is not required on the Hall Effect ignition system, as it has no points and condenser. It would be wise to check the distributor cap and rotor cap for cracks and wear occasionally.

TROUBLESHOOTING

All troubleshooting of the electronic ignition system should generally be left to a qualified technician as, for the most part, substitution of components is the usual method. The ignition coil, however, can be tested using a Digital Volt Ohmmeter (DVOM). The primary resistance should be 0.65 ohms.

If you do not own a DVOM, substitute a known good coil with the same part number and determine if the ignition works. If the ignition system functions normally after replacing the coil, the old coil was faulty. If the ignition system continues to malfunction, the old coil was not the problem in the system.

IGNITION TIMING

▶ See Figures 4 thru 14

Dwell or point gap must be set before the timing is set. Also, the idle speed must be set to specifications.

➡The engine must be warmed up before the timing is set (oil temperature of 122°F–158°F.).

1. Connect a timing light. On most VWs, you won't be able to hook the timing light power leads to the battery, therefore, connect the positive lead from the timing light to terminal #15 on the ignition coil, and hook the negative lead from the test light to ground (intake manifold, etc.). This procedure is for DC lights only and cannot be used on Hall electronic ignition systems.

After hooking up the timing light according to manufacturer's instructions, disconnect the vacuum hose, if so advised by the "Tune-up Specifications" chart, and re-adjust the idle speed if necessary.

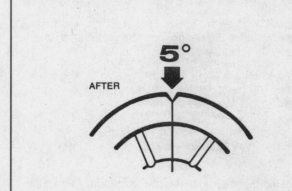

Fig. 5 Ignition timing marks–August 1970 to spring 1973 Type 1 carbureted models, all fuel injected Type 1 models with manual transaxles and 1971 Type 2 models

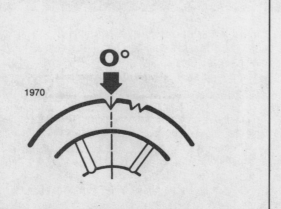

Fig. 4 Ignition timing marks (located on the crankshaft pulley)—1970 (before August) Type 1 models

Fig. 6 Ignition timing marks—spring 1973–1980 Type 1 models with single vacuum hose connected to the distributor

TUNE-UP 2-11

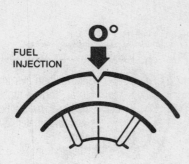

Fig. 7 Ignition timing marks—fuel injected Type 1 models equipped with automatic stick shift transaxles and 1970 Type 2 models

Fig. 10 Ignition timing marks—all 1976–78 Type 2 models and 1979–81 non-California Type 2 models

Fig. 8 Ignition timing marks—all 1972 and 1975 Type 2 models; 1973–74 Type 2 models equipped with automatic transaxles

Fig. 11 Ignition timing marks—1979–81 California Type 2 models

Fig. 9 Ignition timing marks—1973–74 Type 2 carbureted models with manual transaxles

Fig. 12 Ignition timing marks (located on the engine cooling fan)—1970–71 Type 3 models

2-12 TUNE-UP

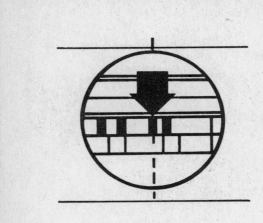

Fig. 13 Ignition timing marks—1972–73 Type 3 models

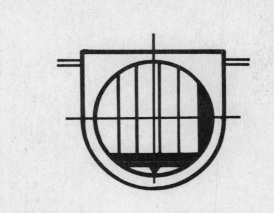

Fig. 14 Ignition timing marks—1971–74 Type 4 models

➡ On Hall electronic ignition systems, the idle stabilizer must be bypassed before the idle can be adjusted. See the illustration.

2. Start the engine and run it at the specified rpm. Aim the timing light at the crankshaft pulley on upright fan engines and at the engine cooling fan on the suitcase engines. The rubber plug in the fan housing will have to be removed before the timing marks on Type 3 and Type 4 engines can be seen.

3. Read the timing and rotate the distributor accordingly.

➡ Rotate the distributor in the opposite direction of normal rotor rotation to advance the timing. Retard the timing by turning the distributor in the normal direction of rotor rotation.

4. It is necessary to loosen the clamp at the base of the distributor before the distributor can be rotated. It may also be necessary to put a small amount of white paint or chalk on the timing marks to make them more visible.

VALVE LASH

Adjustment

➡ 1978–81 Type 2 models have hydraulic lifters that require no adjustment.

Preference should be given to the valve clearance specified on the engine fan housing sticker, if they differ from those in the "Tune-Up Specifications" chart.

➡ The engine must be as cool as possible before adjusting the valves.

➡ If the spark plugs are removed, rotating the crankshaft from position to position will be much easier.

1. Remove the distributor cap and turn the engine until the rotor points to the No. one spark plug wire post in the distributor cap. To bring the piston to exactly top dead center (TDC) on the compression stroke, align the crankshaft timing marks on TDC.

2. Remove the rocker arm covers. At TDC, the pushrods should be down and there should be clearance between the rocker arms and valve stems of both valves of the subject cylinder.

3. With the proper feeler gauge, check the clearance between the adjusting screw and the valve stem of both valves for the No. 1 cylinder (see cylinder numbering diagram). If the feeler gauge slides in snugly without being forced, the clearance is correct. It is better that the clearance is a little loose than a little tight, as a tight adjustment may result in burned exhaust valves. While a lit-

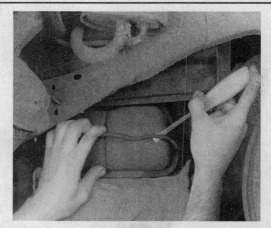

After raising and safely supporting the vehicle, position a shop towel over the heater box and pry the retaining clamp wire down and off of the rocker arm cover . . .

tle looseness is somewhat desirable, too much looseness will cause the tappets to rap loudly as the engine reaches operating temperature. This is because the cylinder barrels and the heads on the engine expand or "grow" when hot, whereas the valve pushrods do not. This growing action increases the valve lash slightly.

4. If the clearance is incorrect, the locknut must be loosened

TUNE-UP 2-13

... then remove the rocker arm cover from the cylinder head

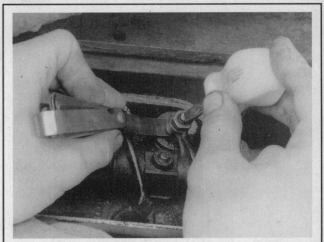
After loosening the locknut, adjust the valve clearance with a screwdriver and a feeler gauge

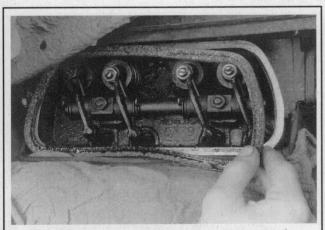
Make certain to remove the old rocker arm cover gasket—a new gasket should be installed whenever the cover is removed from the cylinder head

Once the clearance is correctly set, hold the adjusting screw stationary while tightening the locknut securely

and the adjustment screw turned until the proper clearance is obtained. After tightening down the locknut, it is then advisable to recheck the clearance. It is possible to accidently alter the adjustment when tightening the locknut.

5. The valves are adjusted in a 1-2-3-4 (exact opposite of firing [1-4-3-2] order) sequence. To adjust cylinders 2 through 4, the distributor rotor must be pointed at the appropriate distributor cap post 90° apart from each other. In addition, the crankshaft must be rotated counterclockwise (opposite normal rotation) in 180° degree increments to adjust valves 2, 3 and 4.

➥There should be a red paint mark on the crankshaft 180° opposite of the TDC mark for adjusting valves 2 and 4.

➥Always use new valve cover gaskets.

IDLE SPEED AND MIXTURE ADJUSTMENTS

Carbureted Vehicles

A carburetor adjustment should be performed only after all other variables in a tune-up have been checked and adjusted. This includes checking valve clearance, spark plug gap, breaker point gap and/or dwell angle, and ignition timing. Prior to making any carburetor adjustments, the engine should be brought to operating temperature (122–158°F oil temperature) and you should make sure that the automatic choke is fully open and off of the fast idle cam. Once you have performed all of the preliminary steps, shut off the engine and hook up a tachometer. Connect the hot lead to

2-14 TUNE-UP

the distributor side of the ignition coil and the ground wire to an engine bolt or other good metal to metal connection. Keep the wire clear of the fan.

➡ **An improper carburetor adjustment may have an adverse effect on exhaust emission levels. If any doubt exists, check your state laws regarding the adjusting of emission control equipment.**

SOLEX 30 PICT-3 (1970 TYPE 1 AND TYPE 2 MODELS)

1. Start the engine and bring it to operating temperature. Make sure the car is in neutral.
2. Using the idle speed (bypass) screw, adjust the idle speed to that specified in the "Tune-Up Specifications" chart.

➡ **The bypass screw adjustment is the only adjustment that should be made to 30 PICT-3 carburetor. Do not attempt to adjust the mixture or idle speed by turning the throttle valve adjustment screw, as increased exhaust emissions or poor driveability would result. The bypass screw is the larger adjustment screw.**

SOLEX PICT-3 (1971–74 TYPE 1 MODELS, 1971 TYPE 2 MODELS) AND SOLEX 34 PICT-4 (1973–74 TYPE 1 CALIFORNIA MODELS)

♦ See Figure 15

1. Start the engine and bring it to operating temperature. Make sure the car is in neutral.
2. Shut off the engine. On 1971 models, turn out the throttle valve adjustment screw until it clears the fast idle cam. Then turn in the screw until it makes contact with the fast idle cam. Finally, turn the throttle valve adjusting screw in another one-quarter turn.
3. Slowly turn in the idle mixture (volume control) screw until it bottoms. Then, carefully counting the complete revolutions of the screwdriver, turn it out 2½ to 3 turns.
4. With a tachometer connected to the engine as previously described, start the engine.
5. Using the idle speed (bypass) screw, adjust the idle speed to specifications. Then, using the idle mixture (volume control) screw, adjust until the fastest idle is obtained. Observing the tachometer, turn the volume control screw until the engine speed drops by 20–30 rpm.
6. Finally, using the bypass screw, adjust the idle to specifications.

SOLEX 34 PDSIT-2/3 TWIN CARBURETORS (1972–74) TYPE 2 MODELS

Periodic Adjustment

♦ See Figure 16

1. Start the engine and bring it to operating temperature (122–158°F oil temperature). Make sure the car is in neutral and the parking brake firmly applied.
2. Using the central idle speed adjusting screw (4) on the left carburetor, adjust the idle speed to that listed in the "Tune-Up Specifications" chart.
3a. If a CO meter/exhaust analyzer is available, adjust the carbon monoxide (CO) level to 1–3% CO using the central mixture control screw (5) also on the left carburetor.
3b. If a CO meter is not available, the following procedure is used: First, slowly turn the central mixture control screw (5) in (clockwise) until the engine speed drops noticeably, then turn the screw out (counter-clockwise) until maximum idle speed is at-

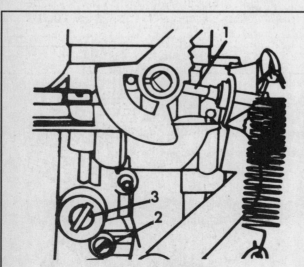

1. Throttle valve adjustment
2. Volume control (mixture) screw
3. By-pass (idle speed) screw

Fig. 15 Solex 34 PICT-3 and 34 PICT-4 idle adjustment locations (30 PICT-3 similar)

1. Synchronizing screw
2. Central idling system left end piece
3. Central idling system left end connecting hose
4. Central idling (idle speed) adjusting screw
5. Central mixture control screw
6. Volume control screw

Fig. 16 Type 2 twin carburetor set-up adjustment locations—34 PICT-2 (left carburetor shown)

TUNE-UP 2-15

tained. Next, turn the screw in once again until rpm drops 20–50 rpm. Finally, turn the screw out ¼ turn.

4. Recheck the idle speed, and adjust as necessary using the central idle speed adjusting screw (4) on the left carburetor.

5. If a satisfactory idle cannot be obtained using this procedure, proceed to "Basic Adjustment."

Basic Adjustment
♦ See Figures 17, 18 and 19

Whenever a carburetor has been removed for service, or if a new or rebuilt carburetor has been installed, a basic carburetor adjustment should be performed.

➡ The throttle valve setting (distance "a") must be 0.004 in.

➡ An exhaust analyzer/CO meter and tachometer is required for this adjustment.

1. Check the synchronization of the carburetors as outlined under "Balancing Multiple Carburetor Installations."
2. Disconnect the throttle linkage rod from the right carburetor.
3. Disconnect the vacuum retard hose from the distributor.
4. Disconnect the cut-off valve wire at the central idling system. Disconnect and plug the left side air pump hose (1973–74 models).
5. Turn the idle volume control screws (6) in on both carburetors until they contact their seats.

> ✱✱ **CAUTION**
>
> **Do not force the screws or the tips may become distorted.**

Then, turn both screws out exactly 2½ turns.

6. Start the engine and bring it to operating temperature (122–158°F oil temperature). Set the idle speed to 500–700 rpm by equally adjusting both volume control screws (6).
7. Disconnect the wire from the electromagnetic idling cut-off valve (smaller of the two valves) at the left carburetor and note the decrease in idle speed. Then, repeat this operation for the right carburetor. The idle speed drop should be equal for both sides. If not, readjust the volume control screws accordingly.
8. Connect the wire for the cut-off valve at the central idling system. Unplug, and connect the left air pump hose. Connect the vacuum retard hose.
9. Take the engine through the upper rpm range for a few quick bursts. Then, adjust the idle speed to specifications as outlined under steps 2–5 of "Periodic Adjustment."

Synchronizing Twin Carburetors
♦ See Figure 20

If a carburetor has been removed or disassembled, or if any part of the linkage has been repaired, the carburetors must be synchronized prior to performing any idle adjustments. To synchronize the carburetors, a special instrument is used to measure air flow, such as the UniSyn® or Auto-Syn®. This commonly available device measures the vacuum created inside the carburetors and provides an index for adjusting the carburetors equally. In order to use the air flow gauge on the 34 PDSIT carburetors, a special diameter adaptor (or a small frozen juice can with both ends removed) must be used.

1. Disconnect the linkage connecting rod from the lower socket

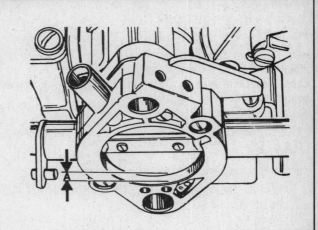

Fig. 17 On 34 PDSIT-2/3 carburetors, the throttle valve closing gap (A) must be 0.004 in.

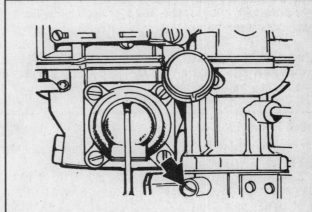

Fig. 18 Volume control screw location (arrow) in the carburetor throttle bodies of 34 PDSIT-2 and PDSIT-3 carburetors

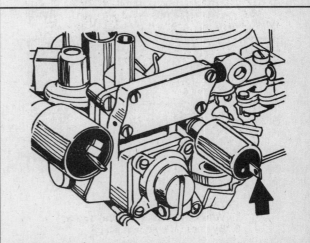

Fig. 19 Terminal connection location (arrow) for the idling cut-off valve of both right and left carburetors

2-16 TUNE-UP

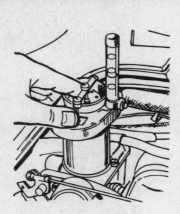

Fig. 20 Using an airflow meter to synchronize the carburetors on a twin carburetor set-up

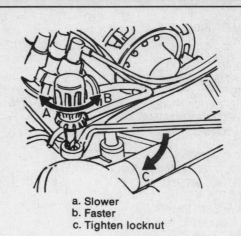

a. Slower
b. Faster
c. Tighten locknut

Fig. 21 Tighten or loosen the idle speed adjusting screw to adjust the idle of Type 3 fuel injected models

of the right carburetor. Without moving the throttle from its closed position, you should be able to snap the linkage connecting rod back onto the socket for the right carburetor. If the ball and socket at the right carburetor do not align perfectly, for 1972 models, adjust with the synchronizing screw (1) at the left carburetor; on 1973–74 models, adjust the length of the connecting rod on the right carburetor until the rod can be snapped onto the socket without disturbing the throttle from its fully closed position. Connect the linkage and hook up a tachometer.

2. Remove the air cleaner ducts from the tops of both carburetors, taking care to leave the central idle system connecting hose (3) and left end piece (2) connected.

3. Start the engine and bring it to operating temperature (122–158°F oil temperature). Make sure the choke flaps are fully open. Using the air flow meter, balance the carburetors using the synchronizing screw (1) on the left carburetor on 1972 models or the connecting rod on the right carburetor on 1973–74 models. Balance the carburetors with the engine running at 2,000–3,000 rpm.

4. Adjust the idle speed to specifications.

Fuel Injected Vehicles

ALL TYPE 3 AND TYPE 4; 1975 AND LATER TYPE 1 AND TYPE 2

Idle Speed
♦ See Figure 21

➡The idle stabilizer on "Hall Effect" electronic ignitions must be bypassed before the idle is set. See electronic ignition section, above, for procedures.

The idle speed is adjusted by a screw located on the left side of the intake air distributor. To adjust the idle speed, loosen the locknut (Type 3 only) and turn the screw with a screwdriver until the idle speed is adjusted to specification. Turning the screw clockwise decreases idle speed; counterclockwise increases idle speed.

On automatic transmission Type 4 and 1975–76 Type 2 models, the idle speed regulator should also be adjusted. With the vehicle idling at 900–1,000 RPM in Park or Neutral, measure "a" in the illustration. It should be 0.020–0.040 in. If not, adjust at arrow. See Chapter 4 for a test for the idle speed regulator.

If turning the screw either in or out does not noticeably affect idle speed, check for the following:

a. Air leaks in the intake manifold system.

b. Air leaks into the crankcase (make sure the oil cap is on correctly).

c. Faulty EGR components (See Chapter 4).

d. If the bypass screw must be turned fully in to lower the idle speed, or if the idle speed is fast prior to adjustment, check the auxiliary air regulator as covered in Chapter 4. If the auxiliary air regulator is OK, check the pressure regulator and deceleration valve as described in Chapter 4.

If, after checking these systems, the bypass screw still makes little change in the idle speed, it is possible that someone has fiddled with the throttle valve adjustment screw. Refer this operation to a competent mechanic.

Idle Mixture

The idle mixture must be adjusted using an exhaust gas analyzer, a tool not usually owned by amateur mechanics; therefore idle mixture adjustment procedures are not given here. The idle mixture does not have to be adjusted unless exhaust emissions are above specified levels, the intake air sensor is replaced or after an engine overhaul.

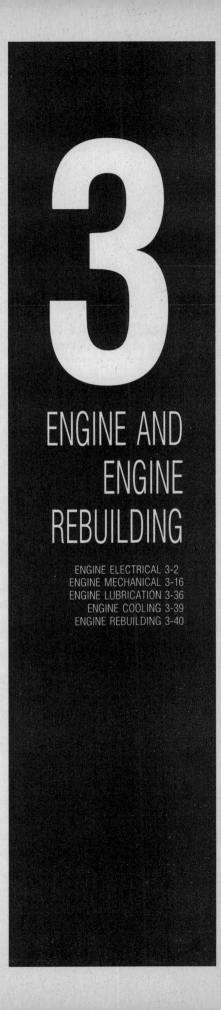

3

ENGINE AND ENGINE REBUILDING

ENGINE ELECTRICAL 3-2
ENGINE MECHANICAL 3-16
ENGINE LUBRICATION 3-36
ENGINE COOLING 3-39
ENGINE REBUILDING 3-40

ENGINE ELECTRICAL 3-2
UNDERSTANDING ELECTRICITY 3-2
   BASIC CIRCUITS 3-2
   TROUBLESHOOTING 3-3
BATTERY, STARTING AND CHARGING SYSTEMS 3-4
   BASIC OPERATING PRINCIPLES 3-4
POINT-TYPE IGNITION SYSTEMS 3-5
IGNITION COIL 3-6
   TESTING 3-6
DISTRIBUTOR 3-7
   REMOVAL & INSTALLATION 3-7
GENERATOR AND ALTERNATOR 3-9
   ALTERNATOR PRECAUTIONS 3-9
   REMOVAL & INSTALLATION 3-9
VOLTAGE REGULATOR 3-13
   REMOVAL & INSTALLATION 3-13
   VOLTAGE ADJUSTMENT 3-13
STARTER 3-13
   STARTER/SEAT BELT INTERLOCK 3-13
   REMOVAL & INSTALLATION 3-13
   SOLENOID REPLACEMENT 3-14
BATTERY 3-14
   REMOVAL & INSTALLATION 3-14
ENGINE MECHANICAL 3-16
DESCRIPTION 3-16
ENGINE 3-21
   REMOVAL & INSTALLATION 3-21
CYLINDER HEAD 3-24
   REMOVAL & INSTALLATION 3-24
   VALVE SEATS 3-27
   OVERHAUL & VALVE GUIDE REPLACEMENT 3-27
ROCKER SHAFTS 3-27
   REMOVAL & INSTALLATION 3-27
INTAKE MANIFOLD 3-28
   REMOVAL & INSTALLATION 3-28
MUFFLERS, TAILPIPES AND HEAT EXCHANGERS 3-28
   REMOVAL & INSTALLATION 3-28
PISTONS AND CYLINDERS 3-30
   REMOVAL & INSTALLATION 3-30
CRANKCASE 3-33
   DISASSEMBLY & ASSEMBLY 3-33
CAMSHAFT AND TIMING GEARS 3-33
   REMOVAL & INSTALLATION 3-33
CRANKSHAFT 3-34
   REMOVAL & INSTALLATION 3-34
CONNECTING RODS 3-35
   REMOVAL & INSTALLATION 3-35
ENGINE LUBRICATION 3-36
OIL STRAINER 3-36
   REMOVAL & INSTALLATION 3-36
OIL COOLER 3-36
   REMOVAL & INSTALLATION 3-36
OIL PUMP 3-36
   REMOVAL & INSTALLATION 3-36
OIL PRESSURE RELIEF VALVE 3-38
   REMOVAL & INSTALLATION 3-38
HYDRAULIC VALVE LIFTERS 3-38
   ADJUSTMENT 3-38
   REMOVAL & INSTALLATION 3-38
   BLEEDING 3-38
ENGINE COOLING 3-39
FAN HOUSING AND FAN 3-39
   REMOVAL & INSTALLATION 3-39
AIR FLAP AND THERMOSTAT 3-39
   ADJUSTMENT 3-39
ENGINE REBUILDING 3-40
CYLINDER HEAD RECONDITIONING 3-42
   IDENTIFY THE VALVES 3-42
   REMOVE THE VALVES AND SPRINGS 3-42
   HOT-TANK THE CYLINDER HEAD 3-42
   DEGREASE THE REMAINING CYLINDER HEAD PARTS 3-43
   DE-CARBON THE CYLINDER HEAD 3-43
   CHECK THE VALVE STEM-TO-GUIDE CLEARANCE (VALVE ROCK) 3-44
   KNURLING THE VALVE GUIDES 3-44
   REPLACING THE VALVE GUIDES 3-44
   RESURFACING (GRINDING) THE VALVE FACE 3-45
   REPLACING VALVE SEAT INSERTS 3-47
   RESURFACING THE VALVE SEATS 3-47
   CHECKING VALVE SEAT CONCENTRICITY 3-48
   LAPPING THE VALVES 3-48
   CHECK THE VALVE SPRINGS 3-48
   INSTALL THE VALVES 3-49
   INSPECT THE ROCKER SHAFTS AND ROCKER ARMS 3-49
   INSPECT THE PUSHRODS AND PUSHROD TUBES 3-49
CRANKCASE RECONDITIONING 3-51
   DISASSEMBLING CRANKCASE 3-51
   HOT TANK THE CRANKCASE 3-51
   INSPECT THE CRANKCASE 3-51
   ALIGN BORE THE CRANKCASE 3-51
   CHECK CONNECTING ROD SIDE CLEARANCE, AND FOR STRAIGHTNESS 3-51
   DISASSEMBLE THE CRANKSHAFT 3-52
   INSPECT THE CRANKSHAFT 3-52
   INSPECT THE CONNECTING RODS 3-52
   CHECK CONNECTING ROD BEARING (OIL) CLEARANCE 3-53
   CHECK MAIN BEARING (OIL) CLEARANCE 3-53
   CLEAN AND INSPECT THE CAMSHAFT 3-53
   CHECK THE CAMSHAFT BEARINGS 3-54
   CHECK THE LIFTERS (TAPPETS) 3-54
   ASSEMBLE THE CRANKSHAFT 3-54
   INSTALLING THE CRANKSHAFT AND CAMSHAFT 3-54
   CHECK TIMING GEAR BACKLASH 3-55
   ASSEMBLING THE CRANKCASE 3-55
   CHECK CRANKSHAFT END-PLAY 3-55
COMPONENT LOCATIONS
UNDER HOOD ENGINE COMPONENTS 3-18
SPECIFICATION CHARTS
STARTER SPECIFICATIONS 3-15
ALTERNATOR, GENERATOR, AND REGULATOR SPECIFICATIONS 3-16
GENERAL ENGINE SPECIFICATIONS 3-20
TORQUE SPECIFICATIONS 3-20
VALVE SPECIFICATIONS 3-21
CRANKSHAFT AND CONNECTING ROD SPECIFICATIONS 3-21
PISTON AND RING SPECIFICATIONS 3-21
HELI-COIL SPECIFICATIONS 3-41

3-2 ENGINE AND ENGINE REBUILDING

ENGINE ELECTRICAL

Understanding Electricity

For any electrical system to operate, there must be a complete circuit. This simply means that the power flow from the battery must make a full circle. When an electrical component is operating, power flows from the battery to the components, passes through the component (load) causing it to function, and returns to the battery through the ground path of the circuit. This ground may be either another wire or a metal part of the vehicle (depending upon how the component is designed).

BASIC CIRCUITS

Perhaps the easiest way to visualize a circuit is to think of connecting a light bulb (with two wires attached to it) to the battery. If one of the two wires was attached to the negative post (−) of the battery and the other wire to the positive post (+), the circuit would be complete and the light bulb would illuminate. Electricity could follow a path from the battery to the bulb and back to the battery. It's not hard to see that with longer wires on our light bulb, it could be mounted anywhere on the vehicle. Further, one wire could be fitted with a switch so that the light could be turned on and off. Various other items could be added to our primitive circuit to make the light flash, become brighter or dimmer under certain conditions, or advise the user that it's burned out.

Ground

Some automotive components are grounded through their mounting points. The electrical current runs through the chassis of the vehicle and returns to the battery through the ground (−) cable; if you look, you'll see that the battery ground cable connects between the battery and the body of the vehicle.

Load

Every complete circuit must include a "load" (something to use the electricity coming from the source). If you were to connect a wire between the two terminals of the battery (DON'T do this, but take our word for it) without the light bulb, the battery would attempt to deliver its entire power supply from one pole to another almost instantly. This is a short circuit. The electricity is taking a short cut to get to ground and is not being used by any load in the circuit. This sudden and uncontrolled electrical flow can cause great damage to other components in the circuit and can develop a tremendous amount of heat. A short in an automotive wiring harness can develop sufficient heat to melt the insulation on all the surrounding wires and reduce a multiple wire cable to one sad lump of plastic and copper. Two common causes of shorts are broken insulation (thereby exposing the wire to contact with surrounding metal surfaces or other wires) or a failed switch (the pins inside the switch come out of place and touch each other).

Switches and Relays

Some electrical components which require a large amount of current to operate also have a relay in their circuit. Since these cir-

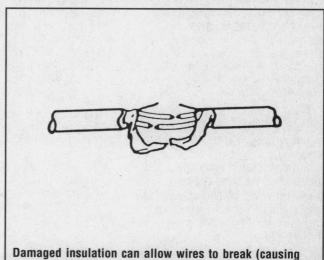

Damaged insulation can allow wires to break (causing an open circuit) or touch (causing a short circuit)

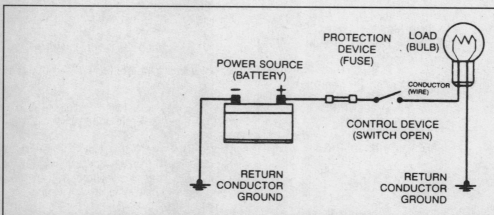

Here is an example of a simple automotive circuit. When the switch is closed, power from the positive battery terminal flows through the fuse, the switch and then the load (light bulb). The light illuminates and the circuit is completed through the return conductor and the vehicle ground. If the light did not work, the tests could be made with a voltmeter or test light at the battery, fuse, switch or bulb socket

cuits carry a large amount of current (amperage or amps), the thickness of the wire in the circuit (wire gauge) is also greater. If this large wire were connected from the load to the control switch on the dash, the switch would have to carry the high amperage load and the dash would be twice as large to accommodate wiring harnesses as thick as your wrist. To prevent these problems, a relay is used. The large wires in the circuit are connected from the battery to one side of the relay and from the opposite side of the relay to the load. The relay is normally open, preventing current from passing through the circuit. An additional, smaller wire is connected from the relay to the control switch for the circuit. When the control switch is turned on, it grounds the smaller wire to the relay and completes its circuit. The main switch inside the relay closes, sending power to the component without routing the main power through the inside of the vehicle. Some common circuits which may use relays are the horn, headlights, starter and rear window defogger systems.

Protective Devices

It is possible for larger surges of current to pass through the electrical system of your vehicle. If this surge of current were to reach the load in the circuit, it could burn it out or severely damage it. To prevent this, fuses, circuit breakers and/or fusible links are connected into the supply wires of the electrical system. These items are nothing more than a built-in weak spot in the system. It's much easier to go to a known location (the fusebox) to see why a circuit is inoperative than to dissect 15 feet of wiring under the dashboard, looking for what happened.

When an electrical current of excessive power passes through the fuse, the fuse blows (the conductor melts) and breaks the circuit, preventing the passage of current and protecting the components.

A circuit breaker is basically a self repairing fuse. It will open the circuit in the same fashion as a fuse, but when either the short is removed or the surge subsides, the circuit breaker resets itself and does not need replacement.

A fuse link (fusible link or main link) is a wire that acts as a fuse. One of these is normally connected between the starter relay and the main wiring harness under the hood. Since the starter is usually the highest electrical draw on the vehicle, an internal short during starting could direct about 130 amps into the wrong places. Consider the damage potential of introducing this current into a system whose wiring is rated at 15 amps and you'll understand the need for protection. Since this link is very early in the electrical path, it's the first place to look if nothing on the vehicle works, but the battery seems to be charged and is properly connected.

TROUBLESHOOTING

Electrical problems generally fall into one of three areas:
• The component that is not functioning is not receiving current.
• The component is receiving power but is not using it or is using it incorrectly (component failure).
• The component is improperly grounded.

The circuit can be can be checked with a test light and a jumper wire. The test light is a device that looks like a pointed screwdriver with a wire on one end and a bulb in its handle. A jumper wire is simply a piece of wire with alligator clips or spe-

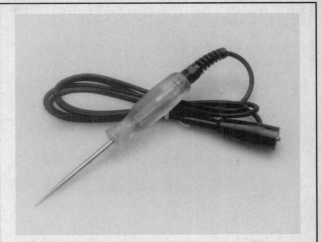
A 12 volt test light is useful when checking parts of a circuit for power

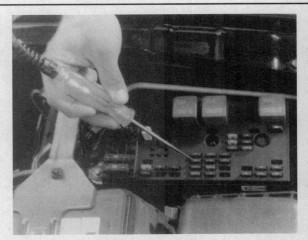
Here, someone is checking a circuit by making sure there is power to the component's fuse

cial terminals on each end. If a component is not working, you must follow a systematic plan to determine which of the three causes is the villain.

1. Turn ON the switch that controls the item not working.

➡ Some items only work when the ignition switch is turned ON.

2. Disconnect the power supply wire from the component.
3. Attach the ground wire of a test light or a voltmeter to a good metal ground.
4. Touch the end probe of the test light (or the positive lead of the voltmeter) to the power wire; if there is current in the wire, the light in the test light will come on (or the voltmeter will indicate the amount of voltage). You have now established that current is getting to the component.
5. Turn the ignition or dash switch **OFF** and reconnect the wire to the component.

If there was no power, then the problem is between the battery and the component. This includes all the switches, fuses, relays and the battery itself. The next place to look is the fusebox; check

3-4 ENGINE AND ENGINE REBUILDING

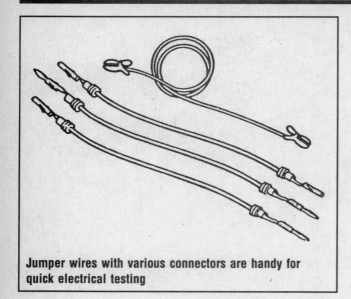

Jumper wires with various connectors are handy for quick electrical testing

carefully either by eye or by using the test light across the fuse clips. The easiest way to check is to simply replace the fuse. If the fuse is blown, and upon replacement, immediately blows again, there is a short between the fuse and the component. This is generally (not always) a sign of an internal short in the component. Disconnect the power wire at the component again and replace the fuse; if the fuse holds, the component is the problem.

※※ WARNING

DO NOT test a component by running a jumper wire from the battery UNLESS you are certain that it operates on 12 volts. Many electronic components are designed to operate with less voltage and connecting them to 12 volts could destroy them. Jumper wires are best used to bypass a portion of the circuit (such as a stretch of wire or a switch) that DOES NOT contain a resistor and is suspected to be bad.

If all the fuses are good and the component is not receiving power, find the switch for the circuit. Bypass the switch with the jumper wire. This is done by connecting one end of the jumper to the power wire coming into the switch and the other end to the wire leaving the switch. If the component comes to life, the switch has failed.

※※ WARNING

Never substitute the jumper for the component. The circuit needs the electrical load of the component. If you bypass it, you will cause a short circuit.

Checking the ground for any circuit can mean tracing wires to the body, cleaning connections or tightening mounting bolts for the component itself. If the jumper wire can be connected to the case of the component or the ground connector, you can ground the other end to a piece of clean, solid metal on the vehicle. Again, if the component starts working, you've found the problem.

A systematic search through the fuse, connectors, switches and the component itself will almost always yield an answer. Loose and/or corroded connectors, particularly in ground circuits, are becoming a larger problem in modern vehicles. The computers and on-board electronic (solid state) systems are highly sensitive to improper grounds and will change their function drastically if one occurs.

Remember that for any electrical circuit to work, ALL the connections must be clean and tight.

➡ **For more information on Understanding and Troubleshooting Electrical Systems, please refer to Section 6 of this manual.**

Battery, Starting and Charging Systems

BASIC OPERATING PRINCIPLES

Battery

The battery is the first link in the chain of mechanisms which work together to provide cranking of the automobile engine. In most modern vehicles, the battery is a lead/acid electrochemical device consisting of six 2v subsections (cells) connected in series so the unit is capable of producing approximately 12v of electrical pressure. Each subsection consists of a series of positive and negative plates held a short distance apart in a solution of sulfuric acid and water.

The two types of plates are of dissimilar metals. This sets-up a chemical reaction, and it is this reaction which produces current flow from the battery when its positive and negative terminals are connected to an electrical accessory such as a lamp or motor. The continued transfer of electrons would eventually convert the sulfuric acid to water, and make the two plates identical in chemical composition. As electrical energy is removed from the battery, its voltage output tends to drop. Thus, measuring battery voltage and battery electrolyte composition are two ways of checking the ability of the unit to supply power. During engine cranking, electrical energy is removed from the battery. However, if the charging circuit is in good condition and the operating conditions are normal, the power removed from the battery will be replaced by the alternator which will force electrons back through the battery, reversing the normal flow, and restoring the battery to its original chemical state.

Starting System

The battery and starting motor are linked by very heavy electrical cables designed to minimize resistance to the flow of current. Generally, the major power supply cable that leaves the battery goes directly to the starter, while other electrical system needs are supplied by a smaller cable. During starter operation, power flows from the battery to the starter and is grounded through the vehicle's frame/body or engine and the battery's negative ground strap.

The starter is a specially designed, direct current electric motor capable of producing a great amount of power for its size. One thing that allows the motor to produce a great deal of power is its tremendous rotating speed. It drives the engine through a tiny pinion gear (attached to the starter's armature), which drives the very large flywheel ring gear at a greatly reduced speed. Another factor allowing it to produce so much power is that only intermittent operation is required of it. Thus, little allowance for air circulation is necessary, and the windings can be built into a very small space.

The starter solenoid is a magnetic device which employs the

ENGINE AND ENGINE REBUILDING

small current supplied by the start circuit of the ignition switch. This magnetic action moves a plunger which mechanically engages the starter and closes the heavy switch connecting it to the battery. The starting switch circuit usually consists of the starting switch contained within the ignition switch, a neutral safety switch or clutch pedal switch, and the wiring necessary to connect these in series with the starter solenoid or relay.

The pinion, a small gear, is mounted to a one way drive clutch. This clutch is splined to the starter armature shaft. When the ignition switch is moved to the **START** position, the solenoid plunger slides the pinion toward the flywheel ring gear via a collar and spring. If the teeth on the pinion and flywheel match properly, the pinion will engage the flywheel immediately. If the gear teeth butt one another, the spring will be compressed and will force the gears to mesh as soon as the starter turns far enough to allow them to do so. As the solenoid plunger reaches the end of its travel, it closes the contacts that connect the battery and starter, then the engine is cranked.

As soon as the engine starts, the flywheel ring gear begins turning fast enough to drive the pinion at an extremely high rate of speed. At this point, the one-way clutch begins allowing the pinion to spin faster than the starter shaft so that the starter will not operate at excessive speed. When the ignition switch is released from the starter position, the solenoid is de-energized, and a spring pulls the gear out of mesh interrupting the current flow to the starter.

Some starters employ a separate relay, mounted away from the starter, to switch the motor and solenoid current on and off. The relay replaces the solenoid electrical switch, but does not eliminate the need for a solenoid mounted on the starter used to mechanically engage the starter drive gears. The relay is used to reduce the amount of current the starting switch must carry.

Charging System

The automobile charging system provides electrical power for operation of the vehicle's ignition system, starting system and all electrical accessories. The battery serves as an electrical surge or storage tank, storing (in chemical form) the energy originally produced by the engine driven generator. The system also provides a means of regulating output to protect the battery from being overcharged and to avoid excessive voltage to the accessories.

The storage battery is a chemical device incorporating parallel lead plates in a tank containing a sulfuric acid/water solution. Adjacent plates are slightly dissimilar, and the chemical reaction of the two dissimilar plates produces electrical energy when the battery is connected to a load such as the starter motor. The chemical reaction is reversible, so that when the generator is producing a voltage (electrical pressure) greater than that produced by the battery, electricity is forced into the battery, and the battery is returned to its fully charged state.

Newer automobiles use alternating current generators or alternators, because they are more efficient, can be rotated at higher speeds, and have fewer brush problems. In an alternator, the field usually rotates while all the current produced passes only through the stator winding. The brushes bear against continuous slip rings. This causes the current produced to periodically reverse the direction of its flow. Diodes (electrical one way valves) block the flow of current from traveling in the wrong direction. A series of diodes is wired together to permit the alternating flow of the stator to be rectified back to 12 volts DC for use by the vehicle's electrical system.

The voltage regulating function is performed by a regulator. The regulator is often built in to the alternator; this system is termed an integrated or internal regulator.

Point-Type Ignition Systems

➡ See Section 2 for electronic ignition information.

There are two basic functions the automotive ignition system must perform: (1) it must control the spark and the timing of the firing to match varying engine requirements; (2) it must increase battery voltage to a point where it will overcome the resistance offered by the spark plug gap and fire the plug.

To accomplish this, an automotive ignition system is divided into two electrical circuits. One circuit, called the primary circuit, is the low voltage circuit. This circuit operates only on battery current and is controlled by the breaker points and the ignition switch. The second circuit is the high voltage circuit, and is called the secondary circuit. This circuit consists of the secondary windings in the coil, the high tension lead between the distributor and the coil (commonly called the coil wire), the distributor cap and rotor, the spark plug leads and the spark plugs.

The coil is the heart of the ignition system. Essentially, a coil is nothing more than a transformer which takes the relatively low voltage available from the battery and increases it to a point where it will fire the spark plug. This increase is quite large, since modern coils produce on the order of about 40,000 volts. The term "coil" is perhaps a misnomer since a coil consists of *two* coils of wire wound about an iron core. These coils are insulated from each other and the whole assembly is enclosed in an oil-filled case. The primary coil is connected to the two primary terminals located on top of the coil and consists of relatively few turns of heavy wire. The secondary coil consists of many turns of fine wire and is connected to the high tension connection on top of the coil. This secondary connection is simply the tower into which the coil wire from the distributor is plugged.

Energizing the coil primary with battery voltage produces current flow through the primary windings. This in turn produces a very large, intense magnetic field. Interrupting the flow of primary current causes the field to collapse. Just as current moving through a wire produces a magnetic field, moving a field across a wire will produce a current. As the magnetic field collapses, its lines of force cross the secondary windings, inducing a current in them. The force of the induced current is concentrated because of the relative shortness of the secondary coil of wire.

The distributor is the controlling element of the system, switching the primary current on and off and distributing the current to the proper spark plug each time a spark is produced. It is basically a stationary housing surrounding a rotating shaft. The shaft is driven at one-half engine speed by the engine's camshaft through the distributor drive gears. A cam which is situated near the top of the shaft has one lobe for each cylinder of the engine. The cam operates the ignition contact points, which are mounted on a plate located on bearings within the distributor housing. A rotor is attached to the top of the distributor shaft. When the bakelite distributor cap is in place, on top of the unit's metal housing, a spring-loaded contact connects the portion of the rotor directly above the center of the shaft to the center connection on top of the distributor. The outer end of the rotor passes very close to the contacts connected to the four high-tension connections around the outside of the distributor cap.

Under normal operating conditions, power from the battery is

3-6 ENGINE AND ENGINE REBUILDING

fed through a resistor or resistance wire to the primary circuit of the coil and is then grounded through the ignition points in the distributor. During cranking, the full voltage of the battery is supplied through an auxiliary circuit routed through the solenoid switch. Current will begin flowing through the primary wiring to the positive connection on the coil, through the primary winding of the coil, through the ground wire between the negative connection on the coil and the distributor, and to ground through the contact points. Shortly after the engine is ready to fire, the current flow through the coil primary will have reached a near maximum value, and an intense magnetic field will have formed around the primary windings. The distributor cam will separate the contact points at the proper time for ignition and the primary field will collapse, causing current to flow in the secondary circuit. A capacitor, known as the "condenser," is installed in the circuit in parallel with the contact points in order to absorb some of the force of the electrical surge that occurs during collapse of the magnetic field. The condenser consists of several layers of aluminum foil separated by insulation. These layers of foil, upon an increase in voltage, are capable of storing electricity, making the condenser a sort of electrical surge tank. Voltages just after the points open may reach 250 V because of the vast amount of energy stored in the primary windings and their magnetic field. A condenser which is defective or improperly grounded will not absorb the shock from the fast-moving stream of electrons when the points open and these electrons will force their way across the point gap, causing burning and pitting.

The very high voltage induced in the secondary windings will cause a surge of current to flow from the coil tower to the center of the distributor, where it will travel along the connecting strip along the top of the rotor. The surge will arc its way across the short gap between the contact on the outer end of the rotor and the connection in the cap for the high-tension lead of the cylinder to be fired. After passing along the high-tension lead, it will travel down the center electrode of the spark plug, which is surrounded by ceramic insulation, and arc its way over to the side electrode, which is grounded through threads which hold the plug in the cylinder head. The heat generated by the passage of the spark will ignite the contents of the cylinder.

Most distributors employ both centrifugal and vacuum advance mechanisms to advance the point at which ignition occurs for optimum performance and economy. Spark generally occurs a few degrees before the piston reaches top dead center (TDC) in order that very high pressures will exist in the cylinder as soon as the piston is capable of using the energy—just a few degrees after TDC. Centrifugal advance mechanisms employ hinged flyweights working in opposition to springs to turn the top portion of the distributor shaft, including the cam and rotor, ahead of the lower shaft. This advances the point at which the cam causes the points to open. A more advanced spark is required at higher engine speeds because the speed of combustion does not increase in direct proportion to increases in engine speed, but tends to lag behind at high revolutions. If peak cylinder pressures are to exist at the same point, advance must be used to start combustion earlier.

Vacuum advance is used to accomplish the same thing when part-throttle operation reduces the speed of combustion because of less turbulence and compression, and poorer scavenging of exhaust gases. Carburetor vacuum below the throttle plate is channeled to a vacuum diaphragm mounted on the distributor. The higher the manifold vacuum, the greater the motion of the diaphragm against spring pressure. A rod between the diaphragm and the plate on which the contact points are mounted rotates the plate on its bearings causing the cam to open the points earlier in relation to the position of the crankshaft.

Ignition Coil

TESTING

Primary Resistance Check (Non-Electronic Ignition)

1. Disconnect all wires from the ignition coil.
2. Connect ohmmeter leads to terminals 15 (positive) and 1 (negative) on the top of the coil.
3. The ohmmeter should read between 1.7 and 2.1 ohms. If much higher, replace the coil.

Secondary Resistance Check (Non-Electronic Ignition)

1. Disconnect all wires from the ignition coil.
2. Connect ohmmeter leads to terminals 1 (negative) and 4 (large, center plug-in terminal).

After disconnecting the negative battery cable, label all of the ignition coil wires . . .

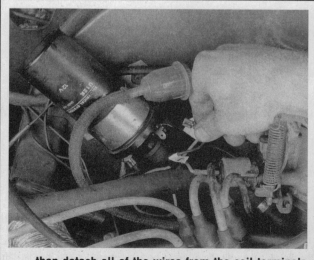

. . . then detach all of the wires from the coil terminals

ENGINE AND ENGINE REBUILDING 3-7

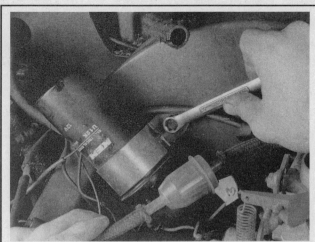

Using a 10mm wrench, remove the ignition coil bracket attaching bolts . . .

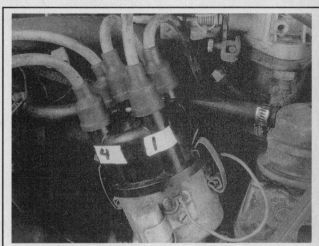

To remove the distributor, label the distributor cap terminals . . .

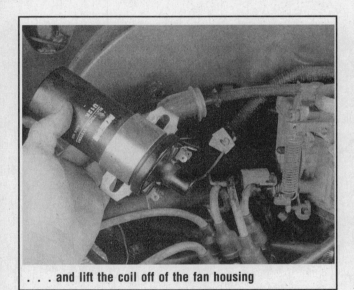

. . . and lift the coil off of the fan housing

. . . and detach the spark plug wires from the cap

3. The ohmmeter should read between 7,000 and 12,000 ohms. If much higher, replace the coil.

Electronic Ignition Coil

Refer to Section 2 under Electronic Ignition for more information on electronic ignition coils.

Distributor

REMOVAL & INSTALLATION

1. Detach the vacuum hoses at the distributor.
2. Disconnect the coil wire and remove the distributor cap.
3. Disconnect the condenser wire.
4. Bring No. 1 cylinder to top dead center (TDC) on the compression stroke by rotating the engine so that the rotor points to the No. 1 spark plug wire tower on the distributor cap and the timing marks are aligned at 0°. Mark the rotor-to-distributor relation-

Unfasten the retaining clamps, then lift the distributor cap off of the distributor housing

3-8 ENGINE AND ENGINE REBUILDING

Position the engine at Top Dead Center (TDC) by turning it until the timing marks align and the distributor rotor points toward the No. 1 terminal position on the cap

Matchmark the distributor housing to the crankcase for proper positioning during installation

To remove the distributor rotor, simply pull it up and off of the distributor shaft . . .

Loosen the distributor retaining clamp nut . . .

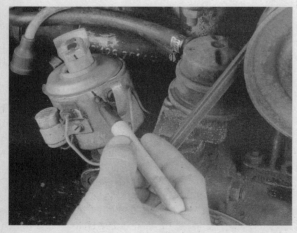
. . . however, to remove the distributor, leave the rotor installed & matchmark the rotor tip to the housing

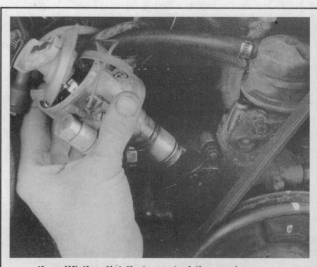
. . . then lift the distributor out of the engine

ENGINE AND ENGINE REBUILDING

ship. Also matchmark the distributor housing-to-crankcase relationship.

5. Unscrew the distributor retaining screw on the crankcase and lift the distributor out.

To install:

6. If the engine has been rotated since the distributor was removed, bring the No. 1 cylinder to TDC on the compression stroke and align the timing marks on 0°. Align the match-marks and insert the distributor into the crankcase. If the matchmarks are gone, have the rotor pointing to the No. 1 spark plug wire tower upon insertion.

7. Replace the distributor retaining screw and reconnect the condenser and coil wires. Reinstall the distributor cap.

8. Retime the engine.

Distributor Driveshaft

1. On carbureted engines, remove the fuel pump.
2. Bring the engine to TDC on the compression stroke of No. 1 cylinder. Align the timing marks at 0°.
3. Remove the distributor.
4. Remove the spacer spring from the driveshaft.
5. Grasp the shaft and turn it slowly to the left while withdrawing it from its bore.
6. Remove the washer found under the shaft.

※※ CAUTION

Make sure that this washer does not fall down into the engine.

To install:

7. Make sure that the engine is at TDC on the compression stroke for No. 1 cylinder with the timing marks aligned at 0°.
8. Replace the washer and insert the shaft into its bore.

➡**Due to the slant of the teeth on the drive gears, the shaft must be rotated slightly to the left when it is inserted into the crankcase.**

9. When the shaft is properly inserted, the offset slot in the drive shaft of Type 1 and 2/1600 engines will be perpendicular to the crankcase joint and the slot offset will be facing the crankshaft pulley. On Type 3, the slot will form a 60° angle with the crankcase joint and the slot offset will be facing the oil cooler. On Type 4 engines, and Type 2/1700, 2/1800 and 2/2000 engines, the slot should be about 12° out of parallel with the center line of the engine and the slot offset should be facing outside the engine.
10. Reinstall the spacer spring.
11. Reinstall the distributor and fuel pump, if removed.
12. Retime the engine.

Generator and Alternator

ALTERNATOR PRECAUTIONS

1. Battery polarity should be checked before any connections, such as jumper cables or battery charger leads, are made. Reversing the battery connections will damage the diodes in the alternator. It is recommended that the battery cables be disconnected before connecting a battery charger.

2. The battery must never be disconnected while the alternator is running.
3. Always disconnect the battery ground lead before working on the charging system, especially when replacing an alternator.
4. Do not short across or ground any alternator or regulator terminals.
5. If electric arc welding has to be done to the car, first disconnect the battery and alternator cables. Never start the car with the welding unit attached.

REMOVAL & INSTALLATION

Types 1 and 2/1600
▶ **See Figure 1**

The generator (alternator) can be removed without removing the engine on these models by loosening the fan housing and lifting it up enough to remove the four generator (alternator) to fan housing bolts.

➡**Although it is possible to remove the generator/alternator with the engine installed in the vehicle, this procedure is much easier with the engine already removed from the car. Access to several of the fan shroud fasteners can be extremely tight.**

1. Disconnect the battery.
2. Disconnect the leads from the generator (alternator) and mark them for reassembly.
3. Remove the air cleaner housing and the carburetor or air flow sensor (fuel injected models). Refer to Section 5 for more information.
4. Slide the accelerator cable out through the fan housing and remove the cable's guide tube.
5. Separate the generator (alternator) pulley halves, noting the number and position of the pulley shims, and remove the belt from the pulley.
6. Remove the retaining strap from the generator (alternator).
7. Remove the cooling air thermostat. See the end of this section for details.
8. Remove the hot air hoses from the fan housing.
9. Remove any wires and hoses which may hinder lifting the fan housing up.
10. On 1971 Type 2s and on all Type 1s except the 1970 model, use a 10mm wrench to remove the bolt that holds the oil cooler cover to the front (flywheel side) of the engine and the bolt that holds the fan housing to the oil cooler flange. Remove the oil cooler cover and flange.
11. Remove the screws at both sides of the fan housing and on 1971 Type 2s and all Type 1s except 1970 models, pry off the clip and disconnect the fan housing flap linkage from the left control flap.
12. Lift up the fan housing enough to remove the four 10mm head bolts that hold the generator (alternator) to the fan housing and remove the generator (alternator) from the vehicle.
13. Remove the fan from the generator (alternator) by unscrewing the special nut and pulling the fan off the keyed generator (alternator) shaft. Note the position of any shims found on Type 2 generators, from chassis number 219000001, as these shims are used to maintain a gap of 0.047 in. between the fan and the fan cover. The Type 1 gap is 0.08 in.

3-10 ENGINE AND ENGINE REBUILDING

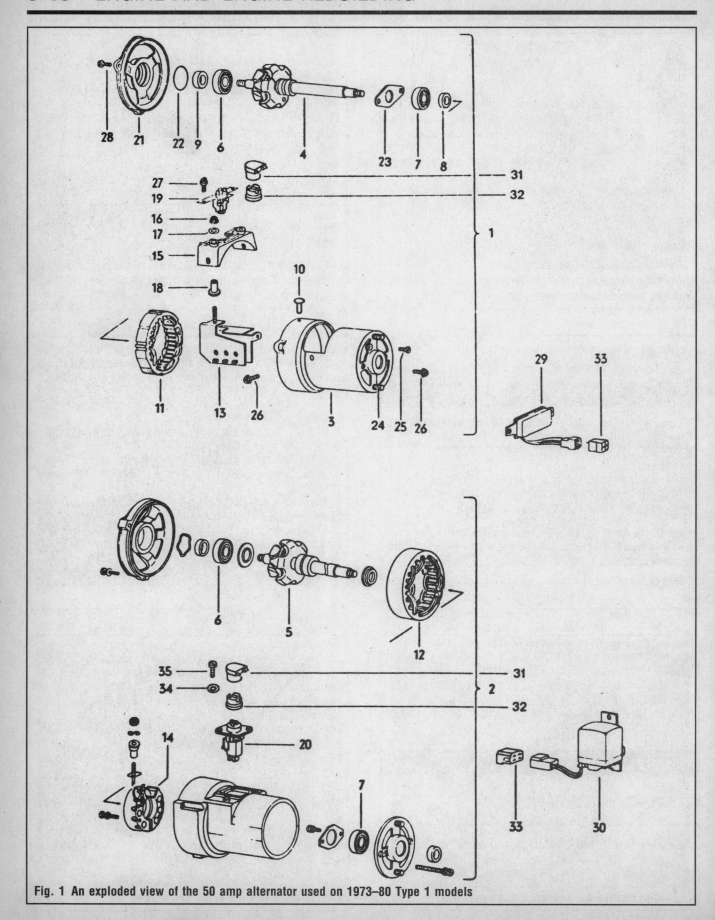

Fig. 1 An exploded view of the 50 amp alternator used on 1973–80 Type 1 models

ENGINE AND ENGINE REBUILDING   3-11

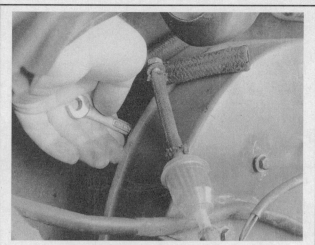

To remove the generator while the engine is in the vehicle, first remove all fan housing mounting bolts

Loosen the 4 generator-to-fan housing bolts with a 10mm wrench . . .

Remove the generator mounting strap adjusting bolt . . .

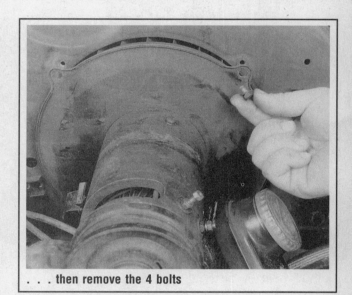

. . . then remove the 4 bolts

. . . then remove the generator mounting strap

Lift the generator out of the fan housing

3-12 ENGINE AND ENGINE REBUILDING

At this point, the generator stand may be removed from the engine— remove the 4 mounting nuts . . .

. . . then lift the generator stand off of the crankcase

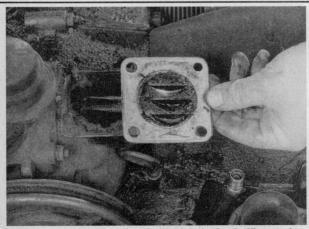

While the generator stand is removed, the baffle can be removed and cleaned—make certain to install it in its original position

14. Reverse the above steps to install. When installing the generator (alternator) the cooling air intake slot in the fan cover must face downward and the generator (alternator) pulley must align with the crankshaft pulley.

Type 2/1800 and 2/2000 (Fuel Injected)

1. Disconnect the negative battery cable.
2. Disconnect the alternator wiring harness at the voltage regulator and starter.
3. Pull out the dipstick and remove the oil filler neck. Disconnect the heater blower from the front of the alternator on 1980–81 Type 2.
4. Loosen the alternator adjusting bolt and remove the drive belt.
5. Remove the right rear engine cover plate and the alternator cover plate.
6. Disconnect the warm air duct at the right side, and remove the heat exchanger bracket and connecting pipe from the blower.
7. Disconnect the cool air intake elbow at the alternator. Remove the attaching bolt and lift out the alternator from above.
8. Reverse the above procedure to install, taking care to ensure that the rubber grommet on the intake cover for the wiring harness is installed correctly. After installation, adjust the drive belt so that the moderate thumb pressure midway on the belt depresses the belt about ½ in.

Type 3

▶ See Figure 2

1. Remove the cooling air intake cover and disconnect the battery.
2. Loosen the fan belt adjustment and remove the fan belt. Removal of the belt is accomplished by removing the nut in the center of the generator pulley and removing the outer pulley half.
3. Remove the two nuts which hold the generator securing strap in place and then remove the strap.
4. Disconnect the generator wiring.
5. Remove the generator.
6. Installation is the reverse of the above. Install the generator so that the mark on the generator housing is in line with the

Fig. 2 When installing the generator on Type 3 models, make sure that the mark on the generator housing is aligned with the notch on the clamping strap

ENGINE AND ENGINE REBUILDING 3-13

notch on the clamping strap. The generator pulley must be aligned with the crankshaft pulley. Make sure that the boot which seals the generator to the air intake housing is properly placed.

Types 4, 2/1700, 2/1800 and 2/2000

The factory procedure recommends removing the engine to remove the alternator. However, it is possible to reach the alternator by first removing the right heater box which will provide access to the alternator.

1. Disconnect the battery.
2. The following is the alternator removal and installation procedure after removing the engine; however, all bolts and connections listed below must be removed, except the engine cooling fan, if the right heater box is removed to gain access to the alternator.
3. Remove the engine.
4. Remove the dipstick, if necessary and the rear engine cover plate.
5. Remove the fan belt.
6. Remove the lower alternator bolt and the alternator cover plate.
7. Disconnect the wiring harness from the alternator.
8. Remove the allen head screws which attach the engine cooling fan, then remove the fan.
9. Remove the rubber elbow from the fan housing.

➡ **This elbow must be in position upon installation because it provides cooling air for the alternator.**

10. Remove the alternator adjusting bracket.

Voltage Regulator

Many 1974 and all later Type 1 VWs with alternators are equipped with integral circuit regulators mounted on the alternators. On all other models, the regulator is mounted separately. Some early model regulators do not have a ground wire; all replacement regulators do. When installing a replacement regulator on an early model vehicle without a ground wire, be sure to attach the ground wire that comes with the replacement regulator.

REMOVAL & INSTALLATION

✳✳ CAUTION

Interchanging the connections on the regulator will destroy the regulator and generator.

Types 1 and 3

The regulator is located under the rear seat on the left side. It is secured to the frame by two screws. Take careful note of the wiring connections before removing the wiring from the regulator. Disconnect the battery before removing the regulator.

Type 2 and Model 14 (Karmann Ghia)

Disconnect the battery. The regulator is located in the engine compartment and is secured in place by two screws. Take careful note of the wiring connections before removing the wiring from the regulator.

Type 4

Turn the engine **OFF**. Disconnect the battery. Make careful note of the wiring connections. The regulator is located near the air cleaner and is mounted either on the air cleaner or on the firewall. It is secured by two screws.

VOLTAGE ADJUSTMENT

Volkswagen voltage regulators are sealed and cannot be adjusted. A malfunctioning regulator must be replaced as a unit.

Starter

The starter motor of the Volkswagen is of the sliding gear-type and is rated at about 0.6, 0.7, or 0.8 horsepower. The motor used in the starter is a series wound-type and draws a heavy current in order to provide the high torque needed to crank the engine during starting. The starter cannot be switched on accidentally while the engine is still running—the device responsible for this safeguard is a nonrepeat switch in the ignition switch. If the engine should stall for any reason, the ignition key must be turned to the "off" position before it is possible to re-start the engine.

The starter is flange-mounted on the right-hand side of the transmission housing. Attached to the starter motor housing is a solenoid which engages the pinion and connects the starter motor to the battery when the ignition key is turned **ON**. When the engine starts, and the key is released from the start position, the solenoid circuit is opened and the pinion is returned to its original position by the return spring. However, if for any reason the starter is not switched OFF immediately after the engine starts, a pinion free-wheeling device stops the armature from being driven so that the starter will not be damaged.

STARTER/SEAT BELT INTERLOCK

All 1974 and some 1975 models are equipped with a starter/seat belt interlock system. This system prevents operation of the starter motor until both front seat occupants buckle up their seat belts. For details, refer to Section 6.

REMOVAL & INSTALLATION

1. Disconnect the battery.
2. Disconnect the wiring from the starter. It will probably be easier if the right-hand rear wheel is removed.

➡ **On fuel injection models, take special note of the terminals from which the various wires are detached (terminal 30, 50, etc.). If the cold start valve wire, which should be connected to terminal 50, is connected to terminal 30 by mistake, the cold start valve will run constantly, causing poor gas mileage, rough idle and flooding.**

3. The starter is held in place by two bolts. Have a helper hold the nut on the top bolt with a wrench in the engine compartment

3-14 ENGINE AND ENGINE REBUILDING

Disconnect the negative battery cable, then use a wrench or socket to remove the starter motor wire retaining nuts

while you remove the bolt from underneath the vehicle. This top bolt is also one of the four main engine to transmission bolts. Remove the lower bolt.

4. Remove the starter from the car.

To install:

5. Before installing the starter, lubricate the outboard bushing with grease. Apply sealing compound to the mating surfaces between the starter and the transmission.

6. Place the long starter bolt in its hole in the starter and locate the starter on the transmission housing. Install the other bolt.

7. Connect the starter wiring and battery cables.

SOLENOID REPLACEMENT

♦ See Figure 3

1. Remove the starter.
2. Remove the nut which secures the connector strip at the end of the solenoid.
3. Take out the two retaining screws on the mounting bracket and withdraw the solenoid after it has been unhooked from its actuating lever.
4. When replacing a defective solenoid with a new one, care should be taken to see that the distance (a) in the accompanying diagram is 19 mm when the magnet is drawn inside the solenoid.
5. Installation is the reverse of removal. In order to facilitate engagement of the actuating rod, the pinion should be pulled out as far as possible when inserting the solenoid.

Battery

The electrical system of the Volkswagen is a negative grounded type. All models except the 1975 and later Type 2 use a 45 amp battery. The 1975 and later Type 2 uses a 54 amp battery. On Type 1 and Type 3 VW models, the battery is located under the right-hand side of the rear seat. On Karmann Ghias and Type 2 models through 1979, the battery is located in the engine compartment. On 1980–81 Type 2s (Vanagon), it is located under the passenger's seat. On Type 4 models, the battery is beneath the driver's seat. On Type 2 Campmobiles equipped with a refrigerator, an additional 45 amp battery is available.

REMOVAL & INSTALLATION

1. Disconnect the battery cables. Note the position of the battery cables for installation. The small diameter battery post is the negative terminal. The negative battery cable is usually black.
2. Undo the battery hold-down strap and lift the battery out of its holder.

✳✳ CAUTION

Do not tilt the battery as acid will spill out.

3. Install the battery in its holder and replace the clamp. Reconnect the battery cables.

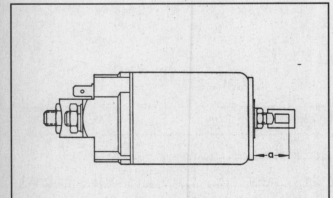

Fig. 3 Before installing a replacement solenoid, make certain that the distance shown (a) is 19mm when the magnet is drawn inside the solenoid

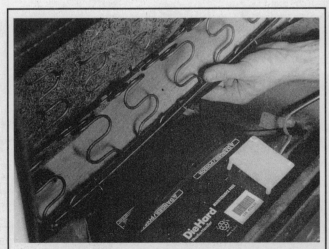

The battery is located under the rear bench seat—the rear seat lifts up and can be removed from the car

ENGINE AND ENGINE REBUILDING 3-15

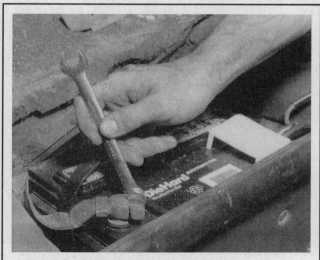
Loosen the negative battery cable terminal nut . . .

The battery must have a shield for the positive terminal, as the rear seat could catch on fire if the terminals contact the metal seat frame

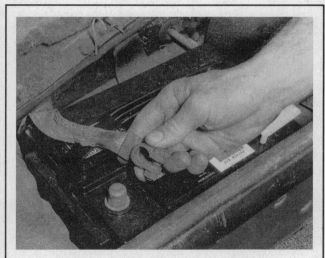
. . . and lift the cable off of the battery terminal

If the battery is to be replaced, remove the hold-down nut and lift the hold-down up and off its mounting stud

Starter Specifications

Starter Number	Lock Test		No-Load Test			Brush Spring Tension (oz)
	Amps	Volts	Amps	Volts	rpm	
111 911 023A	270–290	6	25–40	12	6700–7800	42
311 911 023B	250–300	6	35–45	12	7400–8100	42
003 911 023A	250–300	6	35–50	12	6400–7900	42

3-16 ENGINE AND ENGINE REBUILDING

Alternator, Generator, and Regulator Specifications

Year	Type	Generator Maximum Output (Amps)	Alternator Maximum Output (Amps)	Alternator Stator Winding Resistance (Ohms)	Alternator Exciter Winding Resistance (Ohms)	Regulator Load Current (Amps)	Regulator Regulating Voltage Under Load (Volts)
1970–73	1	30	—	—	—	25 ①	12.5–14.5
1973–80	1	—	50	0.13 ± 0.013	4.0 ± 0.4	25–30	13.8–14.9 ②
1970–71	2/1600	38	—	—	—	25 ①	12.5–14.5
1972–73	2/1700	—	55	0.13 ± 0.013	4.0 ± 0.4	25–30	13.8–14.9 ②
1974–81	2/1800, 2/2000	—	55	0.13 ± 0.013	4.0 ± 0.4	25–30	13.8–14.9 ②
1970–73	3	30	—	—	—	25 ①	12.5–14.5
1971–74	4	—	55	0.13 ± 0.013	4.0 ± 0.4	25–30	13.8–14.9 ②

① @ 2000–2500 generator rpm
② @ 2000 engine rpm
— Not Applicable

ENGINE MECHANICAL

Description

♦ **See Figure 4**

The Volkswagen engine is a flat four cylinder design. This four cycle, overhead valve engine has two pairs of horizontally opposed cylinders. All rear engined VW models are air cooled.

The Type 1 and 2/1600 engine is known as an upright fan engine, that is, the engine cooling fan is mounted vertically on top of the engine and is driven by the generator shaft. The Type 2/1700, Type 2/1800, Type 2/2000, Type 3 and 4 engine, although of the same basic design, i.e. flat four, has the cooling fan driven by the crankshaft and is therefore mounted on the front of the engine. This type of engine is known as the suitcase engine.

Because it is air cooled, the VW engine is slightly noisier than a water cooled engine. This is due to the lack of water jacketing around the cylinders which provides sound deadening on water cooled engines. In addition, air cooled engines tend to run at somewhat higher temperatures, necessitating larger operating clearances to allow more room for the expansion of the parts. These larger operating clearances cause an increase in noise level over a water cooled engine.

The crankshaft of all Volkswagen engines is mounted in a two piece crankcase. The halves are machined to very close tolerances and line bored as a pair and, therefore, should always be replaced in pairs. When fitting them, it is necessary to coat only the mating surfaces with sealing compound and tighten them down to the correct torque. No gasket is used.

The pistons and cylinders are identical on any particular engine. However, it is not possible to interchange pistons and cylinders between engines. The four pistons each have three rings, two compression rings and one oil scraper. Each piston is attached to its connecting rod with a fully floating piston pin.

Each pair of cylinders shares a detachable cylinder head made of light aluminum alloy casting. The cylinder head contains the valves for both cylinders. Shrunk-in valve guides and valve seats are used.

➡ **Complete engine rebuilding procedures are given in the second half of this section.**

ENGINE AND ENGINE REBUILDING 3-17

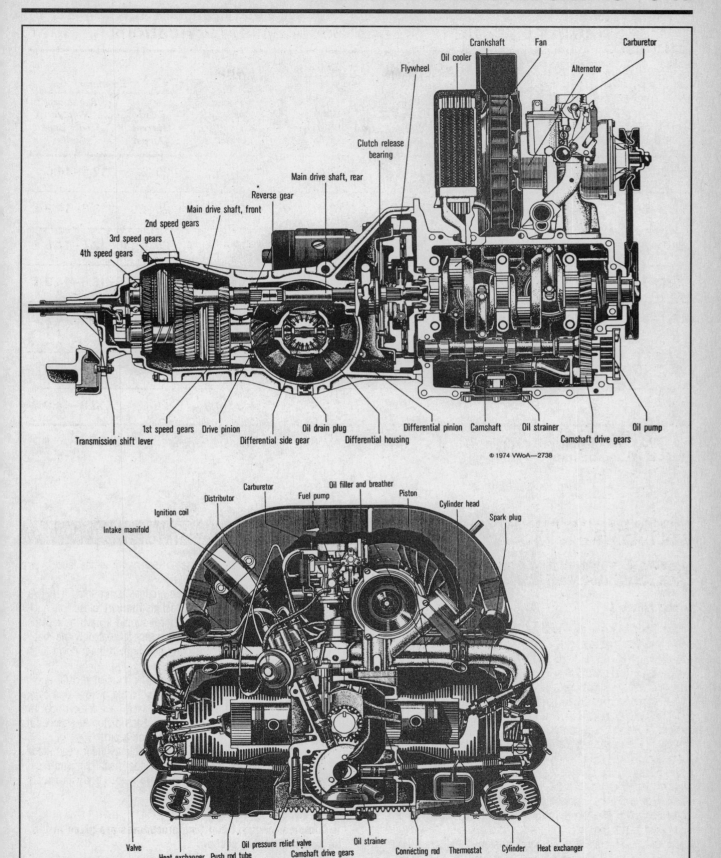

Fig. 4 Longitudinal and frontal cutaway views of a 1600cc engine equipped with a manual transaxle

3-18 ENGINE AND ENGINE REBUILDING

ENGINE COMPONENTS - UPRIGHT 1600cc ENGINE SHOWN

1. Automatic Transaxle Control Valve (ATCV)
2. ATCV vacuum hose (connects to intake air distributor)
3. Heater air feed hose (left side)
4. Exhaust crossover pipe
5. Ignition coil
6. Distributor
7. Oil pressure sender
8. Crankshaft pulley
9. Fuel pump
10. Generator/Alternator stand
11. Intake air distributor
12. Generator/Alternator
13. Oil bath air cleaner assembly
14. Preheated air hose
15. Heater air feed hose (right side)
16. Intake manifold
17. Spark plugs
18. Fan housing shroud
19. Cooling fan (mounted on the end of the generator/alternator shaft inside the fan housing shroud)

ENGINE AND ENGINE REBUILDING

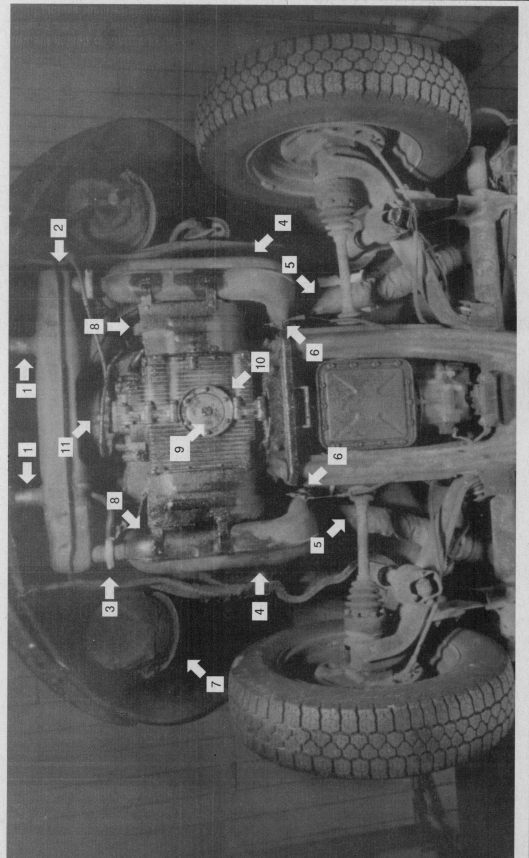

ENGINE COMPONENTS - UPRIGHT 1600cc ENGINE SHOWN

1. Tailpipes
2. Muffler
3. Muffler-to-exhaust manifold connection
4. Exhaust manifold/heater box assembly
5. Heater hoses
6. Heater box control flap cable and connection
7. Vacuum reservoir
8. Under engine air shrouds
9. Engine oil drain plug
10. Engine oil strainer plate
11. Oil and ATF (on automatic models) pump

3-20 ENGINE AND ENGINE REBUILDING

General Engine Specifications

Year	Engine Code	Displacement (cc)	Horsepower @rpm	Torque @rpm (ft lbs)	Bore x Stroke (in.)	Compression Ratio	Oil Pressure @rpm (psi)
1970	B	1584	57/4400	82/3000	3.37 x 2.72	7.5:1 ②	42
1971-72	AE	1584	46/4000	72/2000	3.37 x 2.72	7.3:1	42
1971-74	AK	1584	46/40000	72/2000	3.37 x 2.72	7.3:1	42
1972-74	AH ①, AM	1584	46/4000	72/2000	3.37 x 2.72	7.5:1	42
1972-73	CB	1679	63/4800	81/3200	3.54 x 2.60	7.3:1	42
1973	CD	1679	59/4200	82/3200	3.54 x 2.60	7.3:1	42
1970-73	U	1584	65/4600	87/2800	3.37 x 2.72	7.7:1	42
1972-73	X	1584	52/4000	77/2200	3.37 x 2.72	7.3:1	42
1971	W	1679	85/5000	99.5/3500	3.54 x 2.60	8.2:1	42
1972-74	EA	1679	76/4900	95/2700	3.54 x 2.60	8.2:1	42
1973	EB ①	1679	69/5000	87/2700	3.54 x 2.60	7.3:1	42
1974	EC	1795	72/4800	91/3400	3.66 x 2.60	7.3:1	42
1974	AW	1795	65/4200	92/3000	3.66 x 2.60	7.3:1	42
1975-80	AJ	1584	48/4200	73.1/2800	3.37 x 2.72	7.3:1	42
1975	ED	1795	67/4400	90/2400	3.66 x 2.60	7.3:1	42
1976-81	GD, GE, CV	1970	67/4200	101/3000	3.70 x 2.80	7.3:1	42

① California only
② Type 2—7.7:1

Torque Specifications
(All readings in ft. lbs.)

Vehicle Type		Cylinder Head Nuts	Rod Bearing Bolts	Generator Pulley Nut	Crankshaft Pulley Bolt	Flywheel Bolts	Fan-to-Hub Bolts	Hub-to-Crankshaft Bolts	Crankcase Half Nuts		Drive Plate-to-Crankshaft Bolts	Spark Plugs	Oil Strainer Cover Bolts
									Sealing Nuts	Non-sealing Nuts			
Type 1		23	22-25	40-47	29-36	253 ③	—	—	18	14	—	25	5
Type 2	①	23	22-25	40-47	29-36	253 ③	—	—	18	14	—	25	5
	②	23	24	—	—	80	14	23	23	14	61	22	7-9
Type 3		23	22-25	40-47	94-108	253 ③	—	—	18	14	—	25	5
Type 4		23	24	—	—	80	14	23	23	14	61	22	7-9

① Type 2 models equipped with 1600cc engines.
② Type 2 models equipped with 1700cc, 1800cc or 2000cc engines.
③ These engines are equipped with a single, large gland nut, rather than flywheel bolts.

ENGINE AND ENGINE REBUILDING

Valve Specifications

Engine	Seat Angle (deg.) Intake	Seat Angle (deg.) Exhaust	Face Angle (deg.) Intake	Face Angle (deg.) Exhaust	Valve Seat Width (in.) Intake	Valve Seat Width (in.) Exhaust	Spring Test Pressure (lbs. @ in.)	Valve Guide Inside Diameter (in.) Intake	Valve Guide Inside Diameter (in.) Exhaust	Stem-to-Guide Clearance (in.) Intake	Stem-to-Guide Clearance (in.) Exhaust	Stem Diameter (in.) Intake	Stem Diameter (in.) Exhaust
1600	45	45	44	45	0.05-0.10	0.05-0.10	117.7-134.8 @ 1.22	0.3150-0.3157	①	0.009-0.010	0.009-0.010	0.3125-0.3129	②
1700	30	45	30	45	0.07-0.08	0.078-0.098	168-186 @ 1.14	0.3150-0.3157	0.3534-0.3538	0.018	0.014	0.3125-0.3129	0.3507-0.3511
1800	30	45	30	45	0.07-0.08	0.078-0.098	168-186 @ 1.14	0.3150-0.3157	0.3534-0.3538	0.018	0.014	0.3125-0.3129	0.3507-0.3511
2000	30	45	30	45	0.07-0.08	0.078-0.098	168-186 @ 1.14	0.3150-0.3157	0.3534-0.3538	0.018	0.014	0.3125-0.3129	0.3507-0.3511

① 1975 Type 1 models: 0.353-0.354 in.
All other models: 0.3150-0.3157 in.

② 1975 Type 1 models: 0.350-0.351 in.
All other models: 0.3113-0.3117 in.

Crankshaft and Connecting Rod Specifications
(All measurements are given in inches)

Year	Type Engine	Crankshaft Main Bearing Journal Dia No. 1, 2, 3	Crankshaft Main Bearing Journal Dia No. 4	Main Bearing Oil Clearance No. 1, 3	Main Bearing Oil Clearance No. 2	Main Bearing Oil Clearance No. 4	Crankshaft End-Play	Thrust on No.	Connecting Rods Journal Dia	Connecting Rods Oil Clearance	End-Play
1970-80	1, 2, 3 1600	2.1640-2.1648	1.5379-1.5748	0.0016-0.0004	0.001-0.0003	0.002-0.0004	0.0027-0.0005	1 at flywheel	2.1644-2.1653	0.0008-0.0027	0.004-0.016
1971-81	2, 4 1700, 1800, 2000	2.3609-2.3617	1.5739-1.5748	0.002-0.004	0.0012-0.0035	0.002-0.004	0.0027-0.005	1 at flywheel	2.1644-2.1653 ①	0.0008-0.0027	0.004-0.016

① On 1976 Type 2/2000 models, connecting rod journal diameter is 1.968 in. (50 mm)

Piston and Ring Specifications
(All measurements in inches)

Year	Type Engine Displacement	Piston Clearance	Ring Gap Top Compression	Ring Gap Bottom Compression	Ring Gap Oil Control	Ring Side Clearance Top Compression	Ring Side Clearance Bottom Compression	Ring Side Clearance Oil Control
1970-80	1, 2, 3 1600	0.0016-0.0023	0.012-0.018	0.012-0.018	0.010-0.016	0.0027-0.0039	0.002-0.0027	0.0011-0.0019
1971-81	2, 4 1700, 1800, 2000	0.0016-0.0023	0.014-0.021	0.012-0.022	0.010-0.016	0.0023-0.0035	0.0016-0.0027	0.0008-0.0019

Engine

REMOVAL & INSTALLATION

Types 1, 2, and 3

The Volkswagen engine is mounted on the transaxle, which in turn is attached to the frame. In the Type 1 and 2 models, there are two bolts and two studs attaching the engine to the transaxle. Type 3 engines have an extra mounting at the rear of the engine. Type 3 engines with automatic transaxle have front and rear engine and transaxle mounts. At the front, the gearbox is supported by the rear tubular crossmember; at the rear, a crossmember is bolted to the crankcase and mounted to the body at either end.

When removing the engine from the car, it is recommended that the rear of the car be about 4 ft. off the ground. Remove the engine by bringing it out from underneath the car. Proceed with the following steps to remove the engine.

3-22 ENGINE AND ENGINE REBUILDING

➡An easy way to prevent hooking up your automatic choke wire to the distributor and vice versa when installing the engine is to take the time to mark each wire and the terminal to which it is connected with masking tape, then coding each two pieces of tape 1, 2, 3, etc. When hooking up the wiring, simply match the pieces of tape. Or you can draw a diagram indicating the wire colors and the components to which they attach. Do the same with all vacuum hoses and you should have no problems installing the engine.

1. Disconnect the battery ground cable.
2. Disconnect the generator wiring.
3. Remove the air cleaner. On Type 1 and Type 2 engines, remove the rear engine cover plate. On Type 2/1600 cc engines, remove the rear crossmember.
4. Disconnect the throttle cable and remove the electrical connections to the automatic choke, coil, electromagnetic cut-off-jet, the oil pressure sending unit, the backup light wiring at the right of the engine (Type 1), and all other interfering wiring.

➡Removing the throttle cable on the upright fan engine includes:

   a. Disconnecting the throttle cable from the carburetor;
   b. Removing the spring retainer, the spring cover and the spring;
   c. Pulling the cable out from the transaxle side of the engine and;
   d. Removing the cable guide.

5. Disconnect the fuel hose at the front engine cover plate and seal it to prevent leakage.
6. On Type 3 models, remove the oil dipstick and the rubber boot between the oil filter and the body.
7. Remove the cooling air intake bellows on Type 3 engines after loosening the clip that secures the unit.
8. On Type 3 models, remove the warm air hose.
9. On Type 3 fuel injected engines, remove and plug the pressure line to the left fuel distributor pipe and to the return line on the pressure regulator. Disconnect the fuel injection wiring harness.
10. Raise the car and support it with jackstands.
11. Remove the flexible air hoses between the engine and heat exchangers, disconnect the heater flap cables, remove the electrical heater fan hoses, if equipped, unscrew the two lower engine mounting nuts, and slide a jack under the engine. On Type 2 engines, remove the two bolts from the rubber engine mounts located next to the muffler.
12. On Type 1 Automatic Stick Shift models, disconnect the control valve cable and the manifold vacuum hoses. Disconnect the ATF suction line and plug it. On Type 3 fully automatic transaxle models, disconnect the vacuum hose and the kick-down cable.
13. On all Automatic Stick Shift and fully automatic models, remove the four bolts from the converter drive plate through the holes in the transaxle case. After the engine is removed, hold the torque converter on the transaxle input shaft by using a strap bolted to the bellhousing.
14. Raise the jack until it just contacts the engine and have an assistant hold the two upper mounting bolts so that the nuts can be removed from the bottom. 1971 Type 2s have only an upper right-hand mount bolt.

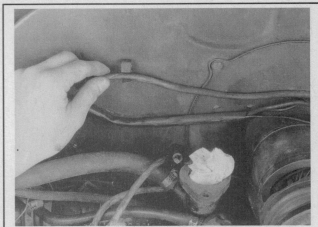

When preparing to drop the engine out of the vehicle, make certain to detach all wires from the engine assembly . . .

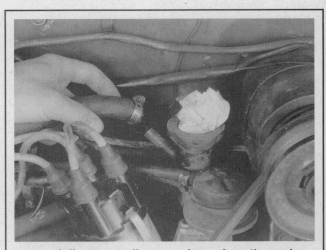

. . . and disconnect all vacuum hoses from the engine assembly as well

Detach any breather hoses from the fan shroud

ENGINE AND ENGINE REBUILDING   3-23

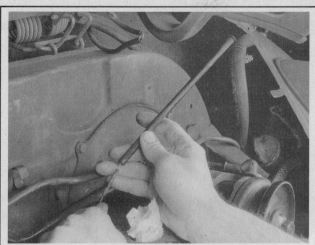
Remove the accelerator cable tube from the fan shroud and pull the accelerator cable through the shroud

With the engine removed from the vehicle, access to the various engine components is much easier

On automatic models, make certain to retain the torque converter in the transaxle using either wire ties (as shown) or bolt-on brackets

➡ 1972 and later Type 2s have two top engine-to-transaxle bolts which also serve as top carrier bolts for the transaxle. The transaxle must be supported when these bolts are removed to prevent damage which could be caused by letting the unit hang. Many later model Type 1 and Type 2 vehicles have a captive left side top engine-to-transaxle nut which is reached from the transaxle side of the unit.

15. When the engine mounts are disconnected and there are no remaining cables or wires left to be disconnected, move the engine toward the back of the car so that the clutch or converter plate disengages from the transaxle.

16. Lower the engine out of the car.

To install:

17. Installation is the reverse of the above. When the engine is lifted into position, the crankshaft should be rotated using the generator pulley so that the clutch plate hub will engage the transaxle shaft splines. Tighten the upper mounting bolts first. Check the clutch, pressure plate, throwout bearing, and pilot bearing for wear.

On Type 3, synthetic washers are used to raise the engine about 3mm when the rear engine mounting is attached and tightened. Use only enough washers in the rear mount so that the engine is lifted no more than 3mm. Care should be used when installing the rear intake housing bellows of the Type 3 engine.

Type 4

♦ See Figures 5, 6 and 7

1. Disconnect the battery.
2. Remove the cooling air bellows, warm air hoses, cooling air intake duct, and air cleaner. On sedans, remove the cooling air fan. On station wagons, remove the dipstick tube rubber boot and the dipstick.
3. Disconnect the fuel injection wiring.
4. Disconnect the coil wires and remove the coil and its bracket.
5. Disconnect the oil pressure switch and the alternator wiring.
6. Disconnect the vacuum hose for the intake air distributor.
7. Disconnect the accelerator cable.

Fig. 5 Remove the 3 torque converter-to-drive plate screws through the access hole in the upper right corner of the drive plate housing—automatic transaxle models

3-24 ENGINE AND ENGINE REBUILDING

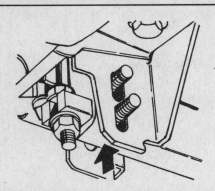

Fig. 6 When installing the engine, position the engine so that the mounting studs are located in the elongated holes, then raise the engine until the studs are at the top of the holes and install the nuts—to adjust the engine, use the threaded rod to the left

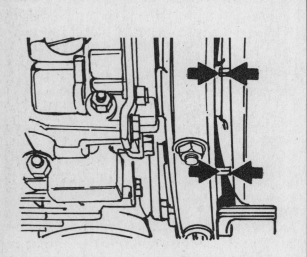

Fig. 7 After installing the engine, make sure that the engine carrier is vertical and parallel to the engine fan housing at the points shown (arrows)

8. Working through the access hole at the upper right corner of the flywheel housing, remove the three screws which secure the torque converter to the drive plate. Remove the ATF oil dipstick and the rubber boot.
9. Remove the two upper engine mounting bolts.
10. Jack up the car and, working beneath the car, remove the muffler shield and the heat exchanger.
11. Disconnect the starter wiring.
12. Remove the heater booster exhaust pipe.
13. Remove the two lower engine mounting nuts.
14. Jack up the engine slightly and remove the four engine carrier screws.

➡ Do not loosen the mountings on the body or the engine transaxle assembly will have to be recentralized in the chassis.

15. Remove the engine from the car.
16. Reverse the removal procedures to install the engine. Install the engine on the lower engine mounting studs and then locate the engine in the engine carrier. When installing the engine in the carrier, lift the engine up so that the four screws are at the top of the elongated holes and tighten them in this position. If it is necessary to raise or lower the engine for adjustment purposes, use the threaded shaft. After the engine is installed, make sure that the rubber buffer is centered in the rear axle carrier. Make sure that the engine carrier is vertical and parallel to the engine fan housing. Readjust it if necessary by moving the brackets on the side members.

Cylinder Head

REMOVAL & INSTALLATION

▶ See Figures 8 and 9

In order to remove the cylinder head from either pair of cylinders, it is necessary to remove the engine.
1. Remove the valve cover and gasket. Remove the rocker arm assembly. Unbolt the intake manifold from the cylinder head. The cylinder head is held in place by eight studs. Since the cylinder head also holds the cylinders in place in the VW engine, and the cylinders are not going to be removed, it will be necessary to hold the cylinders in place after the head is removed.
2. After the rocker arm cover, rocker arm retaining nuts, and rocker arm assembly have been removed, the cylinder head nuts can be removed and the cylinder head lifted off.

To install:
3. When reinstalling the cylinder head, the head should be checked for cracks both in the combustion chamber and in the intake and exhaust ports. Cracked heads must be replaced.
4. Spark plug threads should be checked. New seals should be used on the pushrod tube ends and they should be checked for proper seating.
5. The pushrod tubes should be turned so that the seam faces upward. In order to ensure perfect sealing, used tubes should be stretched slightly before they are reinstalled.

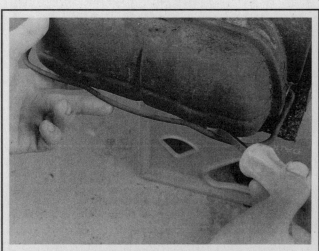

To remove the cylinder head, first pry the rocker arm cover retaining wire down and off of the cover . . .

ENGINE AND ENGINE REBUILDING

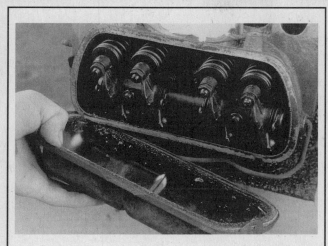
. . . then remove the rocker arm cover from the cylinder head

. . . then remove the shaft from the cylinder head

Make sure to discard the old gasket—a new gasket will be needed upon assembly

Once the rocker arm shaft is removed, the pushrods can be pulled out of their tubes

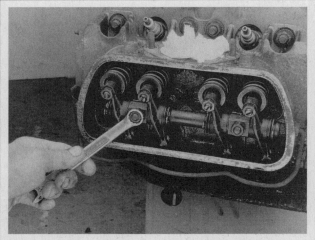
Slowly and alternately remove the rocker arm shaft mounting nuts . . .

Loosen the cylinder head nuts . . .

3-26 ENGINE AND ENGINE REBUILDING

... and slowly slide the cylinder head off of the mounting studs

Remove the lower air shield from between the lower, middle two studs

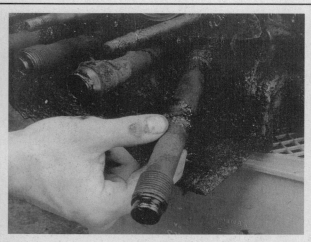

Remove the pushrod tubes from either the cylinder head or the crankcase

Make certain to tighten the cylinder heads to the proper torque value and in the proper sequence to ensure adequate cylinder sealing

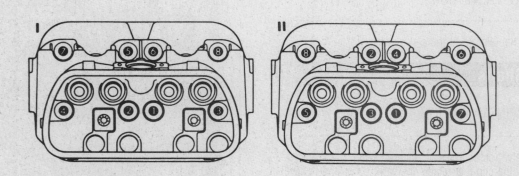

Fig. 8 To install the cylinder heads, first tighten the cylinder head nuts to 7 ft. lbs. following sequence I, then tighten the nuts to 23 ft. lbs. following sequence II—1600cc engines

ENGINE AND ENGINE REBUILDING

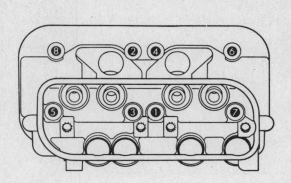

Fig. 9 Make certain to tighten the cylinder head nuts in the sequence shown to ensure proper cylinder sealing—1700cc, 1800cc and 2000cc engines

After removing the rocker arm cover, slowly and alternately remove the rocker arm shaft mounting nuts . . .

6. Install the cylinder head. Using new rocker shaft stud seals, install the pushrods and rocker shaft assembly.

➥**Pay careful attention to the orientation of the shaft as described in the "Rocker Shafts" portion of this section.**

7. Tighten the cylinder head in three stages. Adjust the valve clearance. Using a new gasket, install the rocker cover. It may be necessary to re-adjust the valves after the engine has been run a few minutes and allowed to cool.

VALVE SEATS

On all air-cooled VW engines, the valve seats are shrunk-fit into the cylinder head. This usually involves freezing the seat with a liquid nitrogen or some other refrigerant to about 200°F below zero, and heating up the cylinder head to approximately 400°F. Due to the extreme temperatures required to shrink-fit these items, and because of the extra care needed when working with metals at these extreme temperatures, it is advised that this operation be referred to an experienced repair shop.

OVERHAUL & VALVE GUIDE REPLACEMENT

See "Engine Rebuilding" at the end of this section.

Rocker Shafts

REMOVAL & INSTALLATION

Before the valve rocker assembly can be reached, it is necessary to lever off the clip that retains the valve cover and remove the valve cover. Remove the rocker arm retaining nuts, the rocker arm shaft, and the rocker arms. Remove the stud seals.

Before installing the rocker arm mechanism, be sure that the parts are as clean as possible. Install new stud seals. On Types 1, 2/1600 and 3, install the rocker shaft assembly with the cham-

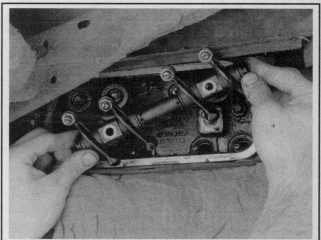

. . . then lift the rocker arm shaft off of the mounting studs

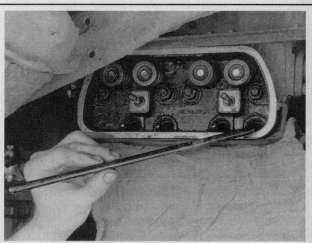

At this point, the pushrods can be removed from the pushrod tubes

3-28 ENGINE AND ENGINE REBUILDING

fered edges of the rocker shaft supports pointing outward and the slots pointing upward. On Type 4 and Type 2/1700, 2/1800 and 2/2000 models, the chamfered edges must point outward and the slots must face downward. The pushrod tube retaining wire must engage the slots in the rocker arm shaft supports as well as the grooves in the pushrod tubes. Tighten the retaining nuts to the proper torque. Use only the copper colored nuts that were supplied with the engine. Make sure that the ball ends of the pushrods are centered in the sockets of the rocker arms. Adjust the valve clearance on models without hydraulic lifters. Install the valve cover using a new gasket.

Intake Manifold

REMOVAL & INSTALLATION

Single Carburetor System

1. Disconnect the battery.
2. Disconnect the generator wiring.
3. Remove the generator. It will be necessary to loosen the fan housing and tilt it back to gain clearance to remove the generator.
4. Disconnect the choke and the accelerator cable.
5. On some models it will be necessary to remove the carburetor from the manifold.
6. Unbolt the manifold from the cylinder head and remove the manifold from the engine.
7. Installation is the reverse of the removal procedure. Always use new gaskets.

Twin Carburetor System

1. Remove the carburetors as outlined in Section 5.
2. Disconnect the tubes from the central idling system mixture distributor.
3. Disconnect all vacuum lines. Label them for purposes of installation.
4. Remove the nuts and bolts retaining the manifolds to the cylinder heads. Carefully lift off each manifold.
5. Installation is the reverse of the removal procedure. Take care to carefully clean all mating surfaces to the carburetors and cylinder heads. Always use new gaskets.

Fuel Injection System

INLET MANIFOLD

1. Remove the air cleaner.
2. Remove the pressure switch which is mounted under the right pair of intake manifold pipes. Disconnect the injector wiring.
3. Remove the fuel injectors by removing the two nuts which secure them in place. On Type 3, do not separate the pair of injectors; they can be removed as a pair and must be left in the injector plate.
4. After removing the intake manifold outer cover plate, remove the two screws which secure the manifold inner cover plate.
5. The manifold may be removed by removing the two nuts and washers which hold the manifold flange to the cylinder head.

To install:

6. Installation is the reverse of the removal procedure. The inner manifold cover should be installed first, but leave the cover loose until the outer cover and manifold are in place. Always use new gaskets. See the following step for proper injector installation.
7. Connect the fuel hoses to the injectors, if removed, after assembling the injectors with the injector retainer plate in place. Make sure that the sleeves are in place on the injector securing studs. Carefully slip the injectors into the manifold and install the securing nuts. Never force the injectors in or out of the manifold. Reconnect the injector wiring.

INTAKE AIR DISTRIBUTOR

The intake air distributor is located at the center of the engines at the junction of the intake manifold pipes.

➡ It is not necessary to remove the distributor if only the manifold pipes are to be removed.

1. Remove the air cleaner and pressure switch which are located under the right pair of manifold pipes.
2. Push the four rubber hoses onto the intake manifold pipes.
3. Remove the accelerator cable and the throttle valve switch.
4. Disconnect the accelerator cable.
5. Disconnect the vacuum hoses leading to the ignition distributor and the pressure sensor and disconnect the hose running to the auxiliary air regulator.
6. Remove those bolts under the air distributor which secure the air distributor to the crankcase and remove the air distributor.
7. Installation is the reverse of removal.

Mufflers, Tailpipes and Heat Exchangers

REMOVAL & INSTALLATION

Muffler

TYPE 1 CARBURETED ENGINE AND 2/1600

1. Working under the hood, disconnect the pre-heater hoses.
2. Remove the pre-heater pipe protection plate on each side of the engine. The plates are secured by three screws.
3. Remove the crankshaft pulley cover plate.
4. Remove the rear engine cover plate from the engine compartment. It is held in place by screws at the center, right, and left sides.
5. Remove the four intake manifold pre-heat pipe bolts. There are two bolts on each side of the engine.
6. Disconnect the warm air channel clamps at the left and right side of the engine.
7. Disconnect the heat exchanger clamps at the left and right side of the engine.
8. Remove the muffler from the engine.
9. Installation is the reverse of the above. Always use new gaskets to install the muffler.

TYPE 1 FUEL INJECTED ENGINE

1. On vehicles with catalytic converters, simply unbolt the muffler from the converter and remove. It may be necessary to remove the tailpipe first to make room.

ENGINE AND ENGINE REBUILDING

2. On vehicles without catalytic converters, unbolt the small EGR pipe (two bolts at each end), then unbolt the muffler from the heat exchangers.

Installation is the reverse of removal.

TYPE 3

The muffler is secured to the heat exchangers with clamps and, on some models, to the body with bolts at the top and at the ends.

TYPES 4, 2/1700, 2/1800 AND 2/2000

The muffler is secured to the left and/or right heat exchangers by three bolts or a muffler clamp. There is a bracket at one end of the muffler. Always use new gaskets when installing a new muffler.

Heat Exchangers

TYPES 1, 2/1600 AND 3

1. Disconnect the air tube at the outlet end of the exchanger.
2. Remove the clamp which secures the muffler to the exchanger.
3. Loosen the clamp which secures the exchanger to the heater hose connection at the muffler.
4. Remove the nuts which secure the exchanger to the forward end of the cylinder head.
5. Remove the heater flap control wire.
6. Reverse the above to install. Always use new gaskets.

TYPES 4, 2/1700, 2/1800 AND 2/2000

1. Disconnect the air hose at the outlet of each exchanger.
2. Disconnect the warm air tube at the outside end of the exchanger.
3. Disconnect the bolts or clamps which secure the exchangers to the muffler.
4. Remove the nuts at each exhaust port which secure the exchanger to the cylinder head.
5. Installation is the reverse of the above. Always use new gaskets.

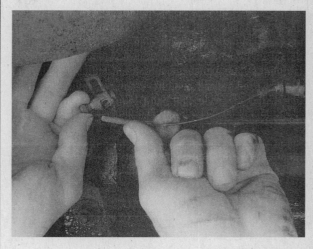

. . . then slide the flap cable out of the retaining fixture

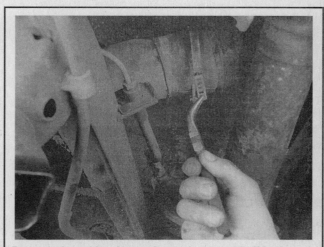

Remove the rubber heater hose clamps and slide the hoses off of the heater boxes

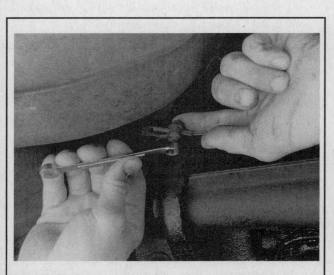

Loosen the heater control flap cable retaining bolt . . .

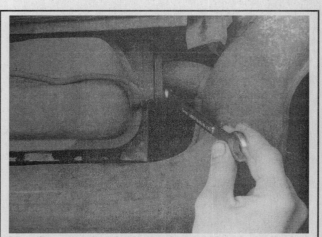

After disconnecting the heater boxes/exhaust manifolds from the muffler assembly, remove the rear exhaust manifold tube-to-cylinder head bolts

3-30 ENGINE AND ENGINE REBUILDING

Tailpipes

TYPES 1 AND 2/1600

Loosen the clamps on the tailpipes and apply penetrating oil. Work the pipe side-to-side while trying to pull the tailpipe out of the muffler.

➡It is often difficult to remove the tailpipes without damaging them.

TYPES 4, 2/1700, 2/1800 AND 2/2000

Remove the bolt which secures the pipe to the muffler. Remove the bolt which secures the pipe to the body and remove the pipe.

Tailpipe and Resonator

TYPE 3

Loosen the clamp at the resonator-to-muffler connection. Remove the bolt at the bend of the tailpipe and remove the resonator and tailpipe from the resonator. Loosen the clamp which secures the tailpipe to the resonator and work them apart.

Pistons and Cylinders

Pistons and cylinders are matched according to their size. When replacing pistons and cylinders, make sure that they are properly sized.

➡See the "Engine Rebuilding" portion of this section for cylinder refinishing.

REMOVAL & INSTALLATION

Cylinder

♦ See Figure 10

1. Remove the engine. Remove the cylinder head, pushrod tubes, and the deflector plate.
2. Slide the cylinder out of its groove in the crankcase and off of the piston. Try not to damage the cooling fins, as they chip easily. Matchmark the cylinders for reassembly. The cylinders must be returned to their original bore in the crankcase. If a cylinder is to be replaced, it must be replaced with a matching piston.

To install:

3. Cylinders should be checked for wear and, if necessary, replaced with another matched cylinder and piston assembly of the same size.
4. Check the cylinder seating surface on the crankcase, cylinder shoulder and gasket for cleanliness and deep scores. When installing the cylinders, a new gasket, if required, should be fitted over the base of the cylinder and worked up to the crankcase-cylinder mating surface. These gaskets are usually made of paper, so be careful. Oil the gasket with regular engine oil after it is installed.
5. The piston, as well as the piston rings and pin must be oiled before reassembly.
6. Be sure that the ring gaps are of the correct dimension. Stagger the ring gaps around the piston, but make sure that the oil ring gap is positioned up when the pistons are in position on the connecting rods.

Slide the cylinder off of the mounting studs and the piston

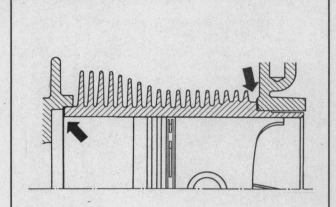
Fig. 10 Prior to installing the cylinders and cylinder heads, make certain that the cylinder sealing surfaces (arrows) are perfectly clean and true

7. Turn the crankshaft until the intended piston is out as far as possible, then fit the ring compressor, oil the cylinder walls and slide the cylinder onto the piston. Make certain that the other exposed piston skirts do not strike the crankcase when the crankshaft is turned. The bottoms of the cylinder barrels have a slight camfer to aid ring fitting. Make sure that the cylinder base gasket is in place on the cylinder barrel.

➡Use a ring compressor that pulls apart or you won't be able to get it off the piston once the cylinder barrel is over the rings. These are available at most automotive stores.

8. Install the deflector plates.
9. Install the pushrod tubes using new gaskets. Install the pushrods. Make sure that the seam in the pushrod tube is facing upward.
10. Install the cylinder head.

ENGINE AND ENGINE REBUILDING

Piston

♦ See Figures 11, 12 and 13

➡ See the "Engine Rebuilding" section for piston ring procedures.

1. Remove the engine. Remove the cylinder head and, after matchmarking the cylinders, remove the cylinders.

➡ You must remove both cylinder barrels from the bank to be worked on even if you are only removing one piston.

2. Matchmark the pistons to indicate the cylinder number and which side points toward the flywheel.
3. Remove the circlips which retain the piston pin.
4. Heating the piston will aid in piston pin removal. To heat the piston, boil a clean rag in water and wrap it around the piston. You can fashion a pin remover by shaving an old piston pin down so that it will slide through the piston. Use it with a mallet

Remove the piston from the connecting rod—the arrow (next to the G) should point toward the flywheel end of the engine during assembly

To remove the pistons, first remove the circlip from both ends of the piston pin bore . . .

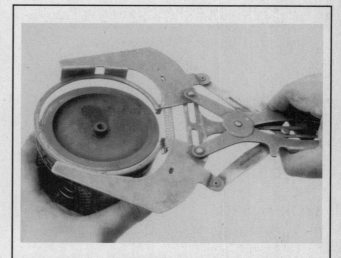
Use a ring expander tool to remove the piston rings

. . . then slide the piston pin out of the piston and connecting rod

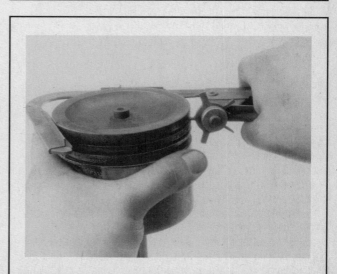
Clean the piston grooves using a ring groove cleaner

ENGINE AND ENGINE REBUILDING

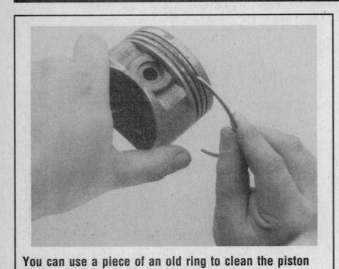

You can use a piece of an old ring to clean the piston grooves, BUT be careful, the ring is sharp

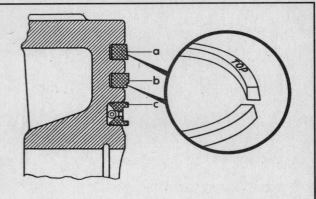

a. Upper compression ring
b. Lower compression ring
c. Oil scraper ring

Fig. 12 The rings must be installed as shown, otherwise they will not produce a proper seal with the cylinder wall—the word TOP must be positioned toward the piston crown

to knock out the pin. Be sure to hold the piston when removing the pin so that the connecting rod is not bent.

5. Remove the piston from the connecting rod.

To install:

6. Before installing the pistons, they should first be cleaned and checked for wear. Remove the old rings. Clean the ring groove with a ring groove cleaner or a broken piece of ring. Clean the piston with solvent but do not use a wire brush or sand paper. Check for any cracks or scuff marks. Check the piston diameter with a micrometer and compare the readings to the specifications. If the running clearance between the piston and cylinder wall is 0.008 in. (0.2mm) or greater, the cylinder and piston

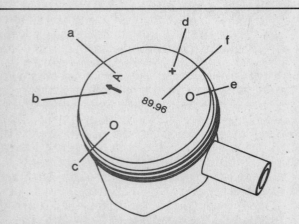

A. Corresponds to the index of the part number—serves as identifying mark
B. Arrow (indented or stamped on) must point toward flywheel
C. Paint spot indicates pistons which are of matching size (blue, pink, green)
D. Weight grading (+ or −)
E. Paint spot indicating weight grading (brown = − weight, grey = + weight)
F. Piston size in mm

Fig. 11 Identification of the various markings on the top of a typical VW piston—make sure that arrow b is oriented toward the front (flywheel end) of the engine

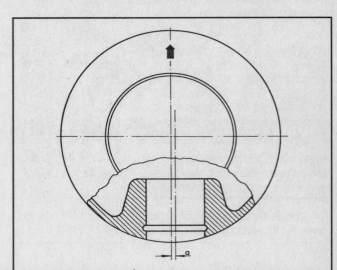

Fig. 13 Since the piston pins are slightly offset from the piston centerline, make certain that the arrow on the piston crown points toward the flywheel end of the engine to ensure proper positioning of the piston pin—dimension (a) should be 0.059 in. (1.5mm)

should be replaced by a set of the same size grading. If the cylinder shows no sign of excessive wear or damage, it is permissible to install a new piston and rings of the appropriate size.

7. Place each ring in turn in its cylinder bore and check the piston ring end-gap. If the gap is too large, replace the ring. If the gap is too narrow, file the end of the ring until the proper gap is obtained.

8. Insert the rings on the piston and check the ring side clearance. If the clearance is too large, replace the piston. Install the rings with the marking "Oben" or "Top" pointing upward.

9. If new rings are installed in a used piston, the ring ridge at the top of the cylinder bore must be removed with a ridge reamer.

ENGINE AND ENGINE REBUILDING 3-33

10. Install a circlip on each piston on the side toward the flywheel, indicated by the arrow on the top of the piston. Install each piston so the arrow points toward the flywheel. If necessary, heat the piston to 167°F (75°C), then install the piston pin and fit the other circlip. Make sure the circlips are seated in their grooves properly.

11. Install the cylinders and the cylinder heads.

Crankcase

DISASSEMBLY & ASSEMBLY

1. Remove the engine.
2. Remove the cylinder heads, cylinders, and pistons.
3. Remove the oil strainer, oil pressure switch, and the crankcase nuts. Remove the flywheel and oil pump. The flywheel is held in place by the bolt, (Type 4 and 1972–81 Type 2 have five bolts), at the center of the flywheel. Matchmark the flywheel so that it can be replaced in the same position.

➡ **On manual transmission models, remove the clutch pressure plate and disc to expose the flywheel bolt(s).**

4. Keep the cam followers in the right crankcase half in position by using a retaining spring.
5. Clean the sludge off of the crankcase and locate *all* of the crankcase retaining nuts. Do not try to separate the halves until you are sure you have removed *all* of the nuts. Use a rubber hammer to break the seal between the crankcase halves.

✱✱ CAUTION

Never insert sharp metal tools, wedges, or any prying device between the crankcase halves. This will ruin the gasket surface and cause serious oil leakage.

6. After the seal between the crankcase halves is broken, remove the right-hand crankcase half, the crankshaft oil seal and the camshaft end plug. The camshaft and crankshaft can now be lifted out of the crankcase half.
7. Remove the cam followers (or lifters) bearing shells, and the oil pressure relief valve.

To assemble:

8. Before starting reassembly, check the crankcase for any damage or cracks.
9. Flush and blow out all ducts and oil passages. Check the studs for tightness. If the tapped holes are worn install a Heli-Coil®.
10. Install the crankshaft bearing dowel pins and bearing shells for the crankshaft and camshaft. Make sure the bearing shells are installed in the proper journal. See Crankshaft Removal & Installation, later in this section, for crankshaft bearing placement.
11. Install the crankshaft and camshaft after the bearings have been well lubricated. When installing the camshaft and the crankshaft, make sure the timing marks are aligned. See Camshaft and Timing Gears Removal & Installation, later in this section, for procedures.
12. Install the oil pressure relief valve.
13. Oil and install the cam followers.

➡ **To keep them from falling out during assembly, liberally coat them with white grease.**

14. Install the camshaft end-plug using sealing compound.
15. Install the thrust washers and crankshaft oil seal. The oil seal must rest squarely on the bottom of its recess in the crankcase. The thrust washers at the flywheel end of the crankshaft are shims used to set the crankshaft end-play.
16. Spread a thin film of sealing compound on the crankcase joining faces and place the two halves together. Tighten the nuts in several stages. Tighten the 8mm nut located next to the 12mm stud of the No. 1 crankshaft bearing first. As the crankcase halves are being tightened, continually check the crankshaft for ease of rotation.

✱✱ CAUTION

Make sure the crankshaft bearings are seated correctly on their dowels or you could crack the bearings, the crankcase or both when tightening the crankcase nuts. See Crankshaft Removal and Installation for instructions.

17. Crankshaft end-play is checked when the flywheel is installed. It is adjusted by varying the number of thickness of the shims located behind the flywheel. Measure the end-play with a dial indicator mounted against the flywheel, and attached firmly to the crankcase.

Camshaft and Timing Gears

REMOVAL & INSTALLATION

Removal of the camshaft requires splitting the crankcases. The camshaft and its bearing shells are then removed from the crankcase halves. Before reinstalling the camshaft, it should be checked for wear on the lobe surfaces and on the bearing surfaces. In addition, the riveted joint between the camshaft timing gear and the camshaft should be checked for tightness. The camshaft should be checked for a maximum run-out of 0.0008 in. The timing gear should be checked for the correct tooth contact and for wear. If the camshaft bearing shells are worn or damaged, new shells should be fitted. The camshaft bearing shells should be installed with the tabs engaging the notches in the crankcase. It is usually a good idea to replace the bearing shells under any circumstances. Before installing the camshaft, the bearing journals and cam lobes should be generously coated with oil. When the camshaft is installed, care should be taken to ensure that the timing gear tooth marked (0) is located between the two teeth of the crankshaft timing gear marked with a center punch. The camshaft end-play is measured at the No. 3 bearing. End-play is 0.0015–0.005 in. (0.04–0.12 mm) and the wear limit is 0.006 in. (0.16 mm).

➡ **Camshaft gears are marked with a −1, 0, +1, etc., along their inner face to denote how much their pitch radius deviates from the standard pitch radius of 0. If your camshaft gear has 0 pitch deviation (i.e., if it is marked on its inner face with a "0"), do not confuse this mark with the zero shaped timing mark on the outer gear face.**

ENGINE AND ENGINE REBUILDING

Crankshaft

REMOVAL & INSTALLATION

Crankshaft Pulley

On the Type 1 and 2/1600, the crankshaft pulley can be removed while the engine is still in the car. However, in this instance it is necessary for the rear cover plate of the engine to be removed. Remove the cover plate after taking out the screws in the cover plate below the crankshaft pulley. Remove the fan belt and the crankshaft pulley securing screw. Using a puller, remove the crankshaft pulley. The crankshaft pulley should be checked for proper seating and belt contact. The oil return thread should be cleaned and lubricated with oil. The crankshaft pulley should be installed in the reverse sequence. Check for oil leaks after installing the pulley.

On the Type 3, the crankshaft pulley can be removed only when the engine is out of the car and the muffler, generator, and cooling air intake housing are removed. After these parts have been removed, take out the plastic cap in the pulley. Remove the crankshaft pulley retaining bolt and remove the pulley.

Type 4 and Type 2/1700, 2/1800 and 2/2000, removal is the same as the Type 3. However, the pulley is secured by three socket head screws and a self locking nut.

Installation for Type 2/1700, 2/1800, 2/2000, 3 and 4 engines is the reverse of removal. When installing, use a new paper gasket between the fan and the crankshaft pulley. If shims are used, do not forget them. Don't use more than two shims. When inserting the pulley, make sure that the pin engages the hole in the fan. Ensure that the clearance between the generator belt and the intake housing is at least 4mm and that the belt is parallel to the housing.

Flywheel

♦ See Figures 14 and 15

➡ In order to remove the flywheel, the crankshaft will have to be prevented from turning. This may be accomplished on Type 1, 2/1600 and Type 3 models by using a 3 or 4 foot length of angle iron or thick stock sheet steel, such as an old fence post. Drill out two holes in the metal bar that correspond to two of the pressure plate retaining bolt holes. The metal bar is installed as per the accompanying illustration.

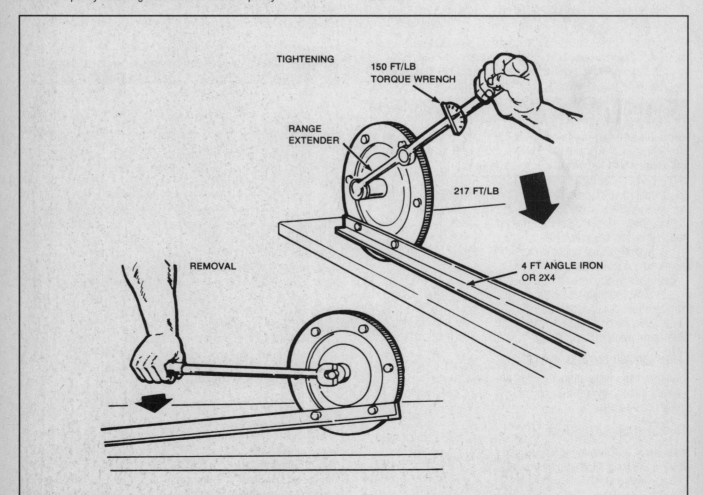

Fig. 14 Because of the large amount of torque holding the flywheel gland nut onto the crankshaft, a removal and installation tool can be fabricated from a 4 ft. length of angle iron stock

ENGINE AND ENGINE REBUILDING

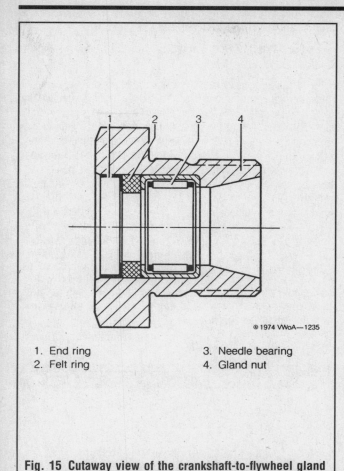

1. End ring
2. Felt ring
3. Needle bearing
4. Gland nut

Fig. 15 Cutaway view of the crankshaft-to-flywheel gland nut—the gland nut also houses a set of needle bearings

TYPES 1, 2/1600 AND 3

The flywheel is attached to the crankshaft with a gland nut and is located by four dowel pins. An oil seal is recessed in the crankcase casting at No. 1 main bearing. A needle bearing, which supports the main driveshaft, is located in the gland nut. Prior to removing the flywheel, it is necessary to remove it, using a 36mm special wrench. Before removing the flywheel, matchmark the flywheel and the crankshaft.

Installation is the reverse of removal. Before installing the flywheel, check the flywheel teeth for any wear or damage. Check the dowel pins for correct fit in the crankshaft and in the flywheel. Adjust the crankshaft end-play and check the needle bearing in the gland nut for wear.

TYPES 2/1700, 2/1800, 2/2000 AND 4

Removal and installation is similar to the Type 1, 2/1600, and 3 except that the flywheel is secured to the crankshaft by five socket head screws.

Crankshaft Oil Seal (Flywheel End)

This seal is removed after removing the flywheel. After the flywheel is removed, inspect the surface on the flywheel joining flange where the seal makes contact. If there is a deep groove or any other damage, the flywheel must be replaced. Remove the oil seal recess and coat it thinly with a sealing compound. Be sure that the seal rests squarely on the bottom of its recess. Make sure that the correct side of the seal is facing outward, that is, the lip of the seal should be facing the inside of the crankcase. Reinstall the flywheel after coating the oil seal contact surface with oil.

➡ Be careful not to damage the seal when sliding the flywheel into place.

Crankshaft

➡ See the "Engine Rebuilding" section for crankshaft refinishing procedures.

Removal of the crankshaft requires splitting the crankcase. After the crankcase is opened, the crankshaft can then be lifted out.

The crankshaft bearings are held in place by dowel pins. These pins must be checked for tightness.

When installing the bearings, make sure that the oil holes in the shells are properly aligned. Be sure that the bearing shells are seated properly on their dowel pins. Bearing shells are available in three undersizes. Measure the crankshaft bearing journals to determine the proper bearing size. Place one half of the No. 2 crankshaft bearing in the crankcase. Slide the No. 1 bearing on the crankshaft so that the dowel pin hole is toward the flywheel and the oil groove faces toward the fan. The No. 3 bearing is installed with the dowel pin hole facing toward the crankshaft web.

To remove the No. 3 main bearing, remove the distributor gear circlip and the distributor drive gear. Mild heat (176°F) must be applied to remove the gear. Next slide the spacer off of the crankshaft. The crankshaft timing gear should now be pressed off the crankshaft after mild heating. When the timing gear is reinstalled, the chamfer must face towards the No. 3 bearing. The No. 3 bearing can then be replaced. When removing and installing the gears on the crankshaft, be careful not to damage the No. 4 bearing journal.

When all of the crankshaft bearings are in place, lift the crankshaft and the connecting rod assembly into the crankcase and align the valve timing marks.

Install the crankcase half and reassemble the engine.

Connecting Rods

REMOVAL & INSTALLATION

The factory suggests you split the crankcase to remove the connecting rods. However, if you're just checking your connecting rod bearings for wear, you can remove them without splitting the case. See the appropriate procedures, located later in this section.

Crankcase Splitting Method

➡ See the "Engine Rebuilding" section for additional information

3-36 ENGINE AND ENGINE REBUILDING

After splitting the crankcase, remove the crankshaft and the connecting rod assembly. Remove the connecting rods, clamping bolts, and the connecting rod caps. Inspect the piston pin bushing. With a new bushing, the correct clearance is indicated by a light finger push fit of the pin at room temperature. Reinsert the new connecting rod bearings after all parts have been thoroughly cleaned. Assemble the connecting rods on the crankshaft, making sure that the rods are oriented properly on the crankshaft. The identification numbers stamped on the connecting rods and connecting rod caps must be on the same side. Note that the marks on the connecting rods are pointing upward, while the rods are pointing toward their respective cylinders. Lubricate the bearing shells before installing them.

Tighten the connecting rod bolts to the specified torque. A slight pre-tension between the bearing halves, which is likely to occur when tightening the connecting rod bolts, can be eliminated by gently striking the side of the bearing cap with a hammer. Do not install the connecting rod in the engine unless it swings freely on its journal. Using a peening chisel, secure the connecting rod bolts in place.

Failure to swing freely on the journal may be caused by improper side clearance, improper bearing clearance or failure to lubricate the rod before assembly.

Non-Crankcase Splitting Method

➡See the "Engine Rebuilding" section for additional information.

Remove the cylinder heads, cylinders and pistons. Put a dab of grease in the end of a socket and, using an extension and ratchet, loosen and carefully remove the connecting rod nuts. The nuts face the piston side of the connecting rod. Turn the crankshaft as necessary to remove both nuts. Have an assistant hold the connecting rod cap from the other side of the engine and gently tap the cap bolts with a brass or plastic drift to separate the caps from the connecting rods. Pull the connecting rod out through the cylinder hole, then remove the cap. Be careful not to drop anything into the crankcase or you may have to split it after all. Reverse procedure to install. Install the connecting rods with the forged marks up: see illustration. Torque to proper specifications and using a small hammer, relieve pre-tension as instructed under Crankcase Splitting Method, above,

➡**You will not be able to stake the nuts using this method, therefore it would be wise to apply Loctite® or an equivalent sealer to the threads.**

ENGINE LUBRICATION

Oil Strainer

REMOVAL & INSTALLATION

The oil strainer can be easily removed by removing the retaining nuts, washers, oil strainer plate, strainer, and gaskets. The Type 2/1700, 2/1800, 2/2000 and Type 4 strainer is secured by a single bolt at the center of the strainer. Once taken out, the strainer must be thoroughly cleaned and all traces of old gaskets removed prior to fitting new ones. The suction pipe should be checked for tightness and proper position. When the strainer is installed, be sure that the suction pipe is correctly seated in the strainer. If necessary, the strainer may be bent slightly. The measurement from the strainer flange to the top of the suction pipe should be 10 mm. The measurement from the flange to the bottom of the strainer should be 6 mm. The cap nuts on Types 1, 2/1600, and 3 must not be overtightened. The Type 4 and Type 2/1700, 2/1800, 2/2000 have a spin-off replaceable oil filter as well as the strainer in the crankcase. The oil filter is located at the left rear corner of the engine.

Oil Cooler

REMOVAL & INSTALLATION

The Type 1 and 2/1600 oil cooler is located under the engine cooling fan housing at the left side of the engine. The Type 3 cooler is located at the same position but is mounted horizontally. The type 4 and Type 2/1700, 2/1800, 2/2000 coolers are mounted near the oil filter, at the left corner of the engine.

The oil cooler may be removed without taking the engine out of the car. On Types 1 and 2/1600, the engine fan housing must be removed. On the Type 3, the cooler is accessible through the left-hand cylinder cover plate. The Type 4 and Type 2/1700, 2/1800, 2/2000 cooler is accessible through the left side engine cowling, working either in the engine compartment or from underneath the car.

The oil cooler can be removed after the three retaining nuts have been taken off. The gaskets should be removed along with the cooler and replaced with new gaskets. If the cooler is leaking, check the oil pressure relief valve. The studs and bracket on the cooler should be checked for tightness. Make certain that the hollow ribs of the cooler do not touch one another. The cooler must not be clogged with dirt. Clean the contact surfaces on the crankcase, install new gaskets, and attach the oil cooler. Types 3 and 4, 2/1700, 2/1800 and 2/2000 have a spacer ring between the crankcase and the cooler at each securing screw. If these rings are omitted, the seals may be squeezed too tightly, resulting in oil stoppage and resultant engine damage. Use double retaining nuts and Loctite® on the cooler studs.

Oil Pump

REMOVAL & INSTALLATION

On Types 1 and 2/1600, the pump can be removed while the engine is in the car, but it is first necessary to remove the cover plate, the crankshaft pulley, and the cover plate under the pulley. On Types 3, 4, 2/1700, 2/1800 and 2/2000, the oil pump can be taken out only after the engine is removed from the car and the air intake housing, the belt pulley fan housing, and fan are dismantled. On the Automatic Stick Shift models, the torque converter oil pump is driven by the engine oil pump.

ENGINE AND ENGINE REBUILDING

Loosen the oil cooler-to-oil cooler mounting flange nuts with a 10mm wrench—this can also be done after the flange is separated from the engine

Lift the oil cooler/oil cooler mounting flange assembly off of the engine . . .

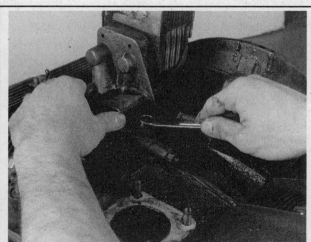
Remove the inboard oil cooler flange-to-engine mounting nut . . .

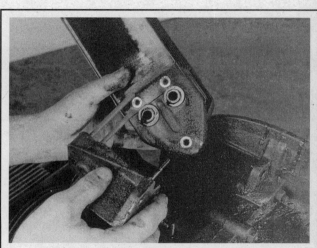
. . . then separate the oil cooler from the mounting flange

. . . then remove the outboard oil cooler flange-to-engine mounting nut—this nut is located under the flange

Make certain to replace all O-ring seals on both the oil cooler and flange before installation

ENGINE AND ENGINE REBUILDING

On Type 1, 2/1600, and 3 remove the nuts from the oil pump cover and then remove the cover and its gasket. Remove the gears and take out the pump with a special extractor that pulls the body out of the crankcase. Care should be taken so as not to damage the inside of the pump housing.

On Type 4, Type 2/1700, 2/1800, and 2/2000 engines, remove the four pump securing nuts and, prying on either side of the pump, pry the pump assembly out of the crankcase. To disassemble the pump, the pump cover must be pressed apart.

Prior to assembly, check the oil pump body for wear, especially the gear seating surface. If the pump body is worn, the result will be loss of oil pressure. Check the driven gear shaft for tightness and, if necessary, peen it tightly into place or replace the pump housing. The gears should be checked for excessive wear, backlash, and end-play. Maximum end-play can be checked using a T-square and a feeler gauge. Check the mating surfaces of the pump body and the crankcase for damage and cleanliness. Install the pump into the crankcase with a new gasket. Do not use any sealing compound. Turn the camshaft several revolutions in order to center the pump body opposite the slot in the camshaft. On Type 1, 2/1600, and 3 the cover may now be installed. On Type 4, Type 2/1700, 2/1800, and 2/2000 models, the pump was installed complete. Tighten the securing nuts.

Oil Pressure Relief Valve

REMOVAL & INSTALLATION

The oil pressure relief valve acts as a safety valve which opens when the oil pressure becomes too great. When the engine is cold, the oil is thick, which makes it easier for the oil pump to move it. This in turn creates greater oil pressure. On a cold engine, the oil pressure relief valve plunger is in its lowest position and allows only some of the oil to travel to lubrication points; the rest is fed back into the crankcase. This prevents the oil pressure from raising high enough to burst the seals in the oil cooler and elsewhere. As the engine warms, the plunger rises because the thinning oil doesn't have as much pressure as it did when cold, and more oil is directed toward the lubrication points.

The oil pressure relief valve is removed by unscrewing the end-plug and removing the gasket ring, spring, and plunger. If the plunger sticks in its bore, it can be removed by screwing a 10mm tap into it.

On 1600cc engines, the valve is located to the left of the oil pump. On Automatic Stick Shift models, it is located on the oil pump housing. On 1700, 1800 and 2000 engines, the valve is located beside the oil filter.

Before installing the valve, check the plunger for any signs of seizure. If necessary, the plunger should be replaced. If there is any doubt about the condition of the spring, it should also be replaced. When installing the relief valve, be careful that you do not scratch the bore. Reinstall the plug with a new gasket.

Type 4 and Type 2/1700, 2/1800, 2/2000 engines have a second oil pressure relief valve located just to the right of, and below the oil filter.

Hydraulic Valve Lifters

ADJUSTMENT

1978–81 Type 2 models are equipped with hydraulic lifters. No valve lash adjustment is required.

REMOVAL & INSTALLATION

The lifters can be removed without removing the engine from the vehicle.

To remove the lifters, remove the rocker shafts with the rockers, withdraw the pushrods and pushrod tubes, then, using a magnetic probe, withdraw the lifters. Matchmark them with their respective cylinders to avoid confusion. Place the lifter with the body down to avoid oil leak-out, thus making bleeding unnecessary before reinstallation.

Installation is the reverse of removal. When the rocker shafts and rockers are installed, set the adjusting screws in the rockers so that the ball shaped part is flush with the surface. Turn the crankshaft until cylinder no. 1 is at TDC compression stroke. Turn the adjusting screws so they just touch the valve stems, then turn them two turns clockwise and tighten the locknut. Adjust the other cylinders in the same fashion.

➡ **Before installing make sure lifter is bled correctly by applying firm thumb pressure on the pushrod socket. A resistance should be felt.**

BLEEDING

Fill a can with engine oil. Remove the lifter lockring, pushrod socket, plunger, ball check valve with spring and plunger spring from the body. Be careful, the spring is very powerful! Place the lifter body in the can of oil so that it is completely submerged. Assemble the lifter except for the lockring and the pushrod socket. The lifter should be assembled while submerged in the oil. Open the valve with a suitable punch or scribe by pushing the punch down through the hole in the plunger to allow the oil to flow out of the lower part of the plunger. Insert the pushrod socket in the lifter. Cut an old Type 1 pushrod in half, then use it in conjunction with a press to slowly force down the pushrod socket until the lockring can be installed. The lifter must be submerged in oil the whole time.

ENGINE AND ENGINE REBUILDING 3-39

ENGINE COOLING

Fan Housing and Fan

REMOVAL & INSTALLATION

Types 1 and 2/1600

FAN HOUSING

1. Remove the two heater hoses and the generator strap.
2. Pull out the lead wire from the coil. Remove the distributor cap and take off the spark plug connectors.
3. Remove the retaining screws that are located on both sides of the fan housing. Remove the rear hood.
4. Remove the outer half of the generator pulley and remove the fan belt.
5. Remove the thermostat securing screw and take out the thermostat.
6. Remove the lower part of the carburetor pre-heater duct.
7. The fan housing can now be removed with the generator. After removal, check the fan housing for damage and for loose air deflector plates.
8. Installation is the reverse of the above.
9. Make sure that the thermostat connecting rod is inserted into its hole in the cylinder head. The fan housing should be fitted properly on the cylinder cover plates so that there is no loss of cooling air.

FAN

1. Remove the generator and fan assembly as described in Generator Removal & Installation.
2. While holding the fan, unscrew the fan retaining nut and take off the fan, spacer washers, and the hub.

To install:

3. Place the hub on the generator shaft, making sure that the woodruff key is securely positioned.
4. Insert the spacer washers. The clearance between the fan and the fan cover is 0.06–0.07 in. Place the fan into position and tighten its retaining nut. Correct the spacing by inserting the proper number of spacer washers. Place any extra washers between the lockwasher and the fan.
5. Reinstall the generator and the fan assembly.

Type 2/1700, 2/1800 and 2/2000

1. On 1973–74 models, the air injection pump and related parts must first be removed. Loosen the air pump and remove the drive belt. Remove the pump and bracket retaining bolts and remove the air pump and retaining brackets. Unbolt and remove the extension shaft and pulley assembly from the fan and fan housing. Using a 12 point allen wrench, loosen the alternator drive belt adjusting bolt. Then, remove the timing scale, fan and crankshaft pulley assembly, and the alternator drive belt.
2. On models without the air injection pump, pry out the alternator cover insert, and, using a 12 point allen wrench, loosen the alternator adjusting bolt. Remove the alternator drive belt, the ignition timing scale and the grille over the fan. Remove the three socket head screws attaching the fan and crankshaft assembly to the crankshaft and remove the fan and pulley.

3. Disconnect the cooling air control cable at the flap control shaft.
4. On models so equipped, pull out the rubber elbow for the alternator from the front half of the fan housing.
5. Remove the four nuts retaining the fan housing to the engine crankcase. The assembled fan housing may then be removed by pulling it to the rear and off the engine. It is not necessary to separate the fan housing halves or remove the alternator to remove the fan housing.
6. Reverse the above procedure to install, taking care to adjust the alternator and air pump drive belts (1973–74 models) so that moderate thumb pressure deflects the belt about ½ in. when applied at a point midway between the longest run. Also, adjust the cooling air control cable as outlined in this section.

Type 3

1. Remove the crankshaft pulley, the rear fan housing half, and the fan.
2. Unhook the linkage and spring at the right-hand air control flap.
3. Remove the screws for the front half of the housing and remove the housing.
4. Install the front half and ensure the correct sealing of the cylinder cover plates.
5. Replace and tighten the two lower mounting screws slightly.
6. Turn the two halves of the fan housing to the left until the crankcase half is contacted by the front lug.
7. Fully tighten the two lower mounting screws.
8. Loosen the nuts at the breather support until it can be moved.
9. Insert and tighten the mounting screws of the upper fan housing half. Tighten the breather support nuts fully.
10. Connect the linkage and spring to the right-hand air control flap.
11. Install the fan and the rear half of the fan housing.

Types 4, 2/1700, 2/1800 and 2/2000

1. Remove the engine. Remove the fan belt.
2. Remove the allen head screws and remove the belt pulley and fan as an assembly.

➥ **It is not necessary to remove the alternator to remove the fan housing.**

3. Remove the spacer and the alternator cover plate.
4. Disconnect the cooling air regulating cable at the shaft.
5. Remove the nuts and remove both halves of the fan housing at the same time.
6. Installation is the reverse of the above.

Air Flap and Thermostat

ADJUSTMENT

Types 1 and 2/1600

1. Loosen the thermostat bracket securing nut and disconnect the thermostat from the bracket.

3-40 ENGINE AND ENGINE REBUILDING

2. Push the thermostat upwards to fully open the air flaps.
3. Reposition the thermostat bracket so that the thermostat contacts the bracket at the upper stop, and then tighten the bracket nut.
4. Reconnect the thermostat to the bracket.

Type 3

1. Loosen the clamp screw on the relay lever.
2. Place the air flaps in the closed position. Make sure that the flaps close evenly. To adjust a flap, loosen its securing screw and turn it on its shaft.
3. With the flaps closed, tighten the clamp screw on the relay lever.

Types 4, 2/1700, 2/1800 and 2/2000

1. Loosen the cable control.
2. Push the air flaps completely closed.
3. Tighten the cable control.

ENGINE REBUILDING

♦ See Figures 16, 17, 18, 19 and 20

This section describes, in detail, the procedures involved in rebuilding a horizontally opposed, air-cooled Volkswagen/Porsche four cylinder engine. It is divided into two sections. The first section, Cylinder Head Reconditioning, assumes that the cylinder head is removed from the engine, all manifolds and sheet metal shrouding is removed, and the cylinder head is on a workbench. The second section, Crankcase Reconditioning, covers the crankcase halves, the connecting rods, crankshaft, camshaft and lifters. It is assumed that the engine is mounted on a work stand (which can be rented), with the cylinder heads, cylinders, pistons, and all accessories removed.

In some cases, a choice of methods is provided. The choice of a method for a procedure is at the discretion of the user. It may be limited by the tools available to a user, or the proximity of a local engine rebuilding or machine shop.

The tools required for the basic rebuilding procedures should, with minor exceptions, be those included in a mechanic's tool kit: An accurate torque wrench (preferably a preset, click type), inside and outside micrometers, electric drill with grinding attachment, valve spring compressor, a set of taps and reamers, a valve lapping tool, and a dial indicator (reading in thousandths of an inch). Special tools, where required, are available from the major tool suppliers (i.e. Zelenda®, Craftsman®, K-D®, Snap-On®). The services of a competent automotive or aviation machine shop must also be readily available.

When assembling the engine, bolts and nuts with no torque specification should be tightened according to size and marking (see chart).

Any parts that will be in frictional contact must be pre-lubricated before assembly to provide protection on initial start-up. Many different pre-lubes are available and each mechanic has his own favorite. However, any product specifically formulated for this purpose, such as Vortex Pre-Lube®, STP®, Wynn's Friction Proofing®, or even a good grade of white grease may be used.

➡Do not use engine oil only, as its viscosity is not sufficient.

Where semi-permanent (locked but removable) installation of nuts or bolts is required, the threads should be cleaned and coated with locking compound. Studs may be permanently installed using a special compound such as Loctite® Stud and Bearing Mount.

Aluminum is used liberally in VW and Porsche engines due to its low weight and excellent heat transfer characteristics. Both the cylinder heads and the crankcase are aluminum alloy castings.

Metric

Bolt Diameter (mm)	Bolt Grade				Wrench Size (mm) Bolt and Nut
	5D	8G	10K	12K	
6	5	6	8	10	10
8	10	16	22	27	14
10	19	31	40	49	17
12	34	54	70	86	19
14	55	89	117	137	22
16	83	132	175	208	24
18	111	182	236	283	27
22	182	284	394	464	32
24	261	419	570	689	36

*—Torque values are for lightly oiled bolts.
CAUTION: Bolts threaded into aluminum require much less torque.

Fig. 16 If no specific torque value is given in the text, tighten the bolts according to the bolt grade of the particular fastener—bolts threaded into aluminum require much less torque

However, a few precautions must be observed when handling aluminum engine parts:

—Never hot-tank aluminum parts, unless the hot-tanking solution is specified for aluminum application (i.e. Oakite® Aluminum Cleaner 164, or ZEP® Hot Vat Aluminum Cleaner). Most hot-tanking solutions are used for ferrous metals only, and "cook" at much higher temperatures than the 175°F used for aluminum cleaners. The result would be a dissolved head or crankcase.

—Always coat threads lightly with engine oil or anti-seize compound before installation, to prevent seizure.

—Never overtorque bolts or spark plugs in aluminum threads. Should stripping occur, threads can be restored using inserts such as the Heli-Coil®, K-D® Insert for Keenserts® kits.

To install a Heli-Coil® insert, tap drill the hole with the stripped threads to the specified size (see chart). If you are performing this operation on a spark plug hole with the head installed, coat the tap with wheel bearing grease to prevent aluminum shavings from falling into the combustion chamber (it will also help if the engine is rotated so that the exhaust valve of the subject cylinder is open, so that when the engine is initially started, if any chips did fall into the engine, they will be blown out the exhaust instead of scoring the cylinder walls, and, if com-

ENGINE AND ENGINE REBUILDING

Heli-Coil Specifications

Thread Size	Heli-Coil Insert Part No.	Insert Length (In.)	Drill Size	Tap Part No.	Insert Tool Part No.	Extracting Tool Part No.
1/2-20	1185-4	3/8	17/64 (.266)	4 CPB	528-4N	1227-6
5/16-18	1185-5	15/32	Q (.332)	5 CPB	528-5N	1227-6
3/8-16	1185-6	9/16	X (.397)	6 CPB	528-6N	1227-6
7/16-14	1185-7	21/32	29/64 (.453)	7 CPB	528-7N	1227-16
1/2-13	1185-8	3/4	33/64 (.516)	8 CPB	528-8N	1227-16

pressed air is available, it may be applied through the spark plug hole and the chips blown out the exhaust port).

➡ **Heli-Coil® tap sizes refer to the size thread being replaced, rather than the actual tap size.**

Using the specified tap, tap the hole for the Heli-Coil®. Place the insert on the proper installation tool (see chart). Apply pressure on the insert while winding it clockwise into the hole, until the top of the insert is one turn below the surface. Remove the installation tool and break the installation tang from the bottom of the insert by moving it up and down. If, for some reason, the Heli-Coil® must be removed, tap the removal tool firmly into the hole, so that it engages the top thread, and turn the tool counterclockwise to extract the insert.

K-D® makes an insert specifically designed for the 14mm spark plugs used in all VW's. The steel insert is 3/8 in. deep and has a lip which will seat the insert automatically to the correct depth. To install the K-D® insert, screw the combination reamer and tap into the damaged hole to ream the hole to the proper size and cut new threads for the insert. Then, screw the insert onto a spark plug, and torque the plug to 15–18 ft. lbs. to seat the insert.

➡ **Apply locking compound to the threads of the insert (cylinder head side) to make the installation permanent.**

Another spark plug insert that has come into favor is the Keenserts® insert. The special features of this type of insert are the locking keys and gas tight sealing ring. The Keenserts® kit consists of a ream and countersink tool, a tap with pilot point, an installation tool (drift), and the inserts. To install a Keenserts® insert, the following procedure is used:

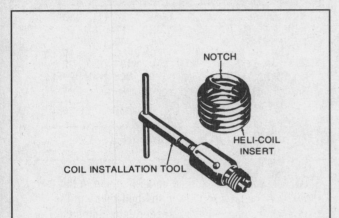

Fig. 17 The Heli-Coil® tool is used to thread the Heli-Coil® into the damaged hole after the hole is drilled oversize

a. Ream and countersink the damaged spark plug hole.
b. Check the countersink depth. It should be 13/16 tool in until the stop comes into full contact with the head.
c. Tap the hole.
d. Select an insert. Mount the insert on the installation tool.
e. Rotate the tool and insert clockwise until the insert bottoms in the hole.
f. Drive the special anti-rotation keys into the head using the installation tool, sleeve, and a hammer.
g. Remove the installation tool. Check that the insert is flush with the cylinder head surface and that all keys have seated at the undercut portion of the insert.
h. To install the sealing ring, place it squarely around the top of the insert. Then, install a flat seated spark plug, with the plug gasket removed, and tighten it to 35 ft. lbs. Remove the plug and check the seating of the ring. This should provide a gas tight seal, flush with the insert top.
i. Finally, install the spark plug with its gasket into the insert, and tighten it to its normal 18 ft. lbs.

To remove a Keenserts® insert, use a 21/32 drill through the center of the insert to a depth of 1/4 in. Remove the locking keys with a punch and remove the insert with an E-Z out® tool.

Snapped bolts or studs may be removed using Vise-Grip® pli-

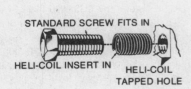

Fig. 18 Once the Heli-Coil® is installed in the hole, a normal bolt can be used again

3-42 ENGINE AND ENGINE REBUILDING

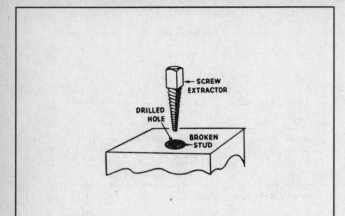

Fig. 19 A broken stud can be removed with a screw extractor after a hole is drilled down the center of the stud

ers. Penetrating oil (e.g. Liquid Wrench®, CRC®) will often aid in breaking the torque of frozen threads. In cases where the stud or bolt is broken off flush with, or below the surface, the following procedure may be used: Drill a hole (using a hardened bit) in the broken stud or bolt, about ½ of its diameter. Select a screw extractor (e.g. E-Z Out®) of the proper diameter, and tap it into the stud or bolt. Slowly turn the extractor counterclockwise to remove the stud or bolt.

One of the problems of small displacement, high-revving engines is that they are prone to developing fatigue cracks and other material flaws because they are highly stressed. One of the more popular procedures for checking metal fatigue and stress is Magnafluxing®. Magnafluxing® coats the part with fine magnetic particles, and subjects the part to a magnetic field. Cracks cause breaks in the magnetic field (even cracks below the surface not visible to the eye), which are outlined by the particles. However, since Magnafluxing® is a magnetic process, it applies only to ferrous metals (crankshafts, flywheels, connecting rods, etc.) It will not work with the aluminum heads and crankcases of these engines which are most prone to cracking.

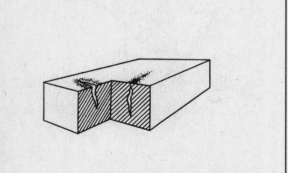

Fig. 20 Magnafluxing an engine component shows surface and sub-surface cracks—Magnflux only works on ferrous (iron or steel) components

Another process of checking for cracks is the Zyglo® process. This process does work with aluminum alloy. First the part is coated with a flourescent dye penetrant. Then the part is subjected to a blacklight inspection, under which cracks glow brightly, both at or below the surface.

A third method of checking for suspected cracks is the use of spot check dye. This method is quicker, and cheaper to perform, although hidden cracks beneath the surface may escape detection. First, the dye is sprayed onto the suspected area and wiped off. Then, the area is sprayed with a developer. The cracks will show up brightly.

If any of the threaded studs for the rocker arms or manifolds become damaged, and they are not broken off below the surface, they may be removed easily using the following procedure. Lock two nuts on the stud and unscrew the stud using the lower nut. It's as easy as that. Then, to make sure that the new stud remains in place, use locking compound on the threads.

Cylinder Head Reconditioning

IDENTIFY THE VALVES

▶ See Figure 21

Keep the valves in order, so that you know which valve (intake and exhaust) goes in which combustion chamber. If the valve faces are not full of carbon, you may number them, front to rear, with a permanent felt tip marker.

REMOVE THE VALVES AND SPRINGS

▶ See Figure 22

Using an appropriate valve spring compressor, compress the valve springs and lift out the keepers with needlenose pliers. Then, slowly release the compressor, and remove the valve, spring and spring retainer. On 1972 and earlier engines, a valve stem seal is used beneath the keepers which can be discarded. Check the keeper seating surfaces on the valve stem for burrs which may scratch the valve guide during installation of the valve. Remove any burrs with a fine file.

This section assumes that the cylinder head is removed for this operation. However, if it is desired to remove the valve springs with the head installed, it will be necessary to screw a compressed air adaptor into the subject spark plug hole and maintain a pressure of 85 psi to keep the valve from dropping down.

Inspect the exhaust valves closely. More often than not, the cause of low compression is a burned exhaust valve. The classic burned valve is cracked on the valve face from the edge of the seat to the stem the way you could cut a pie. Remove all carbon, gum and varnish from the valve stem with a hardwood chisel, or with a wire brush and solvent (i.e. carburetor cleaner, lacquer thinner).

HOT-TANK THE CYLINDER HEAD

Take the head(s) to an engine rebuilding or machine shop and have it (them) hot-tanked to remove grease, corrosion, carbon deposits and scale.

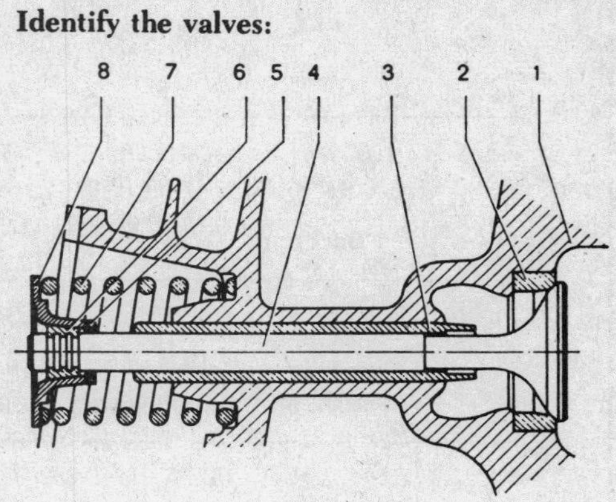

ENGINE AND ENGINE REBUILDING 3-43

Identify the valves:

1. Cylinder head
2. Valve seat insert
3. Valve guide
4. Valve
5. Oil deflector ring (valve stem seal)
6. Valve keeper (key)
7. Valve spring
8. Valve spring cap (retainer)

Fig. 21 Cutaway view of the valve and related components in the cylinder head

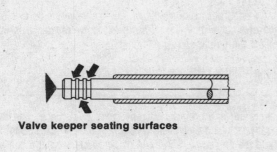

Fig. 22 The tip of the valve stem is manufactured with grooves which hold the valve keepers in place

➡ Make sure that the hot tanking solution is designed to clean aluminum, not to dissolve it.

After hot-tanking, inspect the combustion chambers (around the spark plug hole) and the exhaust ports for cracks. Also, check the plug threads, manifold studs, and rocker arm studs for damage and looseness.

DEGREASE THE REMAINING CYLINDER HEAD PARTS

Using solvent (i.e. Gunk® or Zep® carburetor cleaner), clean the rockers, rocker shafts, valve springs, spring retainers, keepers and the pushrods. You may also use solvent to clean the cylinder head although it will not clean as well as hot-tanking. Also clean the sheet metal shrouding at this time. Do not clean the pushrod tubes in solvent.

DE-CARBON THE CYLINDER HEAD

◆ See Figure 23

Chip carbon away from the combustion chambers and exhaust ports using a chisel made of hardwood. Remove the remaining deposits with a stiff wire brush. You may also use a power brush (drill with wire attachment if you use a very light touch). Remember that you are working with a relatively soft metal (aluminum), and you do not want to grind into the metal. If you have access to a machine shop that works on aluminum heads, ask them about glass-beading the cylinder head.

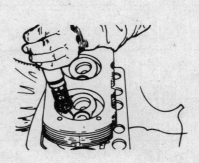

Fig. 23 The combustion chamber can be cleaned of carbon with a rotary wire brush mounted in a drill—work cautiously; the aluminum cylinder heads can be easily damaged

CHECK THE VALVE STEM-TO-GUIDE CLEARANCE (VALVE ROCK)

▶ See Figures 24, 25 and 26

Clean the valve stem with lacquer thinner or carburetor cleaner to remove all gum and varnish. Clean the valve guides using solvent and an expanding wire-type valve guide cleaner or brass bristle brush. Mount a dial indicator to the head so that the gauge pin is at a 90° angle to the valve stem, up against the edge of the valve head. Insert the valve by hand so that the stem end is flush with the end of the guide. Move the valve off its seat, and measure the clearance by rocking the stem back and forth to actuate the dial indicator. Check the figure against specifications. Maximum rock should not exceed the wear limit.

To check whether excessive rock is due to worn valve stems or guides (or both), one of two methods may be used. If a new valve is available, you may recheck the valve rock. If rock is still excessive the guide is at fault. Or, you may measure the old valve stem with a micrometer, and determine if it has passed its wear limit.

	Intake valve guide	Exhaust valve guide	Wear limit
Rock	.008–.009 in. (0.21–0.23 mm)	.011–.013 in. (0.28–0.32 mm)	.031 in. (0.8 mm)
Inside diameter	.3149–.3156 in. (8.00–8.02 mm)		.3172 in. (8.06 mm)

Fig. 24 Valve guide specifications for 1970–74 1600cc engines

	Intake valve guide	Exhaust valve guide	Wear limit
Rock	0.45 mm (.018 in.)		0.9 mm (.035 in.)
Inside diameter	8.00–8.02 mm (.3149–.3156 in.)	8.98–8.99 mm (.3534–.3538 in.)	8.06 or 9.06 mm (.3172 or .3566 in.)

Fig. 25 Valve guide specifications for all 1700cc, 1800cc and 2000cc engines

	Intake valve guide	Exhaust valve guide	Wear limit
Rock	.008–.009 in. (0.21–0.23 mm)	.018 in. (0.45 mm)	.032 or .035 in. (0.8 or 0.9 mm)
Inside diameter	.3149–.3156 in. (8.00–8.02 mm)	.3534–.3538 in. (8.98–8.99 mm)	.3172 or 3566 in. (8.06 or 9.06 mm)

Fig. 26 Valve guide specifications for 1975–80 1600cc engines

In any case, most VW and Porsche mechanics will replace the exhaust valve and guides anyway, since they often wear out inside of 50,000 miles.

VW does not make available oversize valve stems to clean up excessive valve rock. Therefore, if excessive clearance is evident, replace the guides.

KNURLING THE VALVE GUIDES

Knurling is a process whereby metal is displaced and raised, thereby reducing clearance. It is a procedure used in engines where the guides are shrunk in making replacement a costly procedure. Although this operation can be performed on VW and Porsche engines, it is not recommended, since the exhaust guides will eventually need replacement anyway.

REPLACING THE VALVE GUIDES

▶ See Figures 27, 28 and 29

The valve guides are a press fit into the head.

➡If your replacement valve guides do not have a collar at the top, measure the distance the old guides protrude above the head.

Several different methods may be used to remove worn valve guides. One method is to press or tap the guides out of the head using a stepped drift. The problem with this method is the risk of cracking the head. Another method, which reduces this risk, is to first drill out the guide about ⅔ of the length of the guide so that the walls of the guide at the top are paper thin (1/32 in. or so). This relieves most of the tension from the cylinder head guide bore, but still provides a solid base at the bottom of the guide to drift out the guide from the top. A third method of removing guides is to tap threads into the guide and pull it out from the top. After tapping the guide, place an old wrist pin (or some other type of sleeve) over the guide, so that the wrist pin rests squarely on the boss on the cylinder head around the guide. Then, take a long bolt (about 4 or 5 inches long with threads running all the

ENGINE AND ENGINE REBUILDING 3-45

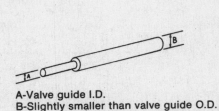

Fig. 27 A valve guide removal tool can be purchased or fabricated, so long as it is shaped as shown

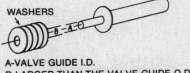

Fig. 29 For valve guides which do not possess a lip around the top edge, use a stack of washers to prevent driving the guide too deeply into the cylinder head

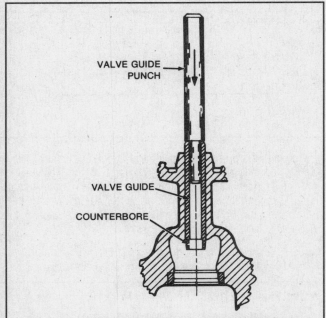

Fig. 28 Using the valve guide removal/installation tool, drive the valve guide into the cylinder head to the specified depth

way up to the bolt head) and thread a nut about halfway up the bolt. Place a washer on top of the wrist pin (see illustration) and thread the bolt into the valve guide until the nut contacts the washer and wrist pin. Finally, screw the nut down against the washer and wrist pin to pull out the guide.

If you are installing the guides without the aid of a press, using only hand tools, it will help to place the new valve guides in the freezer for an hour or so, and the clean, bare cylinder head in the oven at 350–400° F for ½ hour to 45 minutes. Controlling the temperature of the metals in this manner will slightly shrink the valve guides and slightly expand the guide bore in the cylinder head, allowing easier installation and lessening the risk of cracking the head in the process.

Most replacement valve guides, other than those manufactured by VW, have a collar at the top which provides a positive stop to seat the guides in the head. However, VW guides have no such collar. Therefore, on these guides, you will have to determine the height above the cylinder head boss that the guide must extend (about ¼ in.). Then, obtain a stack of washers, their inner diameter slightly larger than the outer diameter of the guide at the top of the guide. If the guide should extend ¼ in., use a ¼ in. thick stack of washers around the guide.

To install the valve guides in the head, use a collared drift, or a special valve guide installation tool of the proper outer diameter (see illustration).

✱✱ CAUTION

If you have heated the head in the oven to aid installation, be extremely careful handling metal of this temperature. Use pot holders, or gloves with thick insulation. Do not set the head down on any surface that may be affected by the heat.

If the replacement guide is collared, drive in the guide until it seats against the boss on the cylinder head. If the guide is not collared, drive in the guide until the installation tool butts against the stack of washers (approx. ¼ in. thick) on the head.

➡ **If you do not heat the head to aid installation, use penetrating lubricant in the guide bore, instead.**

RESURFACING (GRINDING) THE VALVE FACE

◆ See Figures 30, 31, 32 and 33

Using a valve grinding machine, have the valves resurfaced according to specifications (see chart). The valve stem tip should also be squared and resurfaced, by placing the stem in the V-block of the grinder, and turning it while pressing lightly against the grinding wheel.

➡ **After grinding, the minimum valve head margin must be 0.50mm (.020 in.). The valve head margin is the straight surface on the edge of the valve head, parallel with the valve stem.**

3-46 ENGINE AND ENGINE REBUILDING

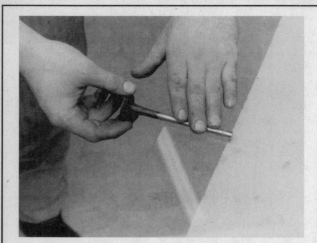

Valve stems may be rolled on a flat surface to check for bends

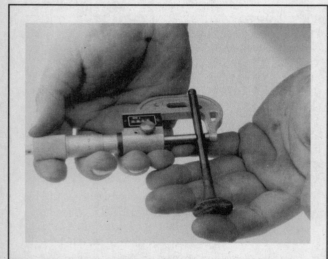

Use a micrometer to check the valve stem diameter

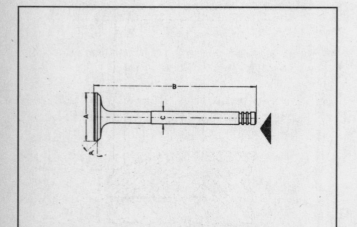

Fig. 30 Inspect the critical valve dimensions—the specific valve values are presented in the accompanying charts

Intake valves: 1600

A	B	C	D
1.259 in. (32.0 mm)	4.4 in. (112 mm)	.3130–3126 in. (7.95–7.94 mm)	44°

Exhaust valves: 1600

A	B	C	D
1.259 in. (32.0 mm)	4.4 in. (112 mm)	1970–'74 .3114–.3118 in. (7.91–7.92 mm)	45°
		1975–'76 .3500–.3510 in. (8.91–8.92 mm)	

Fig. 31 1600cc engine valve specifications—measure the valves to ensure that they meet these specifications

1700, 1800, 2000	Intake valve	Exhaust valve
A (1700)	39.1–39.3 mm dia (1.5394–1.5472 in.)	32.7–33.0 mm dia (1.2874–1.2992 in.)

Fig. 32 Valve specifications for all 1700cc, 1800cc and 2000cc engines—if the valves do not meet these values, machining or replacement is necessary

1700, 1800, 2000	Intake valve	Exhaust valve
A (1800)	41 mm dia (1.614 in.)	34 mm dia (1.338 in.)
A (2000)	37.5 mm dia (1.475 in.)	34 mm dia (1.338 in.)
B	116.8–117.3 mm (4.5984–4.6181 in.)	117.0–117.5 mm (4.6063–4.6260 in.)
C	7.94–7.95 mm dia (.3126–.3130 in.)	8.91–8.92 mm dia (.3508–.3512 in.)
D	29° 30'	45°

Fig. 33 Valve specifications for all 1700cc, 1800cc and 2000cc engines (continued)

ENGINE AND ENGINE REBUILDING

REPLACING VALVE SEAT INSERTS

This operation is not normally performed on VW and Porsche engines due to its expense and special shrink fit of the insert in the head. Usually, if the seat is destroyed, the head is also in bad shape (i.e. cracked, or hammered from a broken valve or piston). Some high-performance engine builders will replace the inserts to accommodate larger diameter valve heads. Otherwise, the operation will usually cost more than replacement of the head. Also, a replacement insert, if not installed correctly, could come out of the head, damaging the engine.

RESURFACING THE VALVE SEATS

▶ See Figures 34 thru 39

Most valve seats can be reconditioned by resurfacing. This is done with a reamer or grinder. First, a pilot is installed in the valve guide (a worn valve guide will allow the pilot to wobble, causing an inaccurate seat cut). When using a reamer, apply steady pressure while rotating clockwise. The seat should clean up in about four complete turns, taking care to remove only as much metal as necessary.

➡ **Never rotate a reamer counterclockwise.**

When using a grinder, lift the cutting stone on and off the seat at approximately two cycles per second, until all flaws are removed.

It takes three separate cuts to recondition a VW or Porsche valve seat. After each cut, check the position of the valve seat using Prussian blue dye (see illustration). First, you cut the center of the seat using a 45° cutter (30° cutter on 1700, 1800 and 2000 cc intake valve seats). Then, you cut the bottom of the seat with a 75° cutter and narrow the top of the seat with a 15° stone. The center of the seat (seat width "a") must be maintained as per the accompanying chart.

Equally as important as the width of the seat is its location in relation to the valve. Using a caliper, measure the distance between the center of the valve face on both sides of a valve. Then,

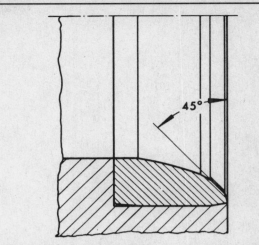

Fig. 35 When cutting the valve seats, first cut a 45 degree contact face—all 1600cc valves, as well as 1700cc, 1800cc and 2000cc exhaust valves

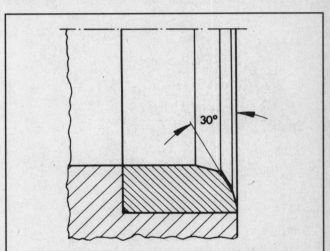

Fig. 36 The first cut for 1700cc, 1800cc and 2000cc intake valves should be at 30 degrees

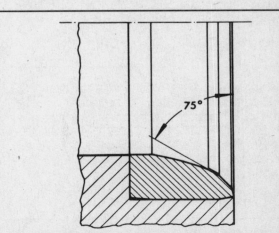

Fig. 37 The second cut (inner cut) on the intake valve seat should be at 75 degrees—1700cc, 1800cc and 2000cc engines

Engine	Intake	Exhaust
1600	.051–.063 in. (1.3–1.6 mm)	.067–.079 in. (1.7–2.0 mm)
1700, 1800, 2000	1.8–2.2 mm (.071–.087 in.)	2.0–2.5 mm (.079–.098 in.)

Fig. 34 The cylinder heads should be machined so that the valve seat width is within these specifications

3-48 ENGINE AND ENGINE REBUILDING

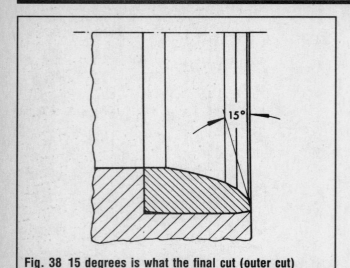

Fig. 38 15 degrees is what the final cut (outer cut) should be

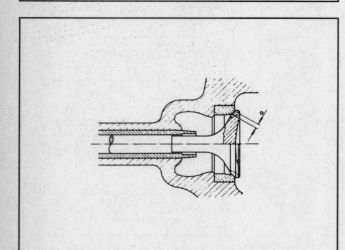

Fig. 39 The valve seat contact width (a) should be sufficient for proper valve sealing

Fig. 40 The valves can be hand-lapped using a special tool, equipped with a suction cup on one end, and a special lapping compound

Prior to lapping, invert the cylinder head, lightly lubricate the valve stem and install the valves in their respective guides. Coat the valve seats with fine Carborundum® grinding compound, and attach the lapping tool suction cup (moistened for adhesion) to the valve head. Then, rotate the tool between your palms, changing direction and lifting the tool often to prevent grooving. Lap the valve until a smooth, polished seat is evident. Finally, remove the tool and thoroughly wash away all traces of grinding compound. Make sure that no compound accumulates in the guides as rapid wear would result.

CHECK THE VALVE SPRINGS

♦ See Figure 41

Place the spring on a flat surface next to a square. Measure the height of the spring and compare that value to that of the other 7 springs. All springs should be the same height. Rotate the spring against the edge of the square to measure distortion. Replace any spring that varies (in both height and distortion) more than 1/16 in.

Use a caliper gauge to check the valve spring free-length

place the caliper on the valve seat, and check that the pointers of the caliper locate in the center of the seat.

CHECKING VALVE SEAT CONCENTRICITY

In order for the valve to seat perfectly in its seat, providing a gas tight seal, the valve seat must be concentric with the valve guide. To check concentricity, coat the valve face with Prussian blue dye and install the valve in its guide. Applying light pressure, rotate the valve 1/4 turn in its valve seat. If the entire valve seat face becomes coated, and the valve is known to be concentric, the seat is concentric.

LAPPING THE VALVES

♦ See Figure 40

With accurately refaced valve seat inserts and new valves, it is not usually necessary to lap the valves. Valve lapping alone is not recommended for use as a resurfacing procedure.

ENGINE AND ENGINE REBUILDING   3-49

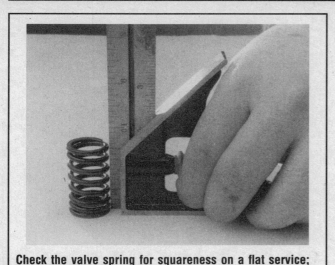

Check the valve spring for squareness on a flat service; a carpenter's square can be used

If you have access to a valve spring tester, you may use the following specifications to check the springs under a load (which is the only specification VW gives).

If any doubt exists as to the condition of the springs, and a spring tester is not available, replace them, they're cheap.

INSTALL THE VALVES

Lubricate the valve stems with white grease (molybdenum-disulphide), and install the valves in their respective guides. Lubricate and install the valve stem seals.

➡ VW has not installed stem seals on new engines since 1972. The reason is, although the seals provide excellent oil control, the guides tend to run "dry" which only hastens their demise. This is especially true for exhaust valves which run at much greater temperatures.

Position the valve springs on the head. The spring is positioned with the closely coiled end facing the head.

Type	Loaded length	Load
1600	1.2 in. (31.0 mm)	117.5–134.9 lb (53.2–61.2 kg)
1700, 1800, 2000	1.141 in. (29.0 mm)	168.0–186.0 lb (76.5–84.5 kg)

Fig. 41 Make certain that the valve springs meet the values shown—if the springs do not match these specifications, they should be replaced

Check the valve stem keys (keepers) for burrs or scoring. The keys should be machined so that the valve may still rotate with the keys held together. Finally, install the spring retainers, compress the springs (using a valve spring compressor), and insert the keys using needlenose pliers or a special tool designed for this purpose.

➡ You can retain the keys with wheel bearing grease during installation.

INSPECT THE ROCKER SHAFTS AND ROCKER ARMS

♦ See Figure 42

Remove the rocker arms, springs and washers from the rocker shaft.

➡ Lay out the parts in the order they are removed.

Inspect the rocker arms for pitting or wear on the valve stem contact point, and check for excessive rocker arm bushing wear where the arm rides on the shaft. If the shaft is grooved noticeably, replace it. Use the following chart to check the rocker arm inner diameter and the rocker shaft outer diameter.

1700, 1800, 2000		1600
.7874–.7882 in. (20.00–20.02 mm)	rocker arms inner diameter (new)	.7086–.7093 in. (18.00–18.02 mm)
.7890 in. (20.04 mm)	wear limit	.710 in. (18.04 mm)
.7854–.7861 in. (19.95–19.97 mm)	rocker arm shaft outer diameter (new)	.7073–.7077 in. (17.97–17.98 mm)
.7846 in. (19.93 mm)	wear limit	.7066 in. (17.95 mm)

Minor scoring may be removed with an emery cloth. If the valve stem contact point of the rocker arm is worn, grind it smooth, removing as little metal as necessary. If it is noticed at this point that the valve stem is worn concave where it contacts the rocker arm, and it is not desired to disassemble the valve from the head, a cap (see illustration) may be installed over the stem prior to installing the rocker shaft assembly.

INSPECT THE PUSHRODS AND PUSHROD TUBES

♦ See Figures 43, 44, 45, 46 and 47

After soaking the pushrods in solvent, clean out the oil passages using fine wire, then blow through them to make sure there

ENGINE AND ENGINE REBUILDING

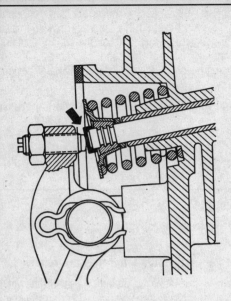

Fig. 42 Inspect the area where the rocker arm rides across the tip of the valve stem—if the stem is worn concave, and the cylinder head cannot be disassembled, a special cap (arrow) can be installed over the valve stem tip

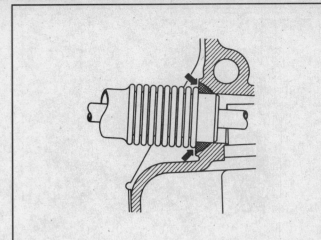

Fig. 44 Make certain to install new silicone seals on both ends of the pushrod tubes

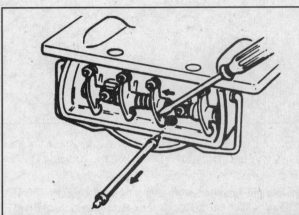

Fig. 45 To replace the pushrod tubes without removing the cylinder heads, first remove the rocker arm cover, push the applicable rocker arm to one side and pull the pushrod out of the tube . . .

are no obstructions. Roll each pushrod over a piece of clean, flat glass. Check for run-out. If a distinct, clicking sound is heard as the pushrod rolls, the rod is bent, necessitating replacement. All pushrods must be of equal length.

Inspect the pushrod tubes for cracks or other damage to the tube that would let oil out and dirt into the engine. The tubes on the 1600 engine are particularly susceptible to damage at the stretchable bellows. Also, on the 1600 engine, the tubes must be maintained at length "a" (see illustration) which is 190–191mm or 7.4–7.52 in. If a tube is too short, it may be carefully stretched, taking care to avoid cracking. However, if the bellows are damaged or if a gritty, rusty sound occurs when stretching the tube, replace it. Always use new seals. When installing tubes in a 1600

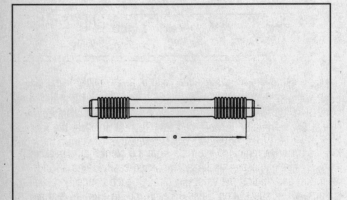

Fig. 43 The pushrod length of a should be 7.40-7.52 in.—if the tubes are installed too short, oil leakage will occur

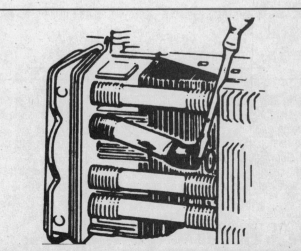

Fig. 46 . . . then pry the damaged tube loose—do not lose the seals from either end of the tube

ENGINE AND ENGINE REBUILDING

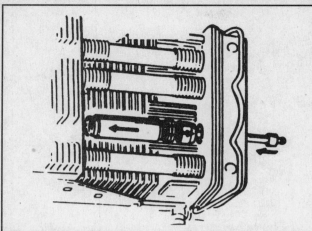

Fig. 47 Using the old seals, position the new tube with the gold end toward the spark plug and tighten until seated

engine, rotate the tubes so that the seams face upwards. When installing tubes in a 1700, 1800 or 2000 engine, make sure the retaining wire for the tubes engages the slots in the supports and rests on the lower edges of the tubes.

If, on an assembled, installed 1600 engine, it is desired to replace a damaged or leaky pushrod tube without pulling the engine, it may be accomplished using a "quick-change" pushrod tube available from several different specialty manufacturers. The special two-piece aluminum replacement tube is installed after removing the valve cover, rocker arm assembly and pushrod of the subject cylinder. The old tube is then pried loose with a screwdriver. Using new seals, the replacement tube is positioned between the head and crankcase, and expanded into place, via a pair of threaded, locking nuts.

Crankcase Reconditioning

DISASSEMBLING CRANKCASE

See "Crankcase Disassembly & Assembly" earlier in this section.

HOT TANK THE CRANKCASE

Using only a hot-tanking solution formulated for aluminum or magnesium alloy, clean the crankcase to remove all sludge, scale, or foreign particles. You may also cold-tank the case, using a strong degreasing solvent, but you will have to use a brush and a lot of elbow grease to get the same results.

After cleaning, blow out all oil passages with compressed air. Remove all old gasket sealing compound from the mating surfaces.

INSPECT THE CRANKCASE

Check the case for cracks using the Zyglo or spot-check method described earlier in this section.

Inspect all sealing or mating surfaces, especially along the crankcase seam, as the crankcase halves are machined in pairs and use no gasket.

Check the tightness of the oil suction pipe. The pipe must be centered over the strainer opening. On 1600 engines, peen over the crankcase where the suction pipe enters the camshaft bearing web.

Check all studs for tightness. Replace any defective studs as mentioned earlier in this section. Check all bearing bores for nicks and scratches. Remove light marks with a file. Deeper scratches and scoring must be removed by align boring the crankshaft bearing bores.

ALIGN BORE THE CRANKCASE

▶ See Figure 48

There are two surfaces on a VW crankcase that take quite a hammering in normal service. One is the main bearing saddles and the other the thrust flange of #1 bearing (at the flywheel end). Because the case is constructed of softer metal than the bearings, it is more malleable. The main bearing saddles are slowly hammered in by the rotation of the heavy crankshaft working against the bearings. This is especially true for an out-of-round crankshaft. The thrust flange of #1 main bearing receives its beating trying to control the end-play of the crankshaft. This beating is more severe in cases of a driver with a heavy clutch foot. Popping the clutch bangs the pressure plate against the clutch disc, against the flywheel, against the crankcase flange, and finally against the thrust flange. All of this hammering leaves its mark on the case, but can be cleaned up by align boring.

Most VW engine rebuilders who want their engines to stay together will align bore the case. This assures proper bearing bore alignment. Then, main bearings with the correct oversize outer diameter (and oversize thrust shoulder on #1) are installed.

Also, as the split crankcase is constructed of light aluminum and magnesium alloy, it is particularly susceptible to warpage due to overheating. Align boring the case will clean up any bearing saddle misalignment due to warpage.

CHECK CONNECTING ROD SIDE CLEARANCE, AND STRAIGHTNESS

Before removing the connecting rods from the crankshaft, check the clearance between the rod and the crank throw using a feeler gauge. Replace any rod exceeding the wear limit. Proper side clearance (also known as end-play or axial play) is .004–.016 in. (0.10–0.40mm).

Also, prior to removing the rods from the crankshaft, check them for straightness. This is accomplished easily using an old wrist pin, and sliding the wrist pin through each connecting rod (small end) in succession. Position each rod, in turn, so that as the pin begins to leave one rod, it is entering the next rod. Any binding indicates a scored wrist pin bushing or misaligned (bent) connecting rod. If the wrist pin absolutely will not slide from one adjacent rod to another, then you've got a really bent rod. Be ready for bent rods on any engine which has dropped a valve and damaged a piston.

3-52 ENGINE AND ENGINE REBUILDING

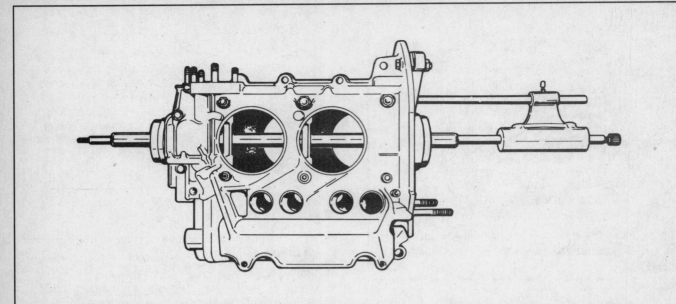

Fig. 48 Because of the importance of the fit between both crankcase halves, make sure to have the case align-bored

DISASSEMBLE THE CRANKSHAFT

Number the connecting rods (1 through 4 from the flywheel side) and matchmark their bearing halves. Remove the connecting rod retaining nuts (do not remove the bolts) from the bit end and remove the rods. Slide off the oil thrower (1600 only) and #4 main bearing. Slide off #1 main bearing from the flywheel end. Remove the snapring (circlip) using snapring pliers. #2 main bearing is the split type, each half of which should remain in its respective crankcase half. Using a large gear puller, or an arbor or hydraulic press, remove the distributor drive gear and crankshaft timing gear and spacer. Don't lose the woodruff key(s).

➡ The 1600 engine has two woodruff keys. The 1700, 1800 and 2000 engines have only one.

Finally, slide off #3 main bearing.

INSPECT THE CRANKSHAFT

Clean the crankshaft with solvent. Run all oil holes through with a brass bristle brush. Blow them through with compressed air. Lightly oil the crankshaft to prevent rusting.

Using a micrometer of known accuracy, measure the crankshaft journals for wear. The maximum wear limit for all journals is .0012 in. (0.03mm). Check the micrometer reading against those specifications listed under "Crankshaft & Connecting Rod Specifications" which appears earlier in this section.

Check the crankshaft run-out. With main bearing journals #1 and #3 supported on V-blocks and a dial gauge set up perpendicular to the crankshaft, measure the run-out at #2 and #4 main bearing journals. Maximum permissible run-out is .0008 in. (0.02mm).

Inspect the crankshaft journals for scratches, ridges, scoring and nicks. All small nicks and scratches necessitate regrinding of the crankshaft at a machine shop. Journals worn to a taper or slightly out-of-round must also be reground. Standard undersizes are .010, .020, .030 in. (0.25, 0.50, 0.75mm).

INSPECT THE CONNECTING RODS

▶ See Figure 49

Check the connecting rods for cracks, bends and burns. Check the rod bolts for damage; replace any rod with a damaged bolt. If

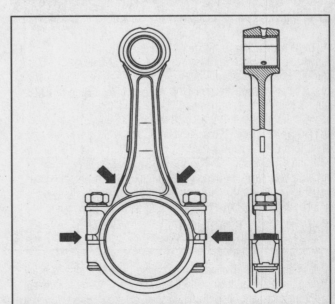

Fig. 49 If it is necessary to balance the crankshaft/connecting rod assembly, metal can be ground off of the connecting rods at the positions shown (arrows)

ENGINE AND ENGINE REBUILDING

possible, take the rods to a machine shop and have them checked for twists and magnafluxed for hidden stress cracks. Also, the rods must be checked for straightness, using the wrist pin method described earlier. If you did not perform this check before removing the rods from the crankshaft, definitely do so before dropping the assembled crankshaft into the case.

Weigh the rods on a gram scale. On 1600 engines, the rods should all weigh within 10 grams (lightest to heaviest); on 1700, 1800 and 2000 engines, within 6 grams. All rods should ideally weigh the same. If not, find the lightest rod and lighten the others to match. Up to 8 grams of metal can be removed from a rod by filing or grinding at the low stress points shown in the illustration.

Check the fit of the wrist pin bushing. At 72° F, the pin should slide through the bushing with only light thumb pressure.

CHECK CONNECTING ROD BEARING (OIL) CLEARANCE

It is always good practice to replace the connecting rod bearings at every teardown. The bearing size is stamped on the back of the inserts. However, if it is desired to reuse the bearings, two methods may be used to determine bearing clearance.

One tedious method is to measure the crankshaft journals using a micrometer to determine what size bearing inserts to use on reassembly (see Crankshaft and Connecting Rod Specifications) to obtain the required 0.0008–0.0027 in. oil clearance.

Another method of checking bearing clearance is the Plastigage® method. This method can only be used on the split-type bearings and not on the ring-type bearings used to support the crankshaft. First, clean all oil from the bearing surface and crankshaft journal being checked. Plastigage® is soluble in oil. Then, cut a piece of Plastigage® the width of the rod bearing and insert it between the journal and bearing insert.

➡ **Do not rotate the rod on the crankshaft.**

Tighten the rod cap nuts to 22–25 ft. lbs. Remove the bearing insert and check the thickness of the flattened Plastigage® using the Plastigage® scale. Journal taper is determined by comparing the width of the Plastigage® strip near its ends. To check for journal eccentricity, rotate the crankshaft 90° and retest. After checking all four connecting rod bearings in this manner, remove all traces of Plastigage® from the journal and bearing. Oil the crankshaft to prevent rusting.

If the oil clearance is .006 in. (0.15mm) or greater, it will be necessary to have the crankshaft ground to the nearest undersize (.010 in.) and use oversize connecting rod bearings.

CHECK MAIN BEARING (OIL) CLEARANCE

It is also good practice to replace the main bearings at every engine teardown as their replacement cost is minimal compared to the replacement cost of a crankshaft or short block. However, if it becomes necessary to reuse the bearings, you may do so after checking the bearing clearance.

Main bearings #1, 3 and 4 are ring-type bearings that slip over the crankshaft. These bearings cannot be checked using the Plasti-

Main Bearing Clearance

	New	Wear limit
Crankshaft bearings 1 + 3 (1600 engine)	0.04–0.10 mm (.0016–.004 in.)	0.18 mm (.007 in.)
Crankshaft bearings 1 + 3 (1700, 1800, 2000 engine)	0.05–0.10 mm (.002–.004 in.)	0.18 mm (.007 in.)
Crankshaft bearing 2 (all models)	0.03–0.09 mm (.001–.0035 in.)	0.17 mm (.0067 in.)
Crankshaft bearing 4 (all models)	0.05–0.10 mm (.002–.004 in.)	0.19 mm (.0075 in.)

gage method. Only the split-type #2 main bearing can be checked using Plastigage. However, since this involves bolting together and unbolting the crankcase halves several times, it is not recommended. Therefore, the main bearings are checked using a micrometer. Use the accompanying chart to determine if the bearing (oil) clearance exceeds its wear limit.

Never reuse a bearing that shows signs of wear, scoring or blueing. If the bearing clearance exceeds its wear limit, it will be necessary to regrind the crankshaft to the nearest undersize and use oversize main bearings.

CLEAN AND INSPECT THE CAMSHAFT

♦ **See Figure 50**

Degrease the camshaft using solvent. Clean out all oil holes and blow through with compressed air. Visually inspect the cam lobes and bearing journals for excessive wear. The edges of the

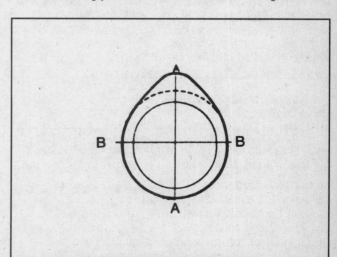

Fig. 50 To compute camshaft lobe lift, subtract measurement B from measurement A

3-54 ENGINE AND ENGINE REBUILDING

camshaft lobes should be square. Slight damage can be removed with silicone carbide oilstone. To check for lobe wear not visible to the eye, measure the camshaft diameter from the tip of the lobe to base (distance A) and then mike the diameter of the camshaft at a 90° angle to the previous measurement (distance B). This will give you camshaft lift. Measure lift for each lobe. If any lobe differs more than .025 in., replace the camshaft.

Check the camshaft for run-out. Place the #1 and #3 journals in V-blocks and rest a dial indicator on #2 journal. Rotate the camshaft and check the reading. Run-out must not exceed 0.0015 in. (0.04mm). Repair is by replacement.

Check the camshaft timing gear rivets for tightness. If any of the gear rivets are loose, or if the gear teeth show a poor contact pattern, replace the camshaft and timing gear assembly. Check the axial (end) play of the timing gear. Place the camshaft in the left crankcase half. The wear limit is .0063 in. (0.16mm). If the end-play is excessive, the thrust shoulder of #3 camshaft bearing is probably worn, necessitating replacement of the cam bearings.

CHECK THE CAMSHAFT BEARINGS

The camshaft bearings are the split-type. #3 camshaft bearing has shoulders on it to control axial play. Since there is no load on the camshaft, the bearings are not normally replaced. However, if the bearings are scored or imbedded with dirt, if the camshaft itself is being replaced, or if the thrust shoulders of #3 bearing are worn (permitting excessive axial play), the bearings should be replaced.

In all cases, clean the bearing saddles and check the oil feed holes for cleanliness. Make sure that the oil holes for the bearing inserts align with those in the crankcase. Coat the bearing surfaces with prelube.

CHECK THE LIFTERS (TAPPETS)

Remove all gum and varnish from the lifters using a toothbrush and carburetor cleaner. The cam following surface of the lifters is slightly convex when new. In service, this surface will wear flat which is OK to reinstall. However, if the cam following surface of the lifter is worn concave, the lifter should be replaced. To check this, place the cam following surface of one lifter against the side of another, using the one lifter as a straightedge. After checking, coat the lifters with oil to prevent rusting.

ASSEMBLE THE CRANKSHAFT

➡ **All dowel pin holes in the main bearings must locate to the flywheel end of the bearing saddles.**

Coat #3 main bearing journal with assembly lubricant. Slide the #3 bearing onto the pulley side of the crankshaft and install the large woodruff key in its recess (the hole in the bearing should be nearest to the flywheel end of the crankshaft). In the meantime, heat both the crankshaft timing gear and distributor drive gears to 176° F in an oil bath. If a hydraulic or an arbor press is available, press on the timing gear, taking care to keep the slot for the woodruff key aligned, the timing marks facing away from the flywheel, and the chamfer in the gear bore facing #3 main bearing journal.

✱✱ CAUTION

Use protective gloves when handling the heated gears.

➡ **Be careful not to scratch the crankshaft journals.**

Or, if a press is not available, you may drive on the gear using a 2 in. diameter length of pipe and a hammer, taking care to protect the flywheel end of the crankshaft with a piece of wood. The woodruff key must lie flat in its recess. Then, slide on the spacer ring and align it with the woodruff key. On 1600 engines, install the smaller woodruff key. Now, press or drive on the distributor drive gear in the same manner as the crankshaft timing gear. Make sure it seats against the spacer ring. Install the snapring (circlip) using snapring pliers. Take care not to scratch #4 main bearing journal. Prelube main bearings #1 and #4 and slide them on the crankshaft. On 1600 engines, install the oil slinger, concave side out.

➡ **Make sure crankshaft timing gear and distributor drive gear fit snugly on the crankshaft once they return to room temperature.**

Install the bearing inserts for the connecting rods and rod caps by pressing in on bearing ends with both thumbs. Make sure the tangs fit in the notches. Don't press in the middle as the inserts may soil or crack. Prelube the connecting rod bearings and journals. Then, install the connecting rods on the crankshaft, making sure the forge marks are up (as they would be installed in the crankcase [3, 1, 4, 2 from flywheel end]), and the rod and bearing cap matchmarks align. Use new connecting rod nuts. After tightening the nuts, make sure that each rod swings freely 180° on the crankshaft by its own weight.

➡ **A slight pretension (binding) of the rod on the crankshaft may be relieved by lightly rapping on the flat side of the big end of the rod with a hammer.**

If the connecting rod nuts are not of the self-locking type (very rare), peen the nuts into the slot on the rods to lock them in place and prevent the possibility of throwing a rod.

INSTALLING THE CRANKSHAFT AND CAMSHAFT

Pencil mark a line on the edge of each ring-type main bearing to indicate the location of the dowel pin hole. Install the lower half of #2 main bearing in the left side of the crankcase so that the shell fits securely over its dowel pin. Prelube the bearing surface.

Lift the crankshaft by two of the connecting rods and lower the assembly into the left crankcase halve. Make sure the other connecting rods protrude through their corresponding cylinder openings. Then, rotate each ring-type main bearing (#1, then #3, then #4) until the pencil marks made previously align with the center of the bearing bore. As each bearing is aligned with its dowel pin, a distinctive click should be heard and the crankshaft should be felt dropping into position. After each bearing is seated, you should not be able to rock any of the main bearings or the crankshaft in the case. Just to be sure, check the bearing installation

ENGINE AND ENGINE REBUILDING

by placing the other half of #2 main bearing over the top of its crankshaft journal. If the upper half rocks, the bearing or bearings are not seated properly on their dowels. Then, install the other half of #2 main bearing in the right crankcase halve. Prelube the bearing surface.

Rotate the crankshaft until the timing marks (twin punch marks on two adjacent teeth) on the timing gear point towards the camshaft side of the case. Lubricate and install the lifters. Coat the lifters for the right half of the case with grease to keep them from falling out during assembly. Coat the camshaft journals and bearing surfaces with assembly lubricant. Install the camshaft so that the single timing mark (0) on the camshaft timing gear aligns (lies between) with the two on the crankshaft timing gear. This is critical as it establishes valve timing.

Install the camshaft end plug using oil-resistant sealer. On cars with manual transmission, the hollow end of the plug faces in towards the engine. On cars equipped with automatic or automatic stick shift transmission, the hollow end faces out towards the front of the car to provide clearance for the torque converter drive plate retaining bolts.

The timing gear mesh is correct if the camshaft does not lift from its bearings when the crankshaft is rotated backwards (opposite normal direction of rotation).

CHECK TIMING GEAR BACKLASH

Mount a dial indicator to the crankcase with its stem resting on a tooth of the camshaft gear. Rotate the gear until all slack is removed, and zero the indicator. Then, rotate the gear in the opposite direction until all slack is removed and record gear backlash. The reading should be between .000 and .002 in. (0.00 and 0.05mm).

ASSEMBLING THE CRANKCASE

▶ See Figure 51

See "Crankcase Assembly & Disassembly" earlier in this section. Use the following installation notes;

a. Make sure all bearing surfaces are prelubed.
b. Always install new crankcase stud seals.
c. Apply only non-hardening oil resistant sealer to all crankcase mating surfaces.
d. Always use new case nuts. Self-sealing nuts must be installed with the red coated side down.
e. All small crankcase retaining nuts are first torqued to 10 ft. lbs., then 14 ft. lbs. All large crankcase retaining nuts are torqued to 20 ft. lbs., then 25 ft. lbs. (except self-sealing large nuts [red plastic insert], which are torqued to a single figure of 18 ft. lbs.). Use a crisscross torque sequence. On 1700, 1800 and 2000 engines, you will have to keep the long case bolt heads from turning.
f. While assembling the crankcase halves, always rotate the crankshaft periodically to check for binding. If any binding occurs, immediately disassemble and investigate the case. Usually, a main bearing has come off its dowel pin, or maybe you forgot to align bore that warped crankcase.

CHECK CRANKSHAFT END-PLAY

▶ See Figure 52

After assembling the case, crankshaft end-play can be checked. End-play is controlled by the thickness of 3 shims located between the flywheel and #1 main bearing flange. End-play is

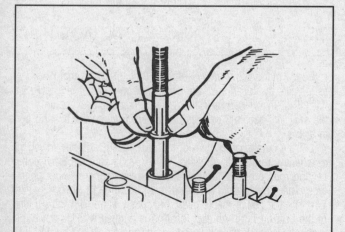

Fig. 51 Before assembling the two crankcase halves, make certain to install new crankcase stud seals

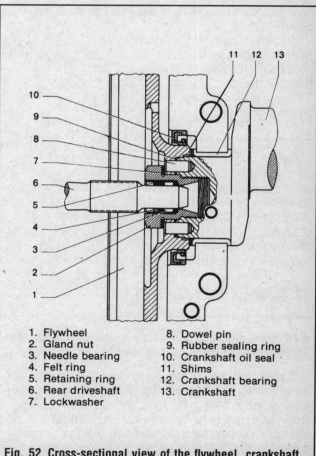

1. Flywheel
2. Gland nut
3. Needle bearing
4. Felt ring
5. Retaining ring
6. Rear driveshaft
7. Lockwasher
8. Dowel pin
9. Rubber sealing ring
10. Crankshaft oil seal
11. Shims
12. Crankshaft bearing
13. Crankshaft

Fig. 52 Cross-sectional view of the flywheel, crankshaft, oil seal and related components used on 1600cc engines

checked with the flywheel installed as follows. Attach a dial indicator to the crankcase with the stem positioned on the face of the flywheel. Move the flywheel in and out and check the reading. End-play should be between .003–.005 in. (0.07–0.13mm). The wear limit is .006 in. (0.15mm).

To adjust end-play, remove the flywheel and reinstall, this time using only two shims. Remeasure the end-play. The difference between the second reading and the .003–.005 in. figure is the required thickness of the third shim. Shims come in the following sizes:

0.24mm—.0095 in.
0.30mm—.0118 in.
0.32mm—.0126 in.
0.34mm—.0134 in.
0.36mm—.0142 in.
0.38mm—.0150 in. (1700, 1800 and 2000 only)

AIR POLLUTION 4-2
NATURAL POLLUTANTS 4-2
INDUSTRIAL POLLUTANTS 4-2
AUTOMOTIVE POLLUTANTS 4-2
   TEMPERATURE INVERSION 4-2
   HEAT TRANSFER 4-3
AUTOMOTIVE EMISSIONS 4-3
EXHAUST GASES 4-3
   HYDROCARBONS 4-3
   CARBON MONOXIDE 4-4
   NITROGEN 4-4
   OXIDES OF SULFUR 4-4
   PARTICULATE MATTER 4-4
CRANKCASE EMISSIONS 4-5
EVAPORATIVE EMISSIONS 4-5
EMISSION CONTROLS 4-5
CRANKCASE VENTILATION
  SYSTEM 4-5
   SERVICING 4-6
EVAPORATIVE EMISSION CONTROL
  SYSTEM 4-6
   SERVICING 4-7
AIR INJECTION SYSTEM 1973–74 4-7
   SERVICING 4-7
EXHAUST GAS RECIRCULATION
  SYSTEM 4-7
   GENERAL DESCRIPTION 4-8
   EGR VALVE SERVICE 4-10
   EGR VALVE INSPECTION 4-10
CATALYTIC CONVERTER SYSTEM 4-12
DECELERATION CONTROL 4-12
   CHECKING THE DECELERATION
    VALVE 4-13
THROTTLE VALVE POSITIONER 4-13
   ADJUSTMENT 4-14
VACUUM DIAGRAMS 4-14

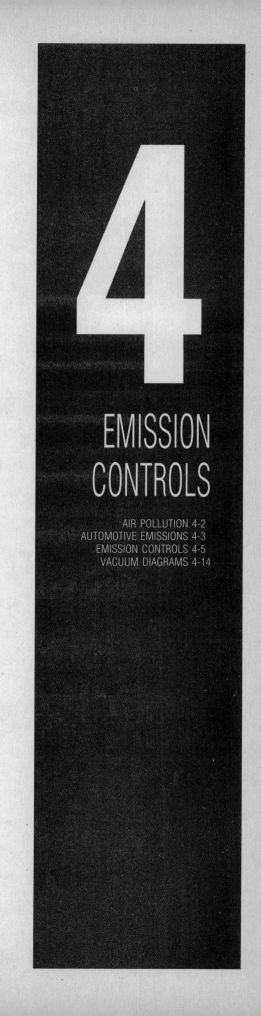

4
EMISSION CONTROLS

AIR POLLUTION 4-2
AUTOMOTIVE EMISSIONS 4-3
EMISSION CONTROLS 4-5
VACUUM DIAGRAMS 4-14

4-2 EMISSION CONTROLS

AIR POLLUTION

The earth's atmosphere, at or near sea level, consists approximately of 78 percent nitrogen, 21 percent oxygen and 1 percent other gases. If it were possible to remain in this state, 100 percent clean air would result. However, many varied sources allow other gases and particulates to mix with the clean air, causing our atmosphere to become unclean or polluted.

Some of these pollutants are visible while others are invisible, with each having the capability of causing distress to the eyes, ears, throat, skin and respiratory system. Should these pollutants become concentrated in a specific area and under certain conditions, death could result due to the displacement or chemical change of the oxygen content in the air. These pollutants can also cause great damage to the environment and to the many man made objects that are exposed to the elements.

To better understand the causes of air pollution, the pollutants can be categorized into 3 separate types, natural, industrial and automotive.

Natural Pollutants

Natural pollution has been present on earth since before man appeared and continues to be a factor when discussing air pollution, although it causes only a small percentage of the overall pollution problem. It is the direct result of decaying organic matter, wind born smoke and particulates from such natural events as plain and forest fires (ignited by heat or lightning), volcanic ash, sand and dust which can spread over a large area of the countryside.

Such a phenomenon of natural pollution has been seen in the form of volcanic eruptions, with the resulting plume of smoke, steam and volcanic ash blotting out the sun's rays as it spreads and rises higher into the atmosphere. As it travels into the atmosphere the upper air currents catch and carry the smoke and ash, while condensing the steam back into water vapor. As the water vapor, smoke and ash travel on their journey, the smoke dissipates into the atmosphere while the ash and moisture settle back to earth in a trail hundreds of miles long. In some cases, lives are lost and millions of dollars of property damage result.

Industrial Pollutants

Industrial pollution is caused primarily by industrial processes, the burning of coal, oil and natural gas, which in turn produce smoke and fumes. Because the burning fuels contain large amounts of sulfur, the principal ingredients of smoke and fumes are sulfur dioxide and particulate matter. This type of pollutant occurs most severely during still, damp and cool weather, such as at night. Even in its less severe form, this pollutant is not confined to just cities. Because of air movements, the pollutants move for miles over the surrounding countryside, leaving in its path a barren and unhealthy environment for all living things.

Working with Federal, State and Local mandated regulations and by carefully monitoring emissions, big business has greatly reduced the amount of pollutant introduced from its industrial sources, striving to obtain an acceptable level. Because of the mandated industrial emission clean up, many land areas and streams in and around the cities that were formerly barren of vegetation and life, have now begun to move back in the direction of nature's intended balance.

Automotive Pollutants

The third major source of air pollution is automotive emissions. The emissions from the internal combustion engines were not an appreciable problem years ago because of the small number of registered vehicles and the nation's small highway system. However, during the early 1950's, the trend of the American people was to move from the cities to the surrounding suburbs. This caused an immediate problem in transportation because the majority of suburbs were not afforded mass transit conveniences. This lack of transportation created an attractive market for the automobile manufacturers, which resulted in a dramatic increase in the number of vehicles produced and sold, along with a marked increase in highway construction between cities and the suburbs. Multi-vehicle families emerged with a growing emphasis placed on an individual vehicle per family member. As the increase in vehicle ownership and usage occurred, so did pollutant levels in and around the cities, as suburbanites drove daily to their businesses and employment, returning at the end of the day to their homes in the suburbs.

It was noted that a smoke and fog type haze was being formed and at times, remained in suspension over the cities, taking time to dissipate. At first this "smog," derived from the words "smoke" and "fog," was thought to result from industrial pollution but it was determined that automobile emissions shared the blame. It was discovered that when normal automobile emissions were exposed to sunlight for a period of time, complex chemical reactions would take place.

It is now known that smog is a photo chemical layer which develops when certain oxides of nitrogen (NOx) and unburned hydrocarbons (HC) from automobile emissions are exposed to sunlight. Pollution was more severe when smog would become stagnant over an area in which a warm layer of air settled over the top of the cooler air mass, trapping and holding the cooler mass at ground level. The trapped cooler air would keep the emissions from being dispersed and diluted through normal air flows. This type of air stagnation was given the name "Temperature Inversion."

TEMPERATURE INVERSION

In normal weather situations, surface air is warmed by heat radiating from the earth's surface and the sun's rays. This causes it to rise upward, into the atmosphere. Upon rising it will cool through a convection type heat exchange with the cooler upper air. As warm air rises, the surface pollutants are carried upward and dissipated into the atmosphere.

When a temperature inversion occurs, we find the higher air is no longer cooler, but is warmer than the surface air, causing the cooler surface air to become trapped. This warm air blanket can extend from above ground level to a few hundred or even a few thousand feet into the air. As the surface air is trapped, so are the pollutants, causing a severe smog condition. Should this stagnant air mass extend to a few thousand feet high, enough air movement with the inversion takes place to allow the smog layer to rise above ground level but the pollutants still cannot dissipate. This inversion can remain for days over an area, with the smog level only rising or lowering from ground level to a few hundred feet high. Meanwhile, the pollutant levels increase, causing eye irritation, respiratory problems, reduced visibility, plant damage and in some cases, even disease.

This inversion phenomenon was first noted in the Los Angeles, California area. The city lies in terrain resembling a basin and with certain weather conditions, a cold air mass is held in the basin while a warmer air mass covers it like a lid.

Because this type of condition was first documented as prevalent in the Los Angeles area, this type of trapped pollution was named Los Angeles Smog, although it occurs in other areas where a large concentration of automobiles are used and the air remains stagnant for any length of time.

HEAT TRANSFER

Consider the internal combustion engine as a machine in which raw materials must be placed so a finished product comes out. As in any machine operation, a certain amount of wasted material is formed. When we relate this to the internal combustion engine, we find that through the input of air and fuel, we obtain power during the combustion process to drive the vehicle. The by-product or waste of this power is, in part, heat and exhaust gases with which we must dispose.

AUTOMOTIVE EMISSIONS

Before emission controls were mandated on internal combustion engines, other sources of engine pollutants were discovered along with the exhaust emissions. It was determined that engine combustion exhaust produced approximately 60 percent of the total emission pollutants, fuel evaporation from the fuel tank and carburetor vents produced 20 percent, with the final 20 percent being produced through the crankcase as a by-product of the combustion process.

Exhaust Gases

The exhaust gases emitted into the atmosphere are a combination of burned and unburned fuel. To understand the exhaust emission and its composition, we must review some basic chemistry.

When the air/fuel mixture is introduced into the engine, we are mixing air, composed of nitrogen (78 percent), oxygen (21 percent) and other gases (1 percent) with the fuel, which is 100 percent hydrocarbons (HC), in a semi-controlled ratio. As the combustion process is accomplished, power is produced to move the vehicle while the heat of combustion is transferred to the cooling system. The exhaust gases are then composed of nitrogen, a diatomic gas (N_2), the same as was introduced in the engine, carbon dioxide (CO_2), the same gas that is used in beverage carbonation, and water vapor (H_2O). The nitrogen (N_2), for the most part, passes through the engine unchanged, while the oxygen (O_2) reacts (burns) with the hydrocarbons (HC) and produces the carbon dioxide (CO_2) and the water vapors (H_2O). If this chemical process would be the only process to take place, the exhaust emissions would be harmless. However, during the combustion process, other compounds are formed which are considered dangerous. These pollutants are hydrocarbons (HC), carbon monoxide (CO), oxides of nitrogen (NOx) oxides of sulfur (SOx) and engine particulates.

The heat from the combustion process can rise to over 4000°F (2204°C). The dissipation of this heat is controlled by a ram air effect, the use of cooling fans to cause air flow and a liquid coolant solution surrounding the combustion area to transfer the heat of combustion through the cylinder walls and into the coolant. The coolant is then directed to a thin-finned, multi-tubed radiator, from which the excess heat is transferred to the atmosphere by 1 of the 3 heat transfer methods, conduction, convection or radiation.

The cooling of the combustion area is an important part in the control of exhaust emissions. To understand the behavior of the combustion and transfer of its heat, consider the air/fuel charge. It is ignited and the flame front burns progressively across the combustion chamber until the burning charge reaches the cylinder walls. Some of the fuel in contact with the walls is not hot enough to burn, thereby snuffing out or quenching the combustion process. This leaves unburned fuel in the combustion chamber. This unburned fuel is then forced out of the cylinder and into the exhaust system, along with the exhaust gases.

Many attempts have been made to minimize the amount of unburned fuel in the combustion chambers due to quenching, by increasing the coolant temperature and lessening the contact area of the coolant around the combustion area. However, design limitations within the combustion chambers prevent the complete burning of the air/fuel charge, so a certain amount of the unburned fuel is still expelled into the exhaust system, regardless of modifications to the engine.

HYDROCARBONS

Hydrocarbons (HC) are essentially fuel which was not burned during the combustion process or which has escaped into the atmosphere through fuel evaporation. The main sources of incomplete combustion are rich air/fuel mixtures, low engine temperatures and improper spark timing. The main sources of hydrocarbon emission through fuel evaporation on most vehicles used to be the vehicle's fuel tank and carburetor float bowl.

To reduce combustion hydrocarbon emission, engine modifications were made to minimize dead space and surface area in the combustion chamber. In addition, the air/fuel mixture was made more lean through the improved control which feedback carburetion and fuel injection offers and by the addition of external controls to aid in further combustion of the hydrocarbons outside the engine. Two such methods were the addition of air injection systems, to inject fresh air into the exhaust manifolds and the installation of catalytic converters, units that are able to burn traces of hydrocarbons without affecting the internal combustion process or fuel economy.

To control hydrocarbon emissions through fuel evaporation, modifications were made to the fuel tank to allow storage of the fuel vapors during periods of engine shut-down. Modifications were also made to the air intake system so that at specific times during engine operation, these vapors may be purged and burned by blending them with the air/fuel mixture.

CARBON MONOXIDE

Carbon monoxide is formed when not enough oxygen is present during the combustion process to convert carbon (C) to carbon dioxide (CO_2). An increase in the carbon monoxide (CO) emission is normally accompanied by an increase in the hydrocarbon (HC) emission because of the lack of oxygen to completely burn all of the fuel mixture.

Carbon monoxide (CO) also increases the rate at which the photo chemical smog is formed by speeding up the conversion of nitric oxide (NO) to nitrogen dioxide (NO_2). To accomplish this, carbon monoxide (CO) combines with oxygen (O_2) and nitric oxide (NO) to produce carbon dioxide (CO_2) and nitrogen dioxide (NO_2). ($CO + O_2 + NO \rightarrow CO_2 + NO_2$).

The dangers of carbon monoxide, which is an odorless and colorless toxic gas are many. When carbon monoxide is inhaled into the lungs and passed into the blood stream, oxygen is replaced by the carbon monoxide in the red blood cells, causing a reduction in the amount of oxygen supplied to the many parts of the body. This lack of oxygen causes headaches, lack of coordination, reduced mental alertness and, should the carbon monoxide concentration be high enough, death could result.

NITROGEN

Normally, nitrogen is an inert gas. When heated to approximately 2500°F (1371°C) through the combustion process, this gas becomes active and causes an increase in the nitric oxide (NO) emission.

Oxides of nitrogen (NOx) are composed of approximately 97–98 percent nitric oxide (NO). Nitric oxide is a colorless gas but when it is passed into the atmosphere, it combines with oxygen and forms nitrogen dioxide (NO_2). The nitrogen dioxide then combines with chemically active hydrocarbons (HC) and when in the presence of sunlight, causes the formation of photo-chemical smog.

Ozone

To further complicate matters, some of the nitrogen dioxide (NO_2) is broken apart by the sunlight to form nitric oxide and oxygen. (NO_2 + sunlight $\rightarrow$ NO + O). This single atom of oxygen then combines with diatomic (meaning 2 atoms) oxygen (O_2) to form ozone (O_3). Ozone is one of the smells associated with smog. It has a pungent and offensive odor, irritates the eyes and lung tissues, affects the growth of plant life and causes rapid deterioration of rubber products. Ozone can be formed by sunlight as well as electrical discharge into the air.

The most common discharge area on the automobile engine is the secondary ignition electrical system, especially when inferior quality spark plug cables are used. As the surge of high voltage is routed through the secondary cable, the circuit builds up an electrical field around the wire, which acts upon the oxygen in the surrounding air to form the ozone. The faint glow along the cable with the engine running that may be visible on a dark night, is called the "corona discharge." It is the result of the electrical field passing from a high along the cable, to a low in the surrounding air, which forms the ozone gas. The combination of corona and ozone has been a major cause of cable deterioration. Recently, different and better quality insulating materials have lengthened the life of the electrical cables.

Although ozone at ground level can be harmful, ozone is beneficial to the earth's inhabitants. By having a concentrated ozone layer called the "ozonosphere," between 10 and 20 miles (16–32 km) up in the atmosphere, much of the ultra violet radiation from the sun's rays are absorbed and screened. If this ozone layer were not present, much of the earth's surface would be burned, dried and unfit for human life.

OXIDES OF SULFUR

Oxides of sulfur (SOx) were initially ignored in the exhaust system emissions, since the sulfur content of gasoline as a fuel is less than $1/10$ of 1 percent. Because of this small amount, it was felt that it contributed very little to the overall pollution problem. However, because of the difficulty in solving the sulfur emissions in industrial pollutions and the introduction of catalytic converter to the automobile exhaust systems, a change was mandated. The automobile exhaust system, when equipped with a catalytic converter, changes the sulfur dioxide (SO_2) into sulfur trioxide (SO_3).

When this combines with water vapors (H_2O), a sulfuric acid mist (H_2SO_4) is formed and is a very difficult pollutant to handle since it is extremely corrosive. This sulfuric acid mist that is formed, is the same mist that rises from the vents of an automobile battery when an active chemical reaction takes place within the battery cells.

When a large concentration of vehicles equipped with catalytic converters are operating in an area, this acid mist may rise and be distributed over a large ground area causing land, plant, crop, paint and building damage.

PARTICULATE MATTER

A certain amount of particulate matter is present in the burning of any fuel, with carbon constituting the largest percentage of the particulates. In gasoline, the remaining particulates are the burned remains of the various other compounds used in its manufacture. When a gasoline engine is in good internal condition, the particulate emissions are low but as the engine wears internally, the particulate emissions increase. By visually inspecting the tail pipe emissions, a determination can be made as to where an engine defect may exist. An engine with light gray or blue smoke emitting from the tail pipe normally indicates an increase in the oil consumption through burning due to internal engine wear. Black smoke would indicate a defective fuel delivery system, causing the engine to operate in a rich mode. Regardless of the color of the smoke, the internal part of the engine or the fuel delivery system should be repaired to prevent excess particulate emissions.

Diesel and turbine engines emit a darkened plume of smoke from the exhaust system because of the type of fuel used. Emission control regulations are mandated for this type of emission and more stringent measures are being used to prevent excess emission of the particulate matter. Electronic components are be-

EMISSION CONTROLS 4-5

ing introduced to control the injection of the fuel at precisely the proper time of piston travel, to achieve the optimum in fuel ignition and fuel usage. Other particulate after-burning components are being tested to achieve a cleaner emission.

Good grades of engine lubricating oils should be used, which meet the manufacturers specification. Cut-rate oils can contribute to the particulate emission problem because of their low flash or ignition temperature point. Such oils burn prematurely during the combustion process causing emission of particulate matter.

The cooling system is an important factor in the reduction of particulate matter. The optimum combustion will occur, with the cooling system operating at a temperature specified by the manufacturer. The cooling system must be maintained in the same manner as the engine oiling system, as each system is required to perform properly in order for the engine to operate efficiently for a long time.

Crankcase Emissions

Crankcase emissions are made up of water, acids, unburned fuel, oil fumes and particulates. These emissions are classified as hydrocarbons (HC) and are formed by the small amount of unburned, compressed air/fuel mixture entering the crankcase from the combustion area (between the cylinder walls and piston rings) during the compression and power strokes. The head of the compression and combustion help to form the remaining crankcase emissions.

Since the first engines, crankcase emissions were allowed into the atmosphere through a road draft tube, mounted on the lower side of the engine block. Fresh air came in through an open oil filler cap or breather. The air passed through the crankcase mixing with blow-by gases. The motion of the vehicle and the air blowing past the open end of the road draft tube caused a low pressure area (vacuum) at the end of the tube. Crankcase emissions were simply drawn out of the road draft tube into the air.

To control the crankcase emission, the road draft tube was deleted. A hose and/or tubing was routed from the crankcase to the intake manifold so the blow-by emission could be burned with the air/fuel mixture. However, it was found that intake manifold vacuum, used to draw the crankcase emissions into the manifold, would vary in strength at the wrong time and not allow the proper emission flow. A regulating valve was needed to control the flow of air through the crankcase.

Testing, showed the removal of the blow-by gases from the crankcase as quickly as possible, was most important to the longevity of the engine. Should large accumulations of blow-by gases remain and condense, dilution of the engine oil would occur to form water, soots, resins, acids and lead salts, resulting in the formation of sludge and varnishes. This condensation of the blow-by gases occurs more frequently on vehicles used in numerous starting and stopping conditions, excessive idling and when the engine is not allowed to attain normal operating temperature through short runs.

Evaporative Emissions

Gasoline fuel is a major source of pollution, before and after it is burned in the automobile engine. From the time the fuel is refined, stored, pumped and transported, again stored until it is pumped into the fuel tank of the vehicle, the gasoline gives off unburned hydrocarbons (HC) into the atmosphere. Through the redesign of storage areas and venting systems, the pollution factor was diminished, but not eliminated, from the refinery standpoint. However, the automobile still remained the primary source of vaporized, unburned hydrocarbon (HC) emissions.

Fuel pumped from an underground storage tank is cool but when exposed to a warmer ambient temperature, will expand. Before controls were mandated, an owner might fill the fuel tank with fuel from an underground storage tank and park the vehicle for some time in warm area, such as a parking lot. As the fuel would warm, it would expand and should no provisions or area be provided for the expansion, the fuel would spill out of the filler neck and onto the ground, causing hydrocarbon (HC) pollution and creating a severe fire hazard. To correct this condition, the vehicle manufacturers added overflow plumbing and/or gasoline tanks with built in expansion areas or domes.

However, this did not control the fuel vapor emission from the fuel tank. It was determined that most of the fuel evaporation occurred when the vehicle was stationary and the engine not operating. Most vehicles carry 5–25 gallons (19–95 liters) of gasoline. Should a large concentration of vehicles be parked in one area, such as a large parking lot, excessive fuel vapor emissions would take place, increasing as the temperature increases.

To prevent the vapor emission from escaping into the atmosphere, the fuel systems were designed to trap the vapors while the vehicle is stationary, by sealing the system from the atmosphere. A storage system is used to collect and hold the fuel vapors from the carburetor (if equipped) and the fuel tank when the engine is not operating. When the engine is started, the storage system is then purged of the fuel vapors, which are drawn into the engine and burned with the air/fuel mixture.

EMISSION CONTROLS

Crankcase Ventilation System

♦ See Figures 1, 2, 3 and 4

All models are equipped with a crankcase ventilation system. The purpose of the crankcase ventilation system is twofold. It keeps harmful vapors from escaping into the atmosphere and prevents the buildup of crankcase pressure. Prior to the 1960's, most cars employed a vented oil filler cap and road draft tube to dispose of crankcase vapor. The crankcase ventilation systems now in use are an improvement over the old method and, when functioning properly, will not reduce engine efficiency.

Type 1 and 2 carbureted engine crankcase vapors are recirculated from the oil breather through a rubber hose to the air cleaner. The vapors then join the air/fuel mixture and are burned in the engine. Fuel injected cars mix crankcase vapors into the air/fuel mixture to be burned in the combustion chambers. Fresh air is forced through the engine to evacuate vapors and recirculate them into the oil breather, intake air distributor, and then to be burned.

4-6 EMISSION CONTROLS

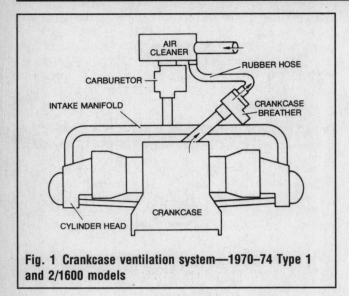

Fig. 1 Crankcase ventilation system—1970–74 Type 1 and 2/1600 models

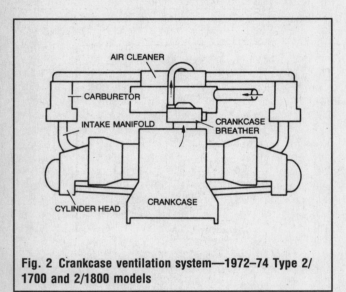

Fig. 2 Crankcase ventilation system—1972–74 Type 2/1700 and 2/1800 models

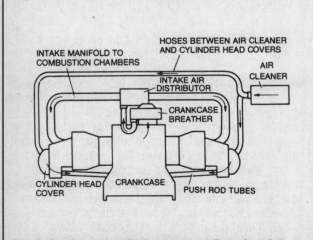

Fig. 3 Crankcase ventilation system—Type 3 and 4 models

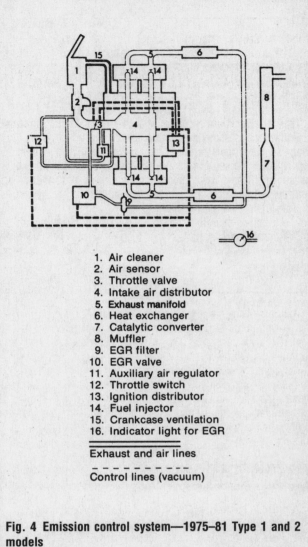

1. Air cleaner
2. Air sensor
3. Throttle valve
4. Intake air distributor
5. Exhaust manifold
6. Heat exchanger
7. Catalytic converter
8. Muffler
9. EGR filter
10. EGR valve
11. Auxiliary air regulator
12. Throttle switch
13. Ignition distributor
14. Fuel injector
15. Crankcase ventilation
16. Indicator light for EGR

Exhaust and air lines

――――――
Control lines (vacuum)

Fig. 4 Emission control system—1975–81 Type 1 and 2 models

SERVICING

The only maintenance required on the crankcase ventilation system is a periodic check. At every tune-up, examine the hoses for clogging or deterioration. Clean or replace the hoses as required.

Evaporative Emission Control System

Required by law since 1971, this system prevents raw fuel vapors from entering the atmosphere. The various systems for different models are similar. They consist of an expansion chamber, activated charcoal filter, and connecting lines. Fuel vapors are vented to the charcoal filter where hydrocarbons are deposited on the element. The engine fan forces fresh air into the filter when the engine is running. The air purges the filter and the hydrocarbons, which are forced into the air cleaner to become part of the air/fuel mixture, are burned.

EMISSION CONTROLS 4-7

SERVICING

Refer to Section 1 for more details.

Air Injection System 1973–74

◆ See Figures, 5, 6 and 7

Type 2 vehicles are equipped with the air injection system, or air pump as it is sometimes called. In this system, an engine driven air pump delivers fresh air to the engine exhaust ports. The additional air is used to promote after-burning of any unburned mixture as it leaves the combustion chamber. In addition, the system supplies fresh air to the intake manifold during gear changes to provide more complete combustion of the air/fuel mixture.

Check the air pump belt tension and examine the hoses for deterioration as a regular part of your tune-up procedure.

SERVICING

The only maintenance required for the system is an air pump belt tension check at 6,000 mile intervals and an air pump filter element replacement at 18,000 mile or 2 year intervals. Refer to Section 1 for the belt tension check.

The air pump filter element is located in a housing adjacent to the pump. To remove the element, loosen the hose clamp and disconnect the filter housing from the pump. Then, loosen the wing nut and draw out the element. Never attempt to clean the old element. Install a new paper element and assemble the filter housing.

Exhaust Gas Recirculation System

In order to control exhaust emissions of oxides of nitrogen (NO_x), an exhaust gas recirculation (EGR) system is employed on 1972 Type 1 and Type 3 models equipped with automatic transaxles and sold in California, on 1973 Type 1 and Type 3 models equipped with automatic transaxles sold nationwide, on all 1973

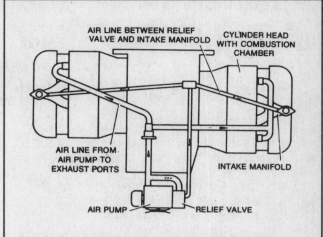

Fig. 6 Air injection (exhaust manifold afterburning) system—1974 Type 2/1800 models

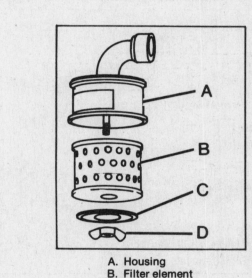

A. Housing
B. Filter element
C. Washer
D. Wingnut

Fig. 7 Exploded view of the Type 2 air pump filter unit

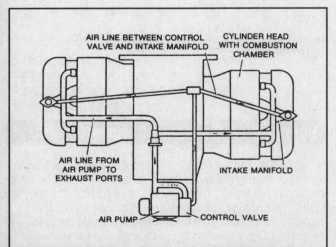

Fig. 5 Air injection (exhaust manifold afterburning) system—1973 Type 2/1700 models

and later Type 2 models, on all 1974 Type 4 models equipped with automatic transaxles, and on all 1974 and later Type 1 models. The system lowers peak flame temperature during combustion by introducing a small (about 10%) percentage of relatively inert exhaust gas into the intake charge. Since the exhaust gas contains little or no oxygen, it cannot react with, nor influence the air/fuel mixture. However, the exhaust gas does (by volume) take up space in the combustion chambers (space that would otherwise be occupied by a heat-producing, explosive air/fuel mixture), and does serve to lower peak combustion chamber temperature. The amount of exhaust gas directed to the combustion chambers is infinitely variable by means of a vacuum operated EGR valve. For system specifics, see the vehicle type breakdown under "General Description."

4-8 EMISSION CONTROLS

GENERAL DESCRIPTION

Type 1
▶ **See Figures 8 and 9**

For 1972, EGR is used only on automatic stick shift models sold in California. Exhaust gas is drawn from the left-hand rear exhaust flange and then cooled in a cooling coil. From here, the gas is filtered in a cyclone filter and finally channelled to the intake manifold, via the EGR valve. The valve permits exhaust gas recirculation during part throttle applications, but not during idling or wide open throttle.

All 1973 models (nationwide) equipped with the automatic stick shift transaxle use an EGR system. As in 1972, the gas is drawn from the left, rear exhaust flange. However, instead of the cooling coil and cyclone filter, a replaceable element type filter is used. The remainder of the system remains unchanged from 1972.

All 1974 Type 1 cars, regardless of equipment, are equipped with the EGR system. The system uses the element type filter and EGR valve which recirculates exhaust gases during part throttle applications as before. However, to improve driveability, all California models use a two-stage EGR valve (one-stage in the 49 state models), and California models equipped with an automatic use an electric throttle valve switch to further limit exhaust gas recirculation to part throttle applications (EGR permitted only between 12° to 72° on a scale of 90° throttle valve rotation).

The EGR system is installed on all 1975–80 models. All applications use the element-type filter and single-stage EGR valve. Recirculation occurs during part throttle applications as before. The system is controlled by a throttle valve switch, which measures throttle position, and an intake air sensor which reacts to engine vacuum. 1977 and later Type 1's destined for California are equipped with a mechanically operated EGR valve. A rod is attached to the throttle valve lever and operates the throttle position. No exhaust gases are recirculated at or near full throttle or at closed throttle. Beginning in 1975, an odometer actuated EGR reminder light (on the dashboard) is used to inform the driver that it is time to service the EGR system. The reminder light measures elapsed mileage and lights at 15,000 mile intervals. A reset button is located behind the switch.

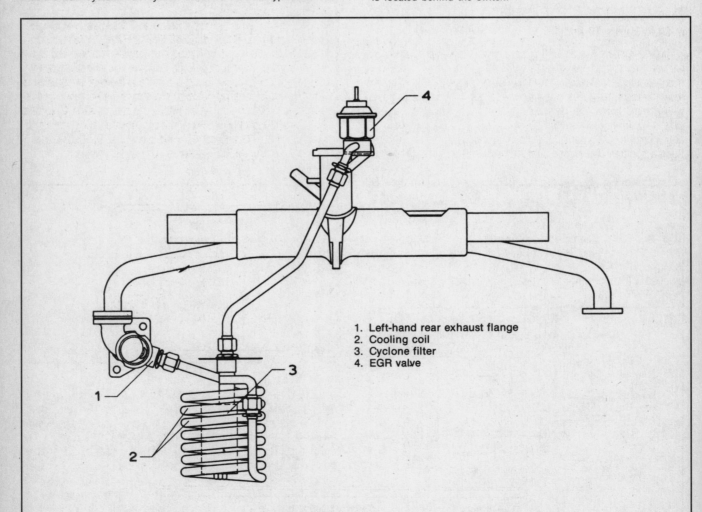

1. Left-hand rear exhaust flange
2. Cooling coil
3. Cyclone filter
4. EGR valve

Fig. 8 Exhaust Gas Recirculation (EGR) system—1972 Type 1 California models equipped with the automatic stick shift transaxle

EMISSION CONTROLS 4-9

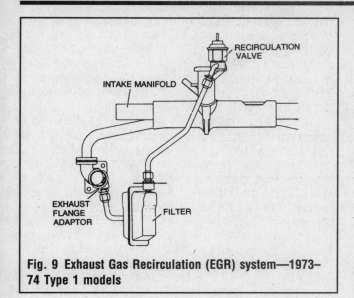

Fig. 9 Exhaust Gas Recirculation (EGR) system—1973–74 Type 1 models

Type 2
▶ See Figures 10 and 11

Type 2 models use an EGR system beginning in 1973. All models use two valves; one at each manifold. Exhaust gas is taken from the muffler, cleaned in a replaceable-type filter, then directed to both intake manifolds, via the EGR valves. On models equipped with manual transaxles, recirculation is vacuum controlled and occurs *both* during part *and* full throttle applications. On models equipped with the automatic, recirculation is controlled both by throttle position and engine compartment (ambient) temperature.

When the ambient temperature exceeds 54° F, a sensor switch (located above the battery) opens, permitting EGR during part throttle applications.

All 1974 Type 2 models use the EGR system, but there are three different systems used. All models use one central EGR valve. Exhaust gas is taken from No. 4 exhaust port, cleaned in an element-type filter, and then directed to both intake manifolds via the single EGR valve. Models equipped with manual transaxles and sold in the 49 states use a single-stage EGR valve which allows recirculation according to the vacuum signal in the left carburetor during part throttle applications. Models equipped with manual transaxles and sold in California use a two-stage EGR valve which recirculates exhaust gases during part throttle openings in two steps. During the first stage, EGR is controlled by the vacuum in the left carburetor. The second stage controls EGR according to the throttle position of the right carburetor. Finally, all models equipped with automatic transaxles (nationwide) use a single-stage EGR valve which controls recirculation according to throttle valve position and engine cooling air temperature. When the cooling system air reaches 185° F, a sensor switch (located between the coil and distributor) opens, permitting EGR during part throttle applications.

All 1975–81 Type 2 models utilize an EGR system. A single-stage EGR valve and element-type filter are used on all applications. Recirculation occurs during part throttle opening and is controlled by engine vacuum, throttle position and engine compartment temperature. At or near full throttle and when the throttle is closed, a solenoid is activated by the throttle valve switch which cuts off the vacuum supply and stops the recirculation of exhaust gases. At 15,000 mile intervals, a dash mounted EGR service reminder light is activated to warn the driver that EGR service is now due. A reset button is located behind the switch.

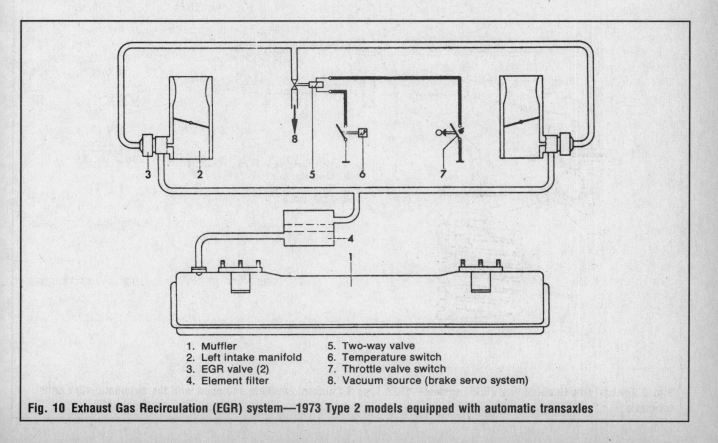

1. Muffler
2. Left intake manifold
3. EGR valve (2)
4. Element filter
5. Two-way valve
6. Temperature switch
7. Throttle valve switch
8. Vacuum source (brake servo system)

Fig. 10 Exhaust Gas Recirculation (EGR) system—1973 Type 2 models equipped with automatic transaxles

4-10 EMISSION CONTROLS

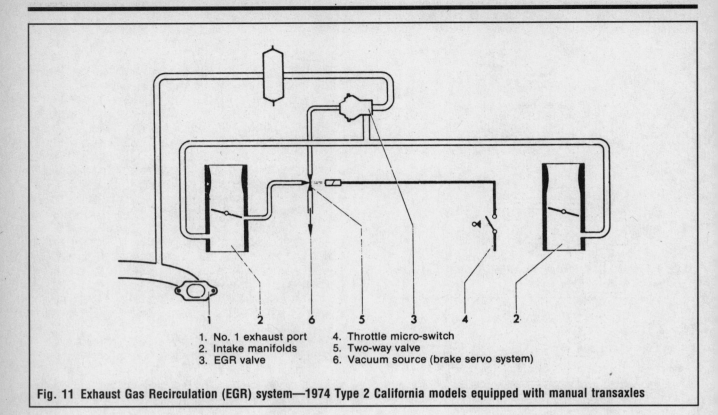

1. No. 1 exhaust port
2. Intake manifolds
3. EGR valve
4. Throttle micro-switch
5. Two-way valve
6. Vacuum source (brake servo system)

Fig. 11 Exhaust Gas Recirculation (EGR) system—1974 Type 2 California models equipped with manual transaxles

Type 3
♦ See Figures 12 and 13

EGR is first used in the 1972 Type 3 models destined for California and equipped with automatic transaxles. Exhaust gas is drawn from the front, right-hand exhaust flange to the EGR valve via a container and cyclone filter. The EGR valve then delivers the exhaust gases to the intake air distributor under part throttle (not full throttle or idling) conditions when the ambient air temperature reaches 65° F *and* only first or second gears are selected.

All 1973 Type 3 models equipped with automatic transaxles use an EGR system. The exhaust gases are cleaned in a replaceable element-type filter in 1973 instead of the cyclone filter and container of the previous year. The EGR valve then delivers the gases according to an electromagnetic valve which permits recirculation above 54° F.

Type 4
♦ See Figure 14

The EGR system appears only on 1974 models (nationwide) equipped with automatic transaxles. On this system, exhaust gas is drawn from the muffler to a single EGR valve via an element type filter. The gases are delivered to the intake air distributor under part throttle conditions.

EGR VALVE SERVICE

The EGR valve should be checked every 15,000 miles and the filter cleaned (cyclone-type) or replaced (element-type).

EGR VALVE INSPECTION

1972–74 Type 1 (Except 1974 California Models)

1. With the engine idling at operating temperature (176° F), pull off the vacuum hose from the EGR valve and push on the black hose from the intake air preheating thermostat instead.
2. If the idle speed drops off sharply or stalls, recirculation is taking place and the valve is OK. If, however, the idle speed does not change, the EGR valve is faulty or a hose is cracked or blocked.
3. Replace the vacuum hoses to their original locations.

1974 Type 1 California and 1974 Type 2 Manual Transaxle California Models

On these 1974 California models, a two-stage EGR valve with a visible pin is used. To check the valve operation, simply make sure that the pin moves in and out relative to engine rpm. If the pin does not move, check the hoses and/or replace the EGR valve.

1975–80 Type 1 Except 1977 and Later California Models

1. Start the engine and pull the electrical connector off the EGR valve vacuum unit (the disk-shaped unit located near the ignition coil).
2. This should make the engine slow down or stall, which means exhaust gases are being recirculated.
3. If no engine speed change occurs, stop the engine, then return the ignition key to the **ON** (not **START**) position.
4. Connect a test light across the terminals of the connector ca-

EMISSION CONTROLS 4-11

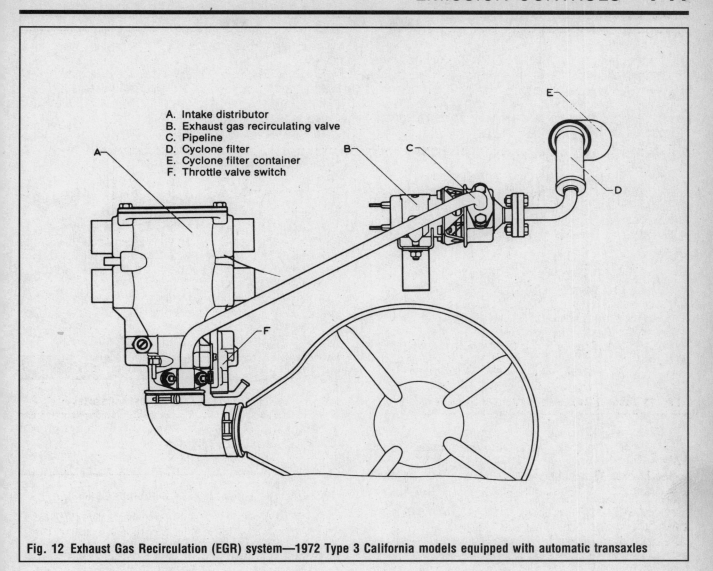

Fig. 12 Exhaust Gas Recirculation (EGR) system—1972 Type 3 California models equipped with automatic transaxles

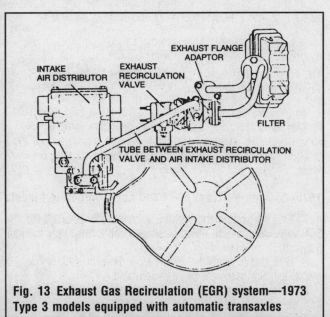

Fig. 13 Exhaust Gas Recirculation (EGR) system—1973 Type 3 models equipped with automatic transaxles

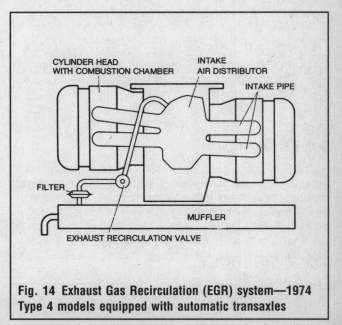

Fig. 14 Exhaust Gas Recirculation (EGR) system—1974 Type 4 models equipped with automatic transaxles

4-12 EMISSION CONTROLS

ble, then move the throttle valve by hand from idle position to about mid-speed range.

5. If the test light goes OFF when the throttle is moved out of idle position, but lights at idle or full throttle position, replace the EGR valve.

6. If the test light does not light at all, there is probably trouble with the wiring or the throttle valve switch, which is located to the left of the alternator.

Type 1 1977 and Later California Models

Remove the E-clip connecting the operating rod from the throttle to the EGR valve. Start the engine and allow it to idle. Manually operate the EGR valve: the engine should slow down or stop when the EGR valve is opened. If no speed change occurs, check the EGR pipe for clogging or replace the EGR valve.

ADJUSTING

1. Idle the engine at 800–950 rpm.
2. Loosen both locknuts on the rod and shorten the length of the rod by turning it.
3. The idle should drop suddenly, indicating the EGR valve has opened. From this position lengthen the rod 1½ turns (1 5/6 turns for 1979–80 models), using the pin in the center of the rod for orientation.
4. Tighten the locknuts.

1973 Type 2

1. Remove the EGR valve.
2. Inspect the valve for cleanliness.
3. Check the valve for freedom of movement by pressing in on the valve pin.
4. Connect the valve to the vacuum hose of another engine or vacuum source and start the engine. At 1,500–2,000 rpm the valve pin should be pulled in and when the speed is reduced it should return to its original position. Replace the EGR valve if it doesn't operate correctly.
5. Replace the washer and install the valve.
6. Repeat this operation on the second valve.

1974 Type 2 With Manual Transaxle (Except California Models)

1. With the engine idling at operating temperature, pull off the vacuum hose at the tee-fitting for the EGR valve and push on the hose from the flow valve of the air pump to the fitting instead.
2. If idle speed drops sharply or stalls, the valve is OK. If the rpm does not change, a hose is blocked or the valve is faulty.
3. Replace the hoses to their original locations.

1975 and Later Type 2

This procedure is the same as 1975–80 Type 1 (Except 1977 and Later California models), except that the EGR valve is located approximately on top of No. 4 cylinder. A long feed pipe connects the EGR valve with the throttle valve housing. The throttle valve switch is located at the bottom of the throttle valve housing and is difficult to see without removing the housing. The switch is black, rectangular and has an electrical connector plugged into it.

1972–73 Type 3

1. Remove the EGR valve.
2. Reconnect the vacuum hose and place the valve on the base.

3. Start the engine. If it doesn't stall, the vacuum line between the valve base and the intake manifold is clogged and must be cleaned.

4. Run the engine at 2,000–3,000 rpm. The closing pin of the EGR valve should pull in 0.15 in. (4 mm) and immediately return to its original position at idle. Replace the EGR valve if it doesn't operate correctly.

5. Install the EGR valve using new seals.

1974 Type 4 Automatic Transmission

1. Run the engine at idle.
2. At the EGR valve, disconnect the hose that runs to the throttle valve housing.
3. Disconnect the hose that runs to the idle speed regulator from the "T" pipe located in front of the intake manifold, and connect the hose from the EGR valve to the "T" pipe. Hold the idle speed regulator plunger back with your hand to prevent it from raising the idle speed.
4. If the engine speed drops or the engine stalls, the EGR valve is working. If nothing happens, either the EGR valve is defective or the lines are clogged.

Catalytic Converter System

All 1975–81 Type 1 and 2 models sold in California and 1977–81 models sold in the other 49 states are equipped with catalytic converters. The converter is installed in the exhaust system, upstream and adjacent to the muffler.

Catalytic converters change noxious emissions of hydrocarbons (HC) and carbon monoxide (CO) into harmless carbon dioxide and water vapor. The reaction takes place inside the converter at great heat using platinum and palladium metals as the catalyst. If the engine is operated on lead-free fuel, they are designed to last 50,000 miles before replacement.

1979–81 Type 2 models sold in California are equipped with an oxygen sensor system installed in the exhaust pipe on the left side of the engine in back of the catalytic converter. At temperatures above 575° F and for various compositions of gases, the sensor sends a signal to the AFC control unit which then corrects fuel injector operating time to insure accurately metered fuel/air mixture and keep the exhaust emissions within the legal limits. The sensor is coupled to a throttle valve switch which disconnects the sensor signal during full throttle. At all other times (except during engine warm up) the sensor is in operation.

The sensor must be replaced every 30,000 miles. A light on the speedometer lights at 30,000 mile intervals to alert you that sensor service is needed. After the service is done, reset the sensor light by pushing the button on the control box which is connected to the speedometer cable and is usually located behind the dash.

Deceleration Control

All 1975 and later Type 1 and Type 2 manual transaxle models, Type 2 automatic transaxle models after chassis No. 226 2 077 583, and Type 3 and 4 California manual transaxle models are equipped with vacuum operated deceleration valves. The deceleration valve prevents an overly rich mixture from reaching the exhaust. During deceleration, the valve opens, allowing air to bypass

EMISSION CONTROLS 4-13

the throttle plate and enter the combustion chamber, thereby leaning the air/fuel mixture. Type 2 automatic transaxle models up to chassis No. 226 2 077 583 and Type 3 and 4 California automatic transaxle models have electrically operated deceleration valves which are activated by the transaxle.

CHECKING THE DECELERATION VALVE

Vacuum Type
◆ See Figure 15

The deceleration valve on the Type 1 is located on the engine compartment hood left hinge mount. It is at the center front of the engine between the fuel injection intake air sensor on the Type 2 (don't confuse it with the EGR canister).

A faulty deceleration valve will cause engine speed to be higher than normal at idle. To test the valve, pinch shut the large fabric covered hose leading into the air cleaner. If the idle speed drops, the valve is faulty and should be replaced.

Electrical Type
◆ See Figure 16

1. From under the vehicle, remove the wire from the automatic transaxle fluid pressure switch which runs to the deceleration valve. The valve is located on the intake air distributor.
2. Turn the ignition **ON,** and ground the disconnected wire against the transmission housing or the chassis frame.

Fig. 16 The electrically operated deceleration valve is attached to only 2 vacuum lines—as compared to 3 on the vacuum operated valve

3. An audible click should be heard. If not, replace the deceleration valve or the automatic transaxle fluid pressure switch.

Throttle Valve Positioner

All 1970–71 Type 1 and 2/1600 models equipped with manual transaxles, and 1972 Type 1 models with manual transaxles sold in California use a throttle valve positioner to hold the throttle butterfly slightly open during deceleration to prevent an excessively

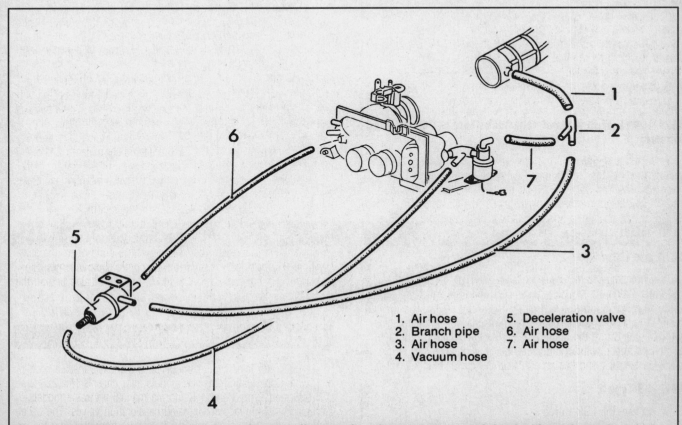

1. Air hose
2. Branch pipe
3. Air hose
4. Vacuum hose
5. Deceleration valve
6. Air hose
7. Air hose

Fig. 15 Identification of the components and vacuum lines used on the vacuum operated deceleration valve system

4-14 EMISSION CONTROLS

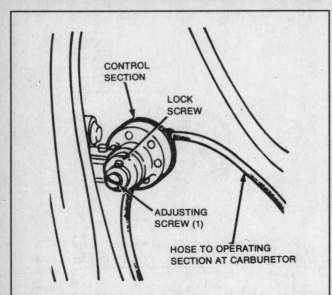

Fig. 17 The throttle valve positioner, located at the left rear of the engine compartment, can be adjusted by loosening the lockscrew and turning the adjusting screw—1970–72 Type 1 and 2/1600 models

rich mixture from reaching the combustion chambers. The throttle valve positioner consists of two parts connected by a hose. The operating part is mounted on the carburetor, connected to the throttle valve arm. It regulates fast idle speed. The control section (altitude corrector) is located at the left side of the engine compartment. It controls throttle valve closing time.

ADJUSTMENT

♦ See Figure 17

➡ The car should first be warmed to operating temperature (122–158°F) for this adjustment. Make sure the choke plate is open.

1. Attach a tachometer (0–3,000 rpm sweep minimum) to the engine with the positive lead to the distributor side of the coil and the negative lead to a good ground. You will also need a stop watch or a good wristwatch with a second hand.

✸✸ CAUTION

Keep yourself, any clothing, jewelry, hair, tools, etc. well clear of the engine belts and pulleys. Make sure you are in a well ventilated area.

2. Start the engine and let it idle in Neutral. Make a check of the fast idle speed by pulling the fast idle lever back so that it contacts the lever stop on the carburetor. The tachometer should read 1,450–1,650 rpm. If the fast idle is not within specifications, turn the adjusting screw (on the lever stop) as required. Disconnect the tachometer.
3. Take the car for a warm-up drive. Recheck the fast idle as in Steps 1 and 2. The tachometer reading should not exceed 1,700 rpm.
4. Now, make a check of the throttle valve closing time. Pull the throttle lever away from the fast idle lever until the tachometer reads 3,000 rpm. While keeping an eye on the second hand of your watch, release the throttle lever and check the time elapsed until the engine reaches idle speed (800–900 rpm). The closing time should be 2.5–4.5 seconds. If the closing time is not within specifications, adjust the control section at the left side of the engine compartment. After loosening the lockscrew, turn the adjusting screw (1) clockwise to increase closing time and counterclockwise to decrease closing time. Recheck the adjustment. If it is within specifications, tighten the lockscrew and disconnect the tachometer.
5. Take the car for another test drive. Once again, recheck the closing time as in Step 4. The closing time now must not exceed 6 seconds. If it does, go back to Step 4.

If, after several attempts, the positioner does not consistently operate correctly, check the hoses for looseness or cracks, then check the diaphragm unit for clogging or diaphragm malfunction.

If the diaphragm unit (the part that is attached to the carburetor) is replaced, the pull rod on the new unit must be adjusted. After installing the new unit, adjust the rod by loosening both locknuts on the rod and turning the middle part to either extend or shorten the length. When the throttle valve is closed, the fast idle lever (inner lever) must not touch either the carburetor body or the throttle valve lever (when the choke is fully open).

VACUUM DIAGRAMS

Following are vacuum diagrams for most of the engine and emissions package combinations covered by this manual. Because vacuum circuits will vary based on various engine and vehicle options, always refer first to the vehicle emission control information label, if present. Should the label be missing, or should vehicle be equipped with a different engine from the vehicle's original equipment, refer to the diagrams below for the same or similar configuration.

If you wish to obtain a replacement emissions label, most manufacturers make the labels available for purchase. The labels can usually be ordered from a local dealer.

EMISSION CONTROLS 4-15

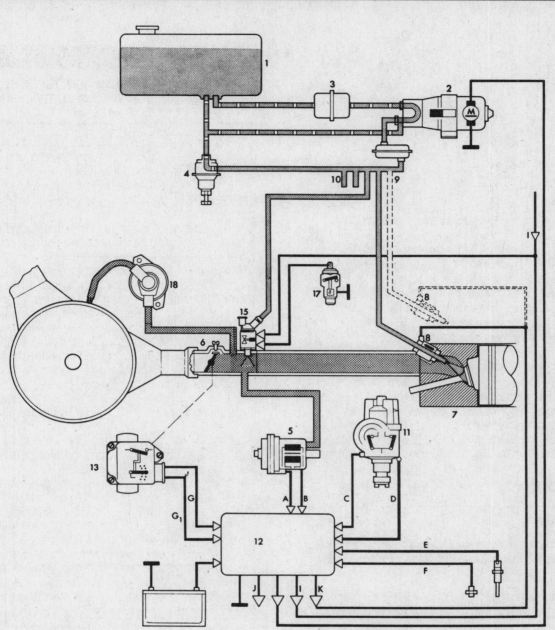

1. Fuel tank
2. Fuel pump
3. Fuel filter
4. Pressure regulator
5. Pressure sensor
6. Intake air distributor
7. Cylinder head
8. Injectors
9. Fuel distributor pipe
10. Fuel distributor pipe with connection for cold starting device
11. Distributor with trigger contacts (distributor contact I, distributor contact II)
12. Control unit
13. Throttle valve switch with acceleration enrichment
15. Cold starting valve
17. Thermostat for cold starting device
18. Auxiliary air regulator
A + B. from pressure sensor (load condition signal)
C + D. from distributor contacts (engine speed and releasing signal)
E + F. from temperature sensors (warmup signal)
G. from throttle valve switch (fuel supply cut-off when coasting)
G1. Acceleration enrichment
I. from starter, terminal 50 solenoid switch (signal for enrichment mixture when starting)
J. to the injectors, cylinders 1 and 4
K. to the injectors, cylinders 2 and 3

Fig. 18 Electronic fuel injection system schematic—1970–74 Type 3 and 4 Federal models (except 1974 Type 4 models equipped with auto) and 1970–71 Type 3 and 4 California models

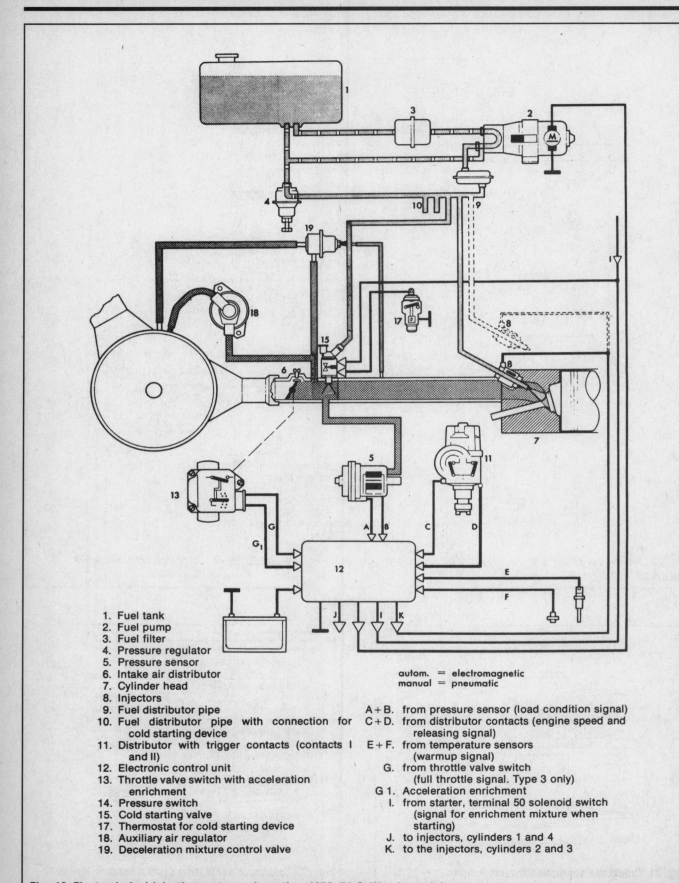

Fig. 19 Electronic fuel injection system schematic—1972–74 California models

EMISSION CONTROLS 4-17

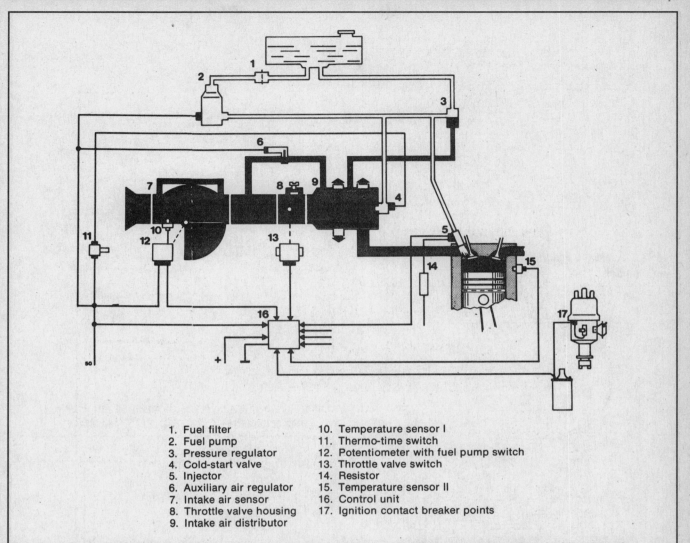

1. Fuel filter
2. Fuel pump
3. Pressure regulator
4. Cold-start valve
5. Injector
6. Auxiliary air regulator
7. Intake air sensor
8. Throttle valve housing
9. Intake air distributor
10. Temperature sensor I
11. Thermo-time switch
12. Potentiometer with fuel pump switch
13. Throttle valve switch
14. Resistor
15. Temperature sensor II
16. Control unit
17. Ignition contact breaker points

Fig. 20 Airflow controlled electronic fuel injection system schematic—1974 Type 4 models equipped with automatic transaxles and all 1975–81 models

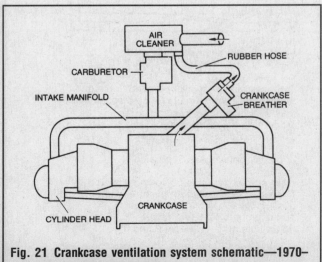

Fig. 21 Crankcase ventilation system schematic—1970–74 Type 1 and 2/1600 models

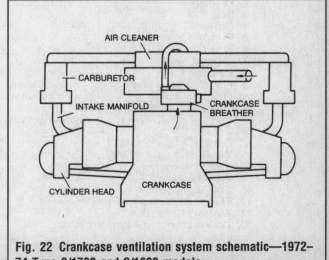

Fig. 22 Crankcase ventilation system schematic—1972–74 Type 2/1700 and 2/1800 models

4-18 EMISSION CONTROLS

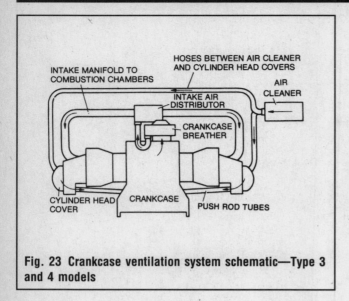

Fig. 23 Crankcase ventilation system schematic—Type 3 and 4 models

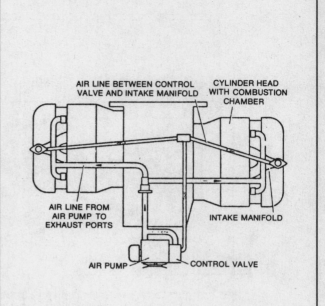

Fig. 25 Air injection (exhaust manifold afterburning) system schematic—1973 Type 2/1700 models

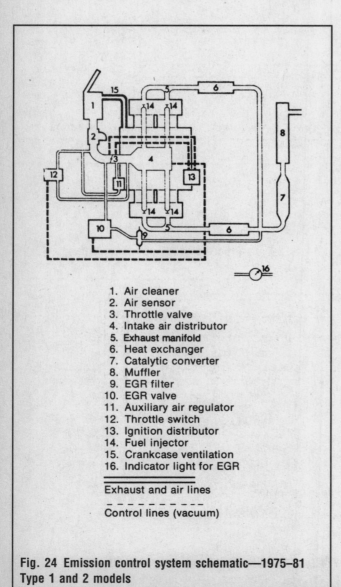

1. Air cleaner
2. Air sensor
3. Throttle valve
4. Intake air distributor
5. Exhaust manifold
6. Heat exchanger
7. Catalytic converter
8. Muffler
9. EGR filter
10. EGR valve
11. Auxiliary air regulator
12. Throttle switch
13. Ignition distributor
14. Fuel injector
15. Crankcase ventilation
16. Indicator light for EGR

——— Exhaust and air lines
– – – Control lines (vacuum)

Fig. 24 Emission control system schematic—1975–81 Type 1 and 2 models

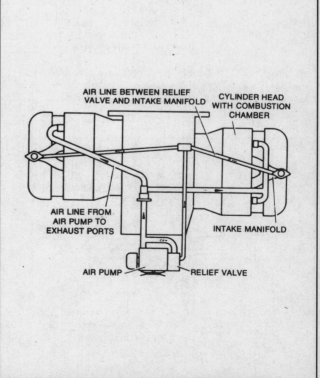

Fig. 26 Air injection (exhaust manifold afterburning) system schematic—1974 Type 2/1800 models

EMISSION CONTROLS 4-19

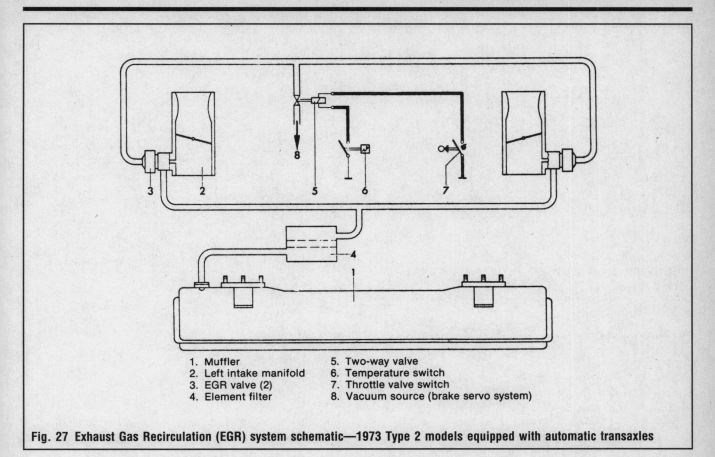

1. Muffler
2. Left intake manifold
3. EGR valve (2)
4. Element filter
5. Two-way valve
6. Temperature switch
7. Throttle valve switch
8. Vacuum source (brake servo system)

Fig. 27 Exhaust Gas Recirculation (EGR) system schematic—1973 Type 2 models equipped with automatic transaxles

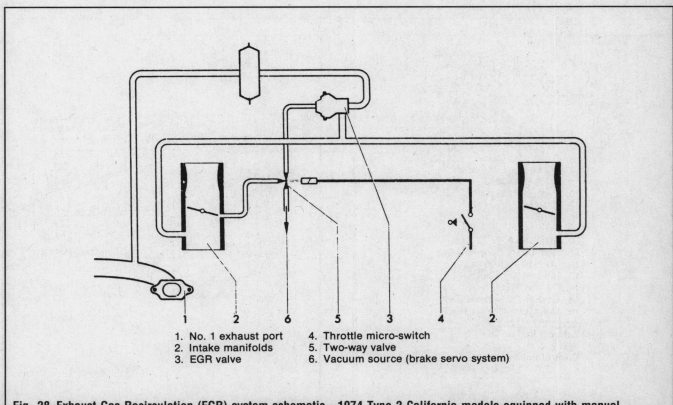

1. No. 1 exhaust port
2. Intake manifolds
3. EGR valve
4. Throttle micro-switch
5. Two-way valve
6. Vacuum source (brake servo system)

Fig. 28 Exhaust Gas Recirculation (EGR) system schematic—1974 Type 2 California models equipped with manual transaxles

4-20 EMISSION CONTROLS

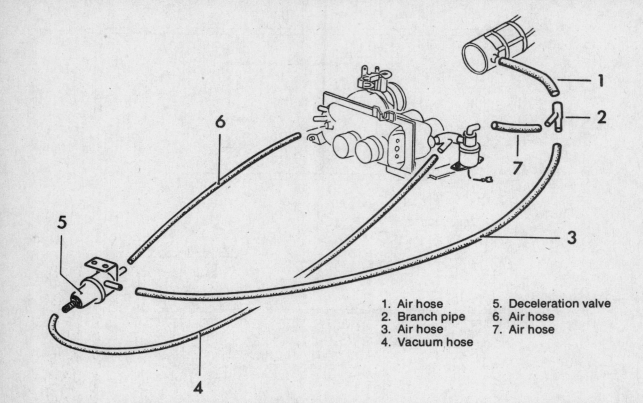

1. Air hose
2. Branch pipe
3. Air hose
4. Vacuum hose
5. Deceleration valve
6. Air hose
7. Air hose

Fig. 29 Vacuum operated deceleration valve system schematic—1975-81 Type 1 and 2 models equipped with manual transaxles, Type 2 models equipped with automatic transaxles after chassis No. 226 2 077 583, and Type 3 and 4 California models equipped with manual transaxles

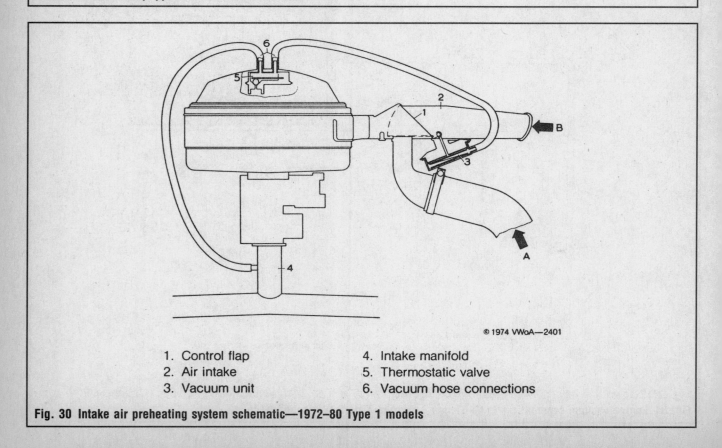

1. Control flap
2. Air intake
3. Vacuum unit
4. Intake manifold
5. Thermostatic valve
6. Vacuum hose connections

Fig. 30 Intake air preheating system schematic—1972-80 Type 1 models

EMISSION CONTROLS 4-21

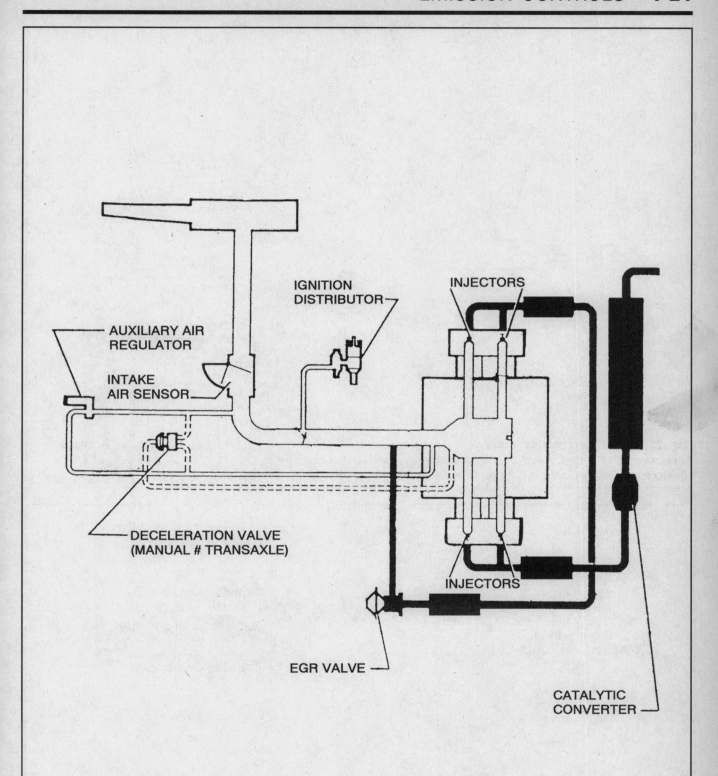

Fig. 31 Engine vacuum schematic—1980–81 non-California Type 2 (Vanagon) models

4-22 EMISSION CONTROLS

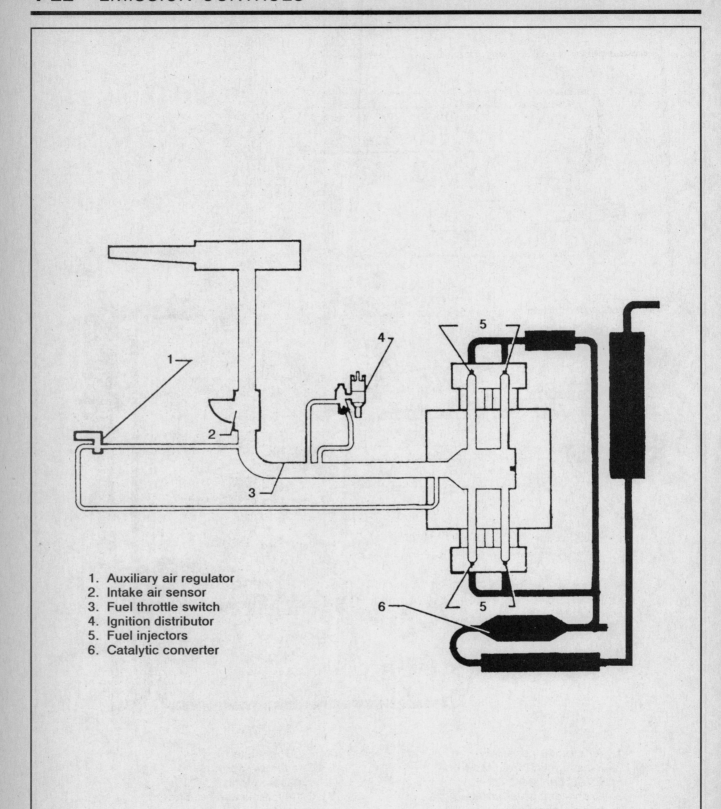

1. Auxiliary air regulator
2. Intake air sensor
3. Fuel throttle switch
4. Ignition distributor
5. Fuel injectors
6. Catalytic converter

Fig. 32 Engine vacuum schematic—1980–81 California Type 2 (Vanagon) models

EMISSION CONTROLS   4-23

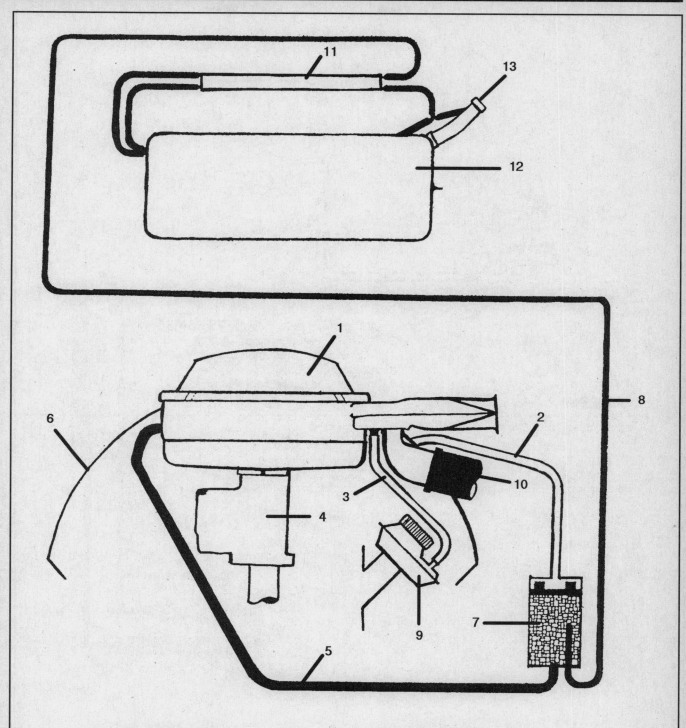

1. Air cleaner assembly
2. Pressure line from fan to EVAP canister
3. Crankcase breather hose to air cleaner assembly
4. Carburetor
5. Breather line from canister to air cleaner assembly
6. Fan shroud
7. EVAP canister
8. Gasoline vapor hose to expansion tank
9. Oil fill/crankcase breather assembly
10. Pre-heated air intake hose
11. Expansion tank
12. Fuel tank
13. Fuel tank fill hose

Fig. 33 Evaporative emissions control (EVAP) system schematic—1970–79 Type 1 and 1970–74 Type 2 models

4-24 EMISSION CONTROLS

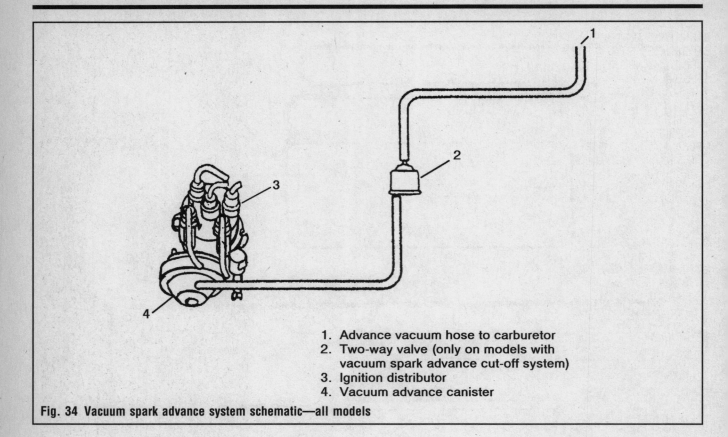

1. Advance vacuum hose to carburetor
2. Two-way valve (only on models with vacuum spark advance cut-off system)
3. Ignition distributor
4. Vacuum advance canister

Fig. 34 Vacuum spark advance system schematic—all models

BASIC FUEL SYSTEM DIAGNOSIS 5-2
CARBURETED FUEL SYSTEM 5-2
UNDERSTANDING THE FUEL
  SYSTEM 5-2
    FUEL TANK 5-2
    FUEL PUMP 5-2
    FUEL FILTERS 5-2
    CARBURETOR 5-2
MECHANICAL FUEL PUMP 5-4
    REMOVAL & INSTALLATION 5-4
    TESTING & ADJUSTING 5-6
CARBURETORS 5-7
    REMOVAL & INSTALLATION 5-7
    ADJUSTMENTS 5-8
    OVERHAUL 5-11
ACCELERATOR CABLE 5-16
    REMOVAL & INSTALLATION 5-16
GASOLINE FUEL INJECTION
  SYSTEM 5-16
DESCRIPTION 5-16
    NON-AIR FLOW CONTROLLED 5-16
    AIR FLOW CONTROLLED 5-16
ELECTRONIC CONTROL (BRAIN)
  BOX 5-17
ELECTRIC FUEL PUMP 5-17
    REMOVAL & INSTALLATION 5-21
    ADJUSTMENTS 5-21
FUEL INJECTORS 5-21
    CHECKING FUEL INJECTOR
      OPERATION 5-22
    CHECKING THE FUEL INJECTOR
      SIGNAL 5-22
    REMOVAL & INSTALLATION 5-22
THROTTLE VALVE SWITCH 5-23
    REMOVAL & INSTALLATION 5-23
    ADJUSTMENT (NON-AIR FLOW
      CONTROLLED ONLY) 5-23
COLD START VALVE 5-23
    CHECKING VALVE LEAKAGE 5-24
    CHECKING VALVE OPERATION 5-24
TRIGGER CONTACTS (NON-AIR FLOW
  CONTROLLED ONLY) 5-24
    REMOVAL & INSTALLATION 5-24
FUEL PRESSURE REGULATOR 5-24
    REMOVAL & INSTALLATION 5-24
    ADJUSTMENT (NON-AIR FLOW
      CONTROLLED ONLY) 5-24
    TESTING 5-25
TEMPERATURE SENSORS I AND II 5-25
    REMOVAL & INSTALLATION 5-25
PRESSURE SENSOR (NON-AIR FLOW
  CONTROLLED ONLY) 5-25
IDLE SPEED REGULATOR 5-25
    TESTING 5-25
AUXILIARY AIR REGULATOR 5-25
    TESTING 5-25
FUEL TANK 5-26
TANK ASSEMBLY 5-26
    REMOVAL & INSTALLATION 5-26
SPECIFICATION CHART
    CARBURETOR SPECIFICATIONS 5-15

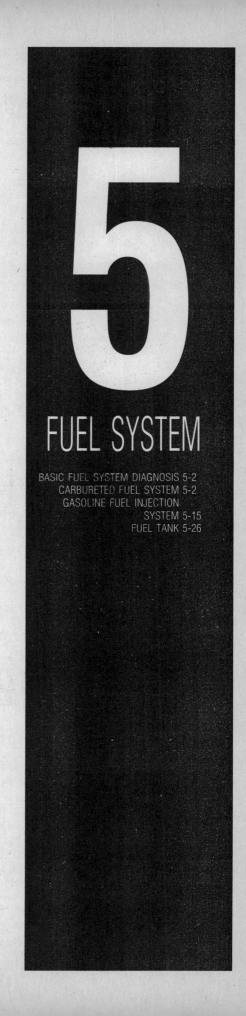

5

FUEL SYSTEM

BASIC FUEL SYSTEM DIAGNOSIS 5-2
CARBURETED FUEL SYSTEM 5-2
GASOLINE FUEL INJECTION
                SYSTEM 5-15
            FUEL TANK 5-26

5-2 FUEL SYSTEM

BASIC FUEL SYSTEM DIAGNOSIS

When there is a problem starting or driving a vehicle, two of the most important checks involve the ignition and the fuel systems. The questions most mechanics attempt to answer first, "is there spark?" and "is there fuel?" will often lead to solving most basic problems. For ignition system diagnosis and testing, please refer to the information on engine electrical components and ignition systems found earlier in this manual. If the ignition system checks out (there is spark), then you must determine if the fuel system is operating properly (is there fuel?).

CARBURETED FUEL SYSTEM

Understanding the Fuel System

An automotive fuel system consists of everything between the fuel tank and the carburetor. This includes the tank itself, all the lines, one or more fuel filters, a fuel pump (mechanical or electric), and the carburetor.

With the exception of the carburetor, the fuel system is quite simple in operation. Fuel is drawn from the tank through the fuel line by the fuel filter, and from there to the carburetor where it is distributed to the cylinders.

FUEL TANK

Normally, fuel tanks are located at the rear of the vehicle, although on most rear-engined cars, they are located at the front. The tank itself also contains a fuel gauge sending unit, and a filler tube. In most tanks, there is also a screen of some sort in the bottom of the tank near the pick-up to filter out impurities. Since the advent of emission controls, tanks are equipped with a control system to prevent fuel vapor from being discharged into the atmosphere. A vent line in the tank which is connected to a filter in the engine compartment. Vapors from the tank are trapped in the filter canister, where they are routed back to the fuel tank, making the system a closed loop. All the fumes are prevented from escaping to the atmosphere. These systems also require the use of a special gas cap which makes an airtight seal.

FUEL PUMP

There are two types of fuel pumps in general use: the mechanical pump and the electric pump. On carbureted engines, mechanical pumps are the more common of the two. Electric pumps are used on all fuel injected cars (and some carburetor-equipped cars) in addition to seeing wide use on a number of imported cars.

Mechanical fuel pumps are usually mounted on the side of the engine block and operated by an eccentric lobe on the engine's camshaft. A pump rocker arm rests against the camshaft eccentric and as the camshaft rotates, causes the rocker arm to rock back and forth. Inside the fuel pump, the rocker arm is connected to a flexible diaphragm. A spring is mounted under the diaphragm to maintain pressure on the diaphragm. As the rocker arm rocks, it pulls the diaphragm down and then releases it. Once the diaphragm is released, the spring pushes it back up. This continual diaphragm motion causes a partial vacuum and pressure in the space above the diaphragm. The vacuum sucks the fuel from the tank and the pressure pushes it toward the carburetor.

As a general rule, mechanical fuel pumps are quite dependable. When trouble does occur, it is usually caused by a cracked or broken diaphragm, which will not draw sufficient fuel. Occasionally, the pump arm or spring will become so worn that the fuel pump can no longer produce an adequate supply of fuel, but this condition can be easily checked. Older fuel pumps are rebuildable, but late-model pumps have a crimped edge and must be replaced if defective.

There are two general types of electric fuel pumps in use today. The impeller type pump uses a vane or impeller which is driven by an electric motor. These pumps are often mounted in the fuel tank, though they are sometimes found below or beside the tank.

The bellows-type pump, is becoming rare. The bellows pump ordinarily is mounted in the engine compartment and contains a flexible metal bellows operated by an electromagnet.

Most electric fuel pumps are not rebuildable and if defective must be replaced. Minor service is usually confined to checking electrical connections and checking for a blown fuse.

FUEL FILTERS

In addition to the screen located in the bottom of the fuel tank, all fuel systems have at least one other filter located somewhere between the fuel tank and the carburetor. On some models, the filter is part of the fuel pump itself, on others it is located in the fuel line, and still others locate the filter in the carburetor inlet or the carburetor body itself.

The fuel filter is usually a paper or bronze element which screens out impurities in the fuel, before it has a chance to reach the carburetor. If you replace the fuel filter, you'll be amazed at the bits of sediment and dirt trapped by the filter.

CARBURETOR

The carburetor is the most complex part of the entire fuel system. Carburetors vary greatly in construction, but they all operate basically the same way; their job is to supply the correct mixture of fuel and air to the engine in response to varying conditions.

Despite their complexity in operation, carburetors function because of a simple physical principle—the venturi principle. Air is drawn into the engine by the pumping action of the pistons. As the air enters the top of the carburetor, it passes through a venturi, which is nothing more than a restriction in the throttle bore. The air speeds up as it passes through the venturi, causing a slight drop in pressure. This pressure drop pulls fuel from the float bowl through a nozzle into the throttle bore, where it mixes

FUEL SYSTEM 5-3

Carburetor Operating Principles

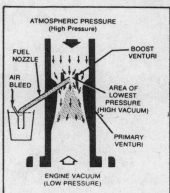

The venturi principle in operation. The pumping action of the pistons creates a vacuum which is amplified by the venturi in the carburetor. This pressure drop will pull fuel from the float bowl through the fuel nozzle. Unfortunately, there is not enough suction present at idle or low speed to make this system work, which is why the carburetor is equipped with an idle and low speed circuit

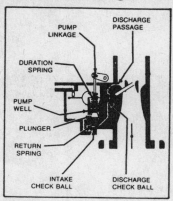

Accelerator pump system. When the throttle is opened, the air flowing through the venturi starts flowing faster almost immediately, but there is a lag in the flow of fuel out of the main nozzle. The result is that the engine runs lean and stumbles. It needs an extra shot of fuel just when the throttle is opened. This shot is provided by the accelerator pump, which is nothing more than a little pump operated by the throttle linkage that shoots a squirt of fuel through a separate nozzle into the throat of the carburetor

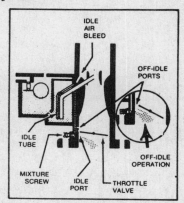

Idle and low-speed system. The vacuum in the intake manifold at idle is high because the throttle is almost completely closed. This vacuum is used to draw fuel into the engine through the idle system and keep it running. Vacuum acts on the idle jet (usually a calibrated tube that sticks down into the main well, below the fuel level) and sucks the fuel into the engine. The idle mixture screw is there to limit the amount of fuel that can go into the engine

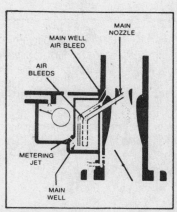

The main metering system may be the simplest system of all, since it is simply the venturi principle in operation. At cruising speeds, the engine sucks enough air to constantly draw fuel through the main fuel nozzle. The main fuel nozzle or jet is calibrated to provide a metering system. The metering system is necessary to prevent an excess amount of fuel flowing into the intake manifold, creating an overly rich mixture

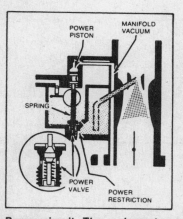

Power circuit. The main metering system works very well at normal engine loads, but when the throttle is in the wide-open position, the engine needs more fuel to prevent detonation and give it full power. The power system provides additional fuel by opening up another passage that leads to the main nozzle. This passageway is controlled by a power valve

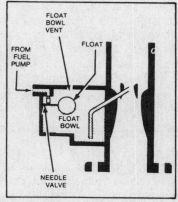

Float circuit. When the fuel pump pushes fuel into the carburetor, it flows through a seat and past a needle which is a kind of shutoff valve. The fuel flows into the float bowl and raises a hinged float so that the float arm pushes the needle into the seat and shuts off the fuel. When the fuel level drops, the float drops and more fuel enters the bowl. In this way, a constant fuel supply is maintained

5-4 FUEL SYSTEM

with the air and forms a fine mist, which is distributed to the cylinders through the intake manifold.

There are six different systems (fuel/air circuits) in a carburetor that make it work; the Flat system, Main Metering system, Idle and Low-Speed system, Accelerator Pump system, Power system, and the Choke system. The way these systems are arranged in the carburetor determines the carburetor's size and shape.

It's hard to believe that the little single-barrel carburetor used on 4 or 6 cylinder engines have all the same basic systems as the enormous 4-barrel's used on V8 engines. Of course, the 4-barrels have more throttle bores ("barrels") and a lot of other hardware you won't find on the little single-barrels. But basically, all carburetors are similar, and if you understand a simple single-barrel, you can use that knowledge to understand a 4-barrel. If you'll study the explanations of the various systems on this page, you'll discover that carburetors aren't as tricky as you thought they were. In fact, they're fairly simple, considering the job they have to do.

It's important to remember that carburetors seldom give trouble during normal operation. Other than changing the fuel and air filters and making sure the idle speed and mixture are OK at every tune-up, there's not much maintenance you can perform on the average carburetor.

Mechanical Fuel Pump

All 1970–74 Type 1 and 2 models utilize a mechanical fuel pump. The pump is located to the left of the generator/alternator on Type 1 and 2/1600 models, and located next to the flywheel on dual carburetor Type 2 models. On Type 1 and 2/1600 models the pump is pushrod operated by an eccentric on the distributor driveshaft. On Type 2/1700 and 2/1800 twin carb models, the pump is operated by a pushrod which rides on a camshaft eccentric.

REMOVAL & INSTALLATION

Types 1 and 2/1600

♦ See Figure 1

1. Disconnect the fuel lines at the pump and plug them to prevent leakage.
2. Remove the two securing nuts.
3. Remove the fuel pump. If necessary, the pushrod, gaskets, and intermediate flange may also be removed.
4. When installing the fuel pump, it is necessary to check the fuel pump pushrod stroke. This is done by measuring the distance that the pushrod projects above the intermediate flange when both gaskets are in place. The rod must project ½ inch. If not, remove or insert enough bottom gaskets under the flange until it does.
5. Fill the cavity in the lower part of the fuel pump housing with grease. Total pushrod length is 4.252 in. for all Type 2/1600 and for Type 1 models equipped with generators. Pushrod length is 3.937 in. for all 1973–74 Type 1 models equipped with alternators. Replace any worn pushrod.
6. Using new gaskets, install the fuel pump and tighten the two securing nuts.
7. Install the fuel hoses.

To remove the mechanical fuel pump on upright engines, disconnect and plug the fuel lines . . .

. . . then remove the two fuel pump attaching bolts

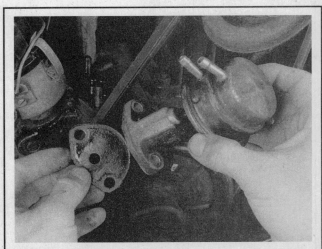
Lift the fuel pump off of the engine case—make sure to remove the old gasket from the engine

FUEL SYSTEM 5-5

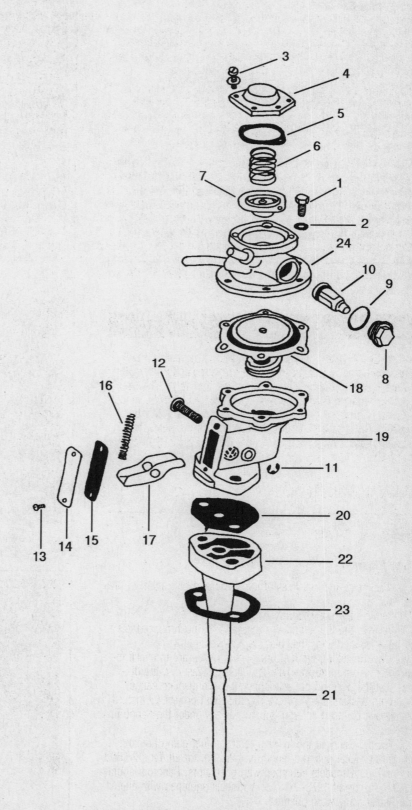

1. Screw
2. Washer
3. Screw
4. Upper pump cover
5. Gasket
6. Cut-off valve spring
7. Cut-off valve diaphragm
8. Fuel filter plug
9. Copper washer
10. Fuel filter
11. Circlip
12. Lever shaft
13. Screw
14. Inspection cover
15. Gasket
16. Operating lever spring
17. Operating lever
18. Pump diaphragm with spring and guide halves attached
19. Pump housing (lower)
20. Gasket
21. Pushrod
22. Intermediate flange
23. Gasket
24. Pump housing (upper)

Fig. 1 Exploded view of the fuel pump used on 1970 Type 1 models

5-6 FUEL SYSTEM

Types 2/1700 and 2/1800 With Twin Carburetors

◆ See Figure 2

1. Remove the engine.
2. Once the engine is removed, remove the upper and lower deflector plates and the carburetor preheater connection to gain access to the pump mounting bolts (adjacent to the flywheel).
3. Disconnect and plug the fuel lines to the carburetors.
4. Remove the two retaining bolts and lift off the fuel pump, gaskets and intermediate flange.
5. Reverse the above procedure to install, using new gaskets. Remember to coat the pushrod and lever with grease.

➡ Prior to installation of the fuel pump, check the action of the camshaft eccentric driven pushrod. Install just the intermediate flange on the engine with 2 gaskets underneath and one on top. Turn the engine over by hand until the pushrod is on the highest point of the camshaft eccentric. Then, measure the distance between the tip of the pushrod and the top gasket surface. Adjust, as necessary, to 0.2 in. by removing or installing gaskets under the intermediate flange. Total pushrod length should be 5.492 in. minimum. Replace, if worn.

TESTING & ADJUSTING

The maximum fuel pump pressure developed by the Type 1 and 2/1600 fuel pump is 3–5 psi at 3,400 rpm for the Type 1 (4,000 rpm for the Type 2). The maximum fuel pump pressure developed by the Type 2 Twin carb fuel pump is 5 psi at 3,800 rpm.

All fuel pumps deliver 400 cc of fuel per minute at 3,800–4,000 rpm.

The only adjustment possible is performed by varying the thickness of the fuel pump flange gaskets. Varying the thickness of the gaskets will change the stroke of the fuel pump pushrod. This ad-

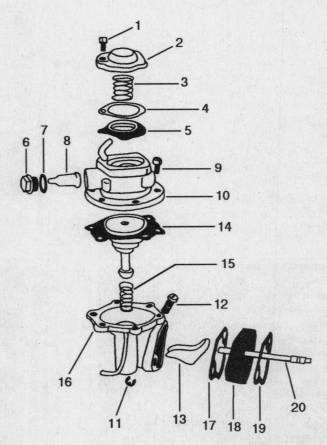

1. Screw
2. Cover
3. Cut-off diaphragm spring
4. Cut-off diaphragm gasket
5. Cut-off diaphragm
6. Plug
7. Washer
8. Filter
9. Screw
10. Pump upper half
11. Circlip
12. Lever shaft
13. Pump lever
14. Diaphragm and gasket
15. Diaphragm spring
16. Pump lower half
17. Gasket
18. Intermediate flange
19. Gasket
20. Pushrod

Fig. 2 Exploded view of the fuel pump used on 1972–74 Type 2 models equipped with the twin carburetor system

FUEL SYSTEM 5-7

justment is not meant to compensate for a pump in bad condition; therefore, do not attempt to vary the height of the pushrod to any great extent.

The fuel pumps used on 1970 Type 1 and 2 models, and 1972–74 Type 2 models may be disassembled for cleaning or repairs. All other mechanical pumps are permanently sealed and must be replaced if found defective.

Carburetors

Carburetors are used on all 1970–74 Type 1 and 2 models. A single downdraft unit is used on all Type 1 models and on 1970–71 Type 2 models. Beginning with the 1972 model year, the Type 2 utilizes twin carburetion. The type of carburetor used (34 PICT-3, etc.) is stamped on the float bowl and should be clearly visible.

REMOVAL & INSTALLATION

Types 1 and 2/1600

1. Remove the air cleaner.
2. Disconnect the fuel hose. Plug it to prevent leakage.
3. Disconnect the vacuum hoses.
4. Remove the automatic choke cable and remove the wire for the electromagnetic pilot jet.
5. Disconnect the accelerator cable at the throttle valve lever.
6. Remove the two nuts securing the carburetor on the intake manifold and then remove the carburetor from the engine.
7. Using a new gasket, install the carburetor on the manifold.
8. Reconnect the fuel and vacuum hoses, the automatic choke cable, and the wiring for the pilot jet.
9. Reconnect the throttle cable and adjust it so that at full throttle there is a gap of 0.04 in. between the throttle lever and its stop on the lower portion of the carburetor body.

➥Open the throttle valve by hand and tighten the adjustment screw, then have an assistant open the throttle and recheck the adjustment.

. . . and disconnect all vacuum lines from the carburetor housing

Label and detach all wiring from the automatic choke assembly and the cut-off solenoid

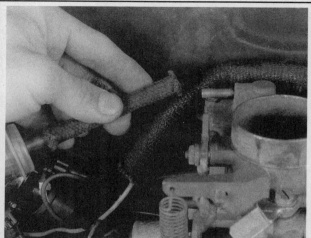
To remove the single carburetor on upright engines, disconnect and plug the fuel line . . .

Loosen the accelerator cable retaining setscrew . . .

5-8 FUEL SYSTEM

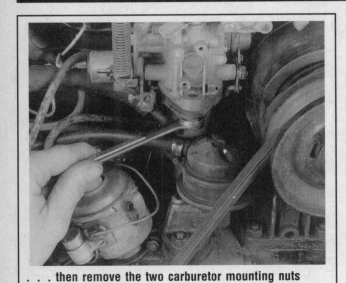

. . . then remove the two carburetor mounting nuts

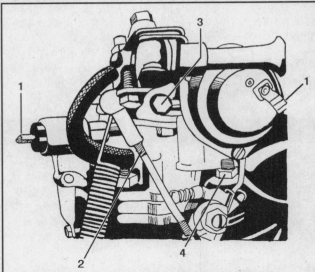

Fig. 3 Identification of the connections for the left carburetor

Lift the carburetor off of the intake air distributor flange and allow the accelerator cable to slide out of the accelerator lever

Type 2/1700 and 2/1800 Twin Carb
♦ See Figure 3

1. Remove the air cleaner.
2. Disconnect and plug the fuel line(s).
3. Disconnect the electrical leads for the automatic choke, pilot jet cut-off valve, and idle mixture cut-off valve.
4. If removing the left carburetor, disconnect the vacuum line (2), the idle mixture line for the central idling system, and the idle air intake line from the top of the carburetor.
5. Remove the linkage cross shaft bracket retaining bolt (3). Disconnect the return spring and the pull rod and release the linkage from both carburetors.
6. Remove the carburetor retaining nuts and remove the carburetor(s).
7. Reverse the above procedure to install, taking care to use new gaskets. After installation, synchronize the carburetors as outlined under "Fuel Adjustments" in Chapter Two.

ADJUSTMENTS

Throttle Linkage
SINGLE AND TWIN CARBURETOR MODELS
♦ See Figure 4

1. Have an assistant hold the accelerator pedal to the floor at wide open throttle. Measure the distance between the throttle valve lever and the stop on the carburetor body. Proper distance (a) is 0.04 in. minimum.
2. To adjust, loosen the cable adjusting screw found in the bottom on the throttle lever.
3. The throttle lever has a rigid cylinder attached to its end. Move the rigid portion in or out of the end of the throttle lever to obtain the proper adjustment and tighten the adjusting screw. The proper adjustment is reached when there is a gap of 0.04 in. (single carb) or 0.04–0.06 in (twin carb) between the throttle valve lever and its stop on the lower portion of the carburetor body. See the note at the end of the Carburetor Removal and Installation procedure.

Float and Fuel Level

➡ The carburetor must be on a level surface to obtain an accurate reading.

A properly assembled carburetor has a preset float level. For the float level to be correct the fiber washer under the needle valve seat must be installed and be of the proper thickness. See the Carburetor Specifications Chart.

The only way to adjust the float level, if it is absolutely necessary, and still retain proper seating of the needle in the needle valve seat, is to vary the thickness of the fiber washer beneath the seat.

Washers are available in thicknesses of 0.50 mm, 0.80 mm, 1.00 mm, and 1.50 mm.

FUEL SYSTEM 5-9

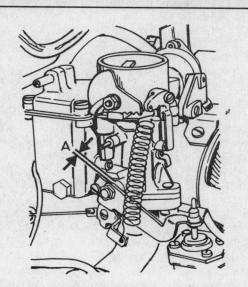

Fig. 4 Full throttle clearance (A) should be 0.04 in. on Type 1 and 2/1600 models, or 0.04–0.06 in. on Type 2/1700 and 2/1800 models

Throttle Valve Gap

TYPE 2 34 PDSIT 2/3 CARBURETORS—CARBURETOR REMOVED

♦ See Figure 5

➡ The choke valve must be closed for proper adjustment.

1. Remove the carburetor from the car.
2. Loosen the two nuts on the automatic choke connecting rod and insert a wire gauge or drill between the throttle valve and the side of the venturi. The gauge should be .024 in. for 1972–73 models and .028 in. for 1974.
3. Move the two nuts up or down on the connecting rod until the throttle valve gap is adjusted and tighten the two nuts (A).

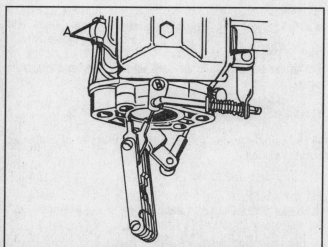

Fig. 5 Adjust the throttle valve gap with the adjusting nuts (A) on 34 PDSIT 2/3 carburetors

TYPE 2 34 PDSIT 2/3 CARBURETORS—CARBURETOR INSTALLED

1. Back out the idle speed screw until the throttle valve is completely closed.
2. Turn the idle speed screw until it just touches the throttle lever.
3. Close the choke valve.
4. Place a 0.09 in. drill or wire gauge between the idle screw and the throttle valve lever. Adjust the two nuts on the automatic choke connecting rod either up or down until the drill can be easily pulled out.
5. It will be necessary to rebalance the carburetors.

Accelerator Pump

ALL CARBURETOR TYPES

♦ See Figures 6, 7 and 8

Improper accelerator pump adjustment is characterized by flat spots during acceleration or a severe hesitation when the throttle is first depressed.

➡ VW now has a special tool available that allows you to check accelerator pump injection quantity without removing the top of the carburetor (air horn). It consists of a measuring glass, and injection pipe and a choke plate retainer. The part number is VW 119.

1. Remove the carburetor from the engine and remove the upper half of the carburetor.
2. Support the carburetor securely in a vise without damaging the carburetor body.
3. Fill the float chamber with gasoline and attach a rubber tube to the injector tube. Place the open end of the tube into a milliliter measuring tube.
4. Move the throttle lever several strokes until all of the air is forced out of the tube. Move the throttle lever an additional ten full strokes and measure the quantity of gas in the measuring

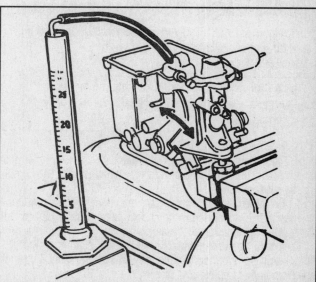

Fig. 6 A special measuring glass is necessary to check the accelerator pump injection quantity

FUEL SYSTEM

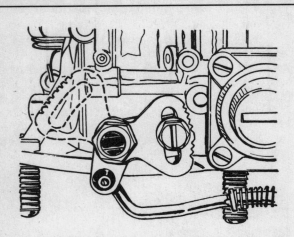

Fig. 7 To adjust the accelerator pump injection quantity, loosen the lockbolt and move the pump lever slightly—the dotted position of the pump lever is for 1971–73 Type 1 models with a generator (Solex 34 PICT-3 carburetor); the other position shown is for 1973–74 Type 1 models equipped with an alternator

Fig. 8 The accelerator pump is adjusted by turning the adjusting screw on 1973–74 Type 1 models sold in California (Solex 34 PICT-4)

tube. Multiply the accelerator pump quantity injected specification by 10 and compare this figure to the amount of gas in the measuring tube.

5. On 30 PICT-3 carburetors, and on 34 PDSIT 2/3 carburetors used on manual transmission Type 2s and on 1972 and some 1973 automatic transmission Type 2s, the injection quantity is decreased by moving the cotter pin on the connecting link to the outer hole, and increased by moving it to the inner hole. On dual carburetors, both cotter pins must be in the same holes on the two connecting links.

6. On some 1973 and all 1974 automatic transmission Type 2s with 34 PDSIT 2/3 carburetors, the cotter pin adjustment has been replaced with a round, threaded adjustment barrel on the end of the connecting rod. Turn the barrel to increase or decrease the injection quantity. The 1974 automatic transmission Type 2 with 34 PDIST 2/3 carburetors also has a thermal valve screwed into the bottom part of the carburetor housing to the side of the accelerator pump. This valve regulates the amount of fuel injected by the accelerator pump according to carburetor body temperature. When temperature is below 70°F, the valve is closed and allows 1.5 cc of fuel to be injected on each stroke. Above 70°F, the valve opens, allowing only about half as much fuel on each stroke, the rest being routed back into the float bowl. When measuring full injection quantity, make sure carburetor body temperature is below 60°F.

7. On 34 PICT-3 carburetors, the injection quantity is decreased by loosening the retaining screw and turning the adjusting lever clockwise, and increased by loosening the retaining screw and turning the adjusting lever counterclockwise. Tighten the adjusting screw after adjusting.

8. On 34 PICT-4 carburetors (used on 1973–74 Type 1 models sold in California) the injection quantity is adjusted by turning the adjusting screw (spring loaded) on the pump operating rod. The 1974 California models of these carburetors also have the partial fuel cutoff mentioned in step 6 for 1974 Type 2 automatic transmission with 34 PDSIT 2/3 carburetor. Make sure carburetor body temperature is below 60°F before checking full injection quantity.

Fast Idle

The fast idle speed is adjusted by means of a screw located at the upper end of the throttle valve arm. This screw rests against a cam with steps cut into its edge.

To adjust the fast idle, start the engine and rotate the cam so that the fast idle screw is resting against the highest step on the fast idle cam. The fast idle speed should be 1,450–1,650 rpm. Turn the fast idle screw either in or out until the proper idle speed is obtained.

On dual carburetor engines it is necessary to adjust the fast idle on only the left carburetor. There is a direct connection between the two carburetors and if the left carburetor is adjusted the right will automatically be adjusted.

Dashpot

1972–74 TYPE 1 WITH MANUAL TRANSAXLE

▶ See Figure 9

➡ The car must be fully warmed up and the choke plate open.

1. Check that the distance (a) between the tip of the dashpot and the throttle lever is 0.040 in. with the dashpot and the throttle lever is 0.040 in. with the dashpot plunger fully retracted and the throttle fully closed on the warm running position of the fast idle cam.

2. To adjust, loosen the two locknuts on the dashpot mounting bracket, and raise or lower the dashpot as needed.

1972–74 TYPE 2 TWIN CARB

➡ This adjustment is required only if the dashpot has been removed or the linkage disassembled.

1. Check that the distance between the tip of the plunger and the tab on the linkage is 0.0015 in. while holding the dashpot plunger in the retracted position.

FUEL SYSTEM 5-11

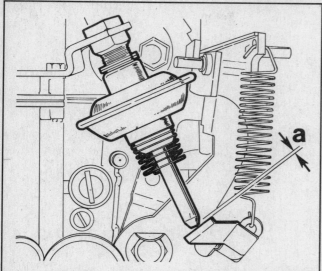

Fig. 9 Adjust the dashpot on 1972–74 Type 1 models so that a is 0.040 in.

2. To adjust, loosen the two locknuts on the dashpot mounting bracket, and raise or lower the dashpot as required.

Automatic Choke

All carburetor equipped models have automatic choke valves which are controlled by the ignition switch. When the engine is running, current is applied to the bi-metal spring inside the choke case causing it to heat up and bend. One end of the bi-metal is attached to a lever on the choke valve: as the bi-metal bends it slowly opens the choke valve. To check the choke valve operation, with the engine cold, remove the air cleaner. The choke valve should be fully closed. Start the engine and watch the choke valve: as normal operating temperature is reached the choke should slowly open. If the choke valve does not open, give it a slight tap. If it pops open, the valve is sticking: spray it with choke cleaner. If the valve still doesn't open, check to see if current is reaching the choke unit when the engine is initially turned on. If it is, the automatic choke should be checked or replaced. The wire to the automatic choke is usually hooked in parallel with the electromagnetic cutoff valve and originates at terminal 15 of the ignition coil.

Electromagnetic Cutoff Valve (Anti-Dieseling Solenoid)

This valve prevents the engine from "running on" or dieseling after the ignition key is turned **OFF**. The unit consists of a small plunger connected to a solenoid. While the engine is running, current is applied to the valve which causes the solenoid to pull the plunger back. When the plunger is pulled back it allows fuel to flow into the idling system of the carburetor; when the plunger is let forward it cuts off fuel flow. See the exploded views of the carburetors to locate the electromagnetic cutoff valve.

This unit is usually trouble free, however, if it does go faulty it can cut off the entire idling system, causing the car to stall when idling but to run if the accelerator pedal is used. The unit simply unscrews. The wiring on the valve is usually hooked in parallel with the automatic choke and originates at terminal 15 of the ignition coil. Type 2 models with dual carburetors have an extra cutoff valve which cuts off the central idling system as well as one regular cutoff valve on each carburetor.

OVERHAUL

▸ See Figures 10, 11 and 12

Efficient carburetion depends greatly on careful cleaning and inspection during overhaul, since dirt, gum, water, or varnish in or on the carburetor parts are often responsible for poor performance.

Overhaul your carburetor in a clean, dust-free area. Carefully disassemble the carburetor, referring often to the exploded views. Keep all similar and look-alike parts segregated during disassembly and cleaning to avoid accidental interchange during assembly. Make a note of all jet sizes.

When the carburetor is disassembled, wash all parts (except diaphragms, electric choke units, pump plunger, and any other plastic, leather, fiber, or rubber parts) in clean carburetor solvent. Do not leave parts in the solvent any longer than is necessary to sufficiently loosen the deposits. Excessive cleaning may remove the special finish from the float bowl and choke valve bodies, leaving these parts unfit for service. Rinse all parts in clean solvent and blow them dry with compressed air to allow them to air dry. Wipe clean all cork, plastic, leather, and fiber parts with a clean, lint-free cloth.

Blow out all passages and jets with compressed air and be sure that there are no restrictions or blockages. Never use wire or similar tools to clean jets, fuel passages, or air bleeds. Clean all jets and valves separately to avoid accidental interchange.

Check all parts for wear or damage. If wear or damage is found, replace the defective parts. Especially check the following:

1. Check the float needle and seat for wear. If wear is found, replace the complete assembly.
2. Check the float hinge pin for wear and the float(s) for dents or distortion. Replace the float if fuel has leaked into it.
3. Check the throttle and choke shaft bores for wear or an out-of-round condition. Damage or wear to the throttle arm, shaft, or shaft bore will often require replacement of the throttle body. These parts require a close tolerance of fit; wear may allow air leakage, which could affect starting and idling.

➡ Throttle shafts and bushings are not included in overhaul kits. They can be purchased separately.

4. Inspect the idle mixture adjusting needles for burrs or grooves. Any such condition requires replacement of the needle, since you will not be able to obtain a satisfactory idle.
5. Test the accelerator pump check valves. They should pass air one way but not the other. Test for proper seating by blowing and sucking on the valve. Replace the valve if necessary. If the valve is satisfactory, wash the valve again to remove breath moisture.
6. Check the bowl cover for warped surfaces with a straightedge.
7. Closely inspect the valves and seats for wear and damage, replacing as necessary.
8. After the carburetor is assembled, check the choke valve for freedom of operation.

Carburetor overhaul kits are recommended for each overhaul. These kits contain all gaskets and new parts to replace those

FUEL SYSTEM

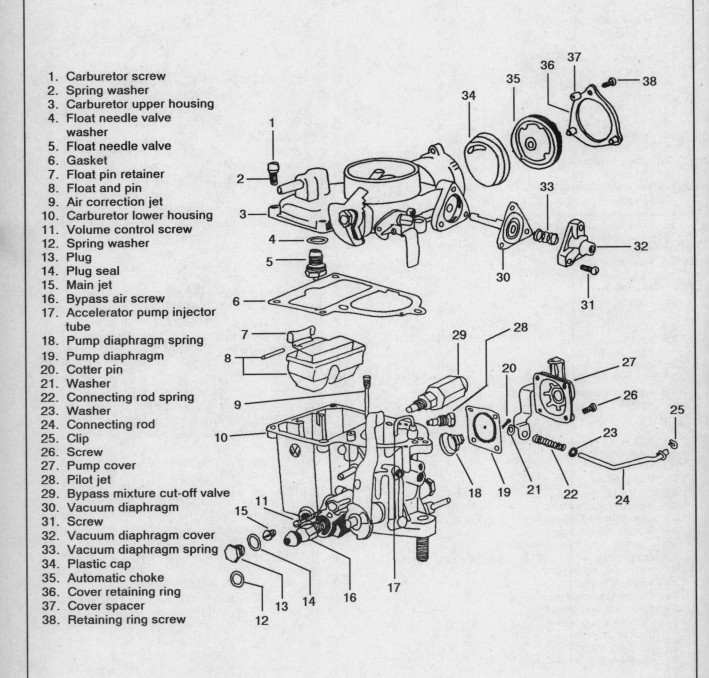

1. Carburetor screw
2. Spring washer
3. Carburetor upper housing
4. Float needle valve washer
5. Float needle valve
6. Gasket
7. Float pin retainer
8. Float and pin
9. Air correction jet
10. Carburetor lower housing
11. Volume control screw
12. Spring washer
13. Plug
14. Plug seal
15. Main jet
16. Bypass air screw
17. Accelerator pump injector tube
18. Pump diaphragm spring
19. Pump diaphragm
20. Cotter pin
21. Washer
22. Connecting rod spring
23. Washer
24. Connecting rod
25. Clip
26. Screw
27. Pump cover
28. Pilot jet
29. Bypass mixture cut-off valve
30. Vacuum diaphragm
31. Screw
32. Vacuum diaphragm cover
33. Vacuum diaphragm spring
34. Plastic cap
35. Automatic choke
36. Cover retaining ring
37. Cover spacer
38. Retaining ring screw

Fig. 10 Exploded view of the Solex 30 PICT-3 carburetor used on 1970 Type 1 and 2/1600 models

FUEL SYSTEM 5-13

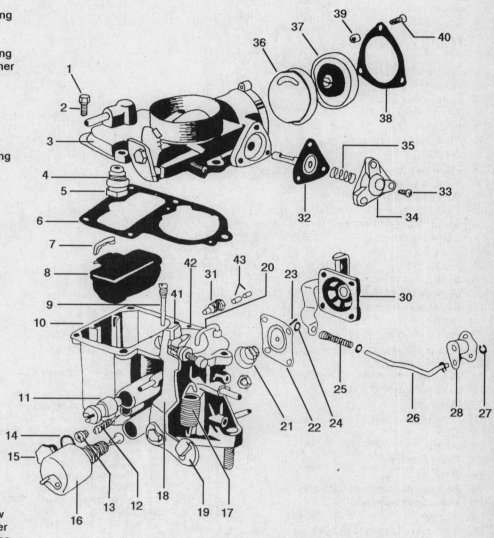

1. Upper carburetor housing screw
2. Spring washer
3. Upper carburetor housing
4. Float needle valve washer
5. Float needle valve
6. Gasket
7. Float pin retainer
8. Float and pin
9. Air correction jet and emulsion tube
10. Lower carburetor housing
11. Bypass screw
12. Volume control screw
13. Main jet
14. Plug washer
15. Plug
16. Bypass air cut-off valve
17. Return spring
18. Fast idling lever
19. Throttle valve lever and stop screw
20. Accelerator pump injection pipe
21. Diaphragm spring
22. Accelerator pump diaphragm
23. Cotter pin
24. Washer
25. Connecting rod and spring
26. Connecting rod
27. Clip
28. Bell crank lever
29. Countersunk head screw
30. Pump cover
31. Pilot jet
32. Vacuum diaphragm
33. Countersunk head screw
34. Vacuum diaphragm cover
35. Vacuum diaphragm spring
36. Plastic cap
37. Insert with spring and heater element
38. Cover retaining ring
39. Retaining ring spacer
40. Retaining ring screw
41. Pilot air hole
42. Auxiliary air hole
43. Auxiliary fuel jet and plug

Fig. 11 Exploded view of the Solex 34 PICT-3 carburetor used on 1971–74 Type 1 and 1971 Type 2/1600 models

5-14 FUEL SYSTEM

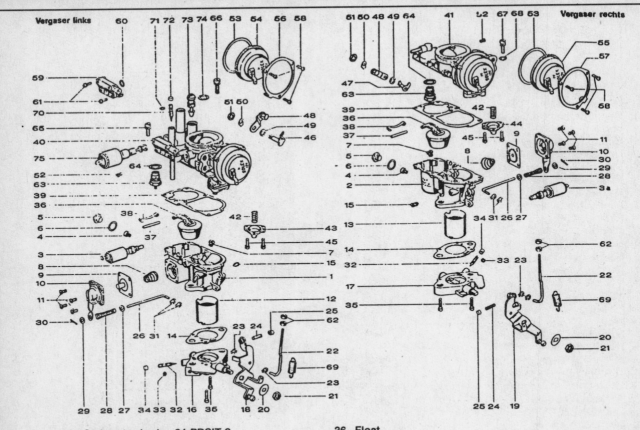

1. Carburetor body—34 PDSIT-2
2. Carburetor body—34 PDSIT-3
3. Electromagnetic idling cutoff valve—34 PDSIT-2
3a. Electromagnetic idling cutoff valve—34 PDSIT-3
4. Main jet
5. Main jet cover plug
6. Main jet cover plug seal
7. Air correction jet
8. Pump diaphragm spring
9. Pump diaphragm
10. Accelerator pump cover
11. Screws
12. Venturi—34 PDSIT-2
13. Venturi—34 PDSIT-3
14. Throttle body gasket
15. Venturi setscrew
16. Throttle body—34 PDSIT-2
17. Throttle body—34 PDSIT-3
18. Throttle arm—34 PDSIT-2
19. Throttle arm—34 PDSIT-3
20. Special washer
21. Nut
22. Connecting rod
23. Circlip
24. Throttle valve opening adjusting screw
25. Plug
26. Connecting link
27. Washer
28. Connecting link
29. Washer
30. Cotter pin
31. Circlip
32. Idle mixture screw
33. O-ring
34. Plug
35. Throttle body screws
36. Float
37. Float pin
38. Float pin retainer
39. Gasket
40. Carburetor upper part (air horn with idle mixture enrichment—34 PDSIT-2
41. Carburetor upper part (air horn)—34 PDSIT-3
42. Vacuum diaphragm spring
43. Vacuum diaphragm cover—34 PDSIT-2
44. Vacuum diaphragm cover—34 PDSIT-3
45. Screws
53. Choke heating element gasket
54. Choke heating element—34 PDSIT-2
55. Choke heating element—34 PDSIT-3
56. Choke cover retaining ring—34 PDSIT-2
57. Choke cover retaining ring—34 PDSIT-3
58. Screws
59. Idle mixture enrichment unit
60. O-ring
61. Screws
62. Connecting rod locknuts
63. Float valve
64. Float valve washer
65. Screws
66. Screws
67. Screws
68. Washer
69. Throttle return spring
70. Idle mixture screw
71. O-ring
72. Plug
73. Idle speed adjusting screw
74. O-ring
75. Central idling system electromagnetic cutoff valve

Vergaser links—left carburetor (34 PDSIT-2)
Vergaser rechts—right carburetor (34 PDSIT-3)

Fig. 12 Exploded view of the Solex 34 PDSIT 2/3 carburetors used on 1972–74 2/1700 and 2/1800 models

FUEL SYSTEM

Carburetor Specifications
(All measurements are given in metric units)

Year	Type Common Designation	Engine Code	Carburetor (Solex)	Venturi Diameter (mm)	Main Jet	Air Correction Jet	Pilot Jet	Aux. Fuel Jet	Aux. Air Jet	Power Fuel Jet	Needle Valve Washer Thickness (mm)	Accelerator Pump Injection Quantity (cc stroke)	Throttle Valve Gap (mm)	Fuel Level (mm)
1970	1/1600 & 2/1600	B	30 PICT-3	24	x112.5	125Z ① 140Z ②	65	45.0	130	100/100	1.5	1.05-1.35	—	19.5-20.5
1971	1/1600	AE	34 PICT-3	26	x130	75Z/80Z	g60	47.5	90	100/100	0.5	1.45-1.75	—	17-19
	2/1600	AE	34 PICT-3	26	x125	60Z	g57.5	42.5	90	95/95	0.5	1.45-1.75	—	17-19
1972	1/1600	AE	34 PICT-3	26	x130	75Z/80Z	g60	47.5	90	100/100	0.5	1.45-1.75	—	17-19
	1/1600	AH ③	34 PICT-3	26	x127.5/x130	75Z/80Z	g55	42.5	90	100	0.5	1.30-1.60	—	17-19
	2/1700 ④	CB	Left 34 PDSIT-2 Right 34 PDSIT-3	26	x137.5	155	55	45.0	0.7	—	0.5	0.8-1.0	0.6	12-14
				26	x137.5	155	55	—	—	—	0.5	0.8-1.0	0.6	12-14
1973	1/1600	AK	34 PICT-3	26	x127.5/x127.5	75Z/80Z	g55	42.5	90	100	0.5	1.30-1.60	—	17-19
		AH	34 PICT-4	26	x112.5	75Z/70Z	g55	42.5	90	100	0.5	1.30-1.60	—	17-19
		AM ③	34 PICT-4	26	x112.5	75Z/70Z	g55	42.5	90	100	0.5	1.30-1.60	—	17-19
	2/1700 ④	CB ⑤	Left 34 PDSIT-2 Right 34 PDSIT-3	26	x130	140	55	45.0	0.7	—	1.0	0.6-0.8	0.6	12-14
				26	x130	140	55	—	—	—	1.0	0.6-0.8	0.6	12-14
		CD ⑥	Left 34 PDSIT-2 Right 34 PDSIT-3	26	x132.5	155	50	45.0	0.7	—	1.0	0.7-1.2	0.6	12-14
				26	x132.5	155	50	—	—	—	1.0	0.7-1.2	0.6	12-14
1974	1/1600	AK	34 PICT-3	26	x127.5/x127.5	75Z/80Z	g55	42.5	90	100	0.5	1.30-1.60	—	17-19
		AH	34 PICT-4	26	x127.5	75Z ⑤ 70Z ⑥	g55	42.5	90	100	0.5	1.10 ⑦	—	17-19
		AM ③	34 PICT-4	26	x112.5	75Z/70Z	g55	42.5	90	100	0.5	1.30-1.60	—	17-19
	2/1800	AW	Left 34 PDSIT-2 Right 34 PDSIT-3	26	x130	175	52.5	45.0	0.7	—	1.0	1.5 ⑧	0.7	12-14
				26	x130	175	52.5	—	—	—	1.0	1.5 ⑧	0.7	12-14
1975-81						Air flow controlled by electronic fuel injection								

① — Type 1 models equipped with manual transaxles.
② — Type 1 models with automatic transaxles and all Type 2 models.
③ — California models only.
④ — Twin carburetor induction system.
⑤ — Manual transaxle models only.
⑥ — Automatic transaxle models only.
⑦ — Engine above 75°F; 1.77cc, engine below 75°F.
⑧ — Engine below 70°F; 0.7cc, engine above 70°F.

5-16 FUEL SYSTEM

that deteriorate most rapidly. Failure to replace all parts supplied with the kit (especially gaskets) can result in poor performance later.

Some carburetor manufacturers supply overhaul kits of three basic types: minor repair; major repair; and gasket kits. Basically, they contain the following:

Minor Repair Kits:
- All gaskets
- Float needle valve
- Volume control screw
- All diaphragms
- Spring for the pump diaphragm

Major Repair Kits:
- All jets and gaskets
- All diaphragms
- Float needle valve
- Volume control screw
- Pump ball valve
- Main jet carrier
- Float

Gasket Kits:
- All gaskets

After cleaning and checking all components, reassemble the carburetor, using new parts and referring to the exploded view. When reassembling, make sure that all screws and jets are tight in their seats, but do not overtighten as the tips will be distorted. Tighten all screws gradually in rotation. Do not tighten needle valves into their seats; uneven jetting will result. Always use new gaskets. Be sure to adjust the float level when reassembling.

Accelerator Cable

REMOVAL & INSTALLATION

1. Disconnect the cable from the accelerator pedal.
2. Disconnect the cable from the throttle lever.
3. Pull the cable from the accelerator pedal end and then remove it from the car.
4. Grease the cable before sliding it into its housing.
5. Slide the cable into its housing and push it through its guide tubes. It may be necessary to raise the car and start the cable into the segments of guide tube found under the car.
6. Install one cable end into the accelerator cable. Slip the other end into the throttle valve lever and adjust the cable.

➡**Make sure that the rubber boot at the rear end of the cable is properly seated so that water will not enter the guide tubes.**

GASOLINE FUEL INJECTION SYSTEM

Description

NON-AIR FLOW CONTROLLED

▶ See Figures 13, 14, 15, 16 and 17

The Bosch Electronic fuel injection system used on all Type 3 models, and on 1971–74 Type 4 models (except 1974 models equipped with an automatic), consists of two parts. One part consists of the actual injection components: the injectors, the fuel pump, pressure regulator, and related wiring and hoses. The second part consists of the injection controls and engine operating characteristics sensors: a manifold vacuum sensor that monitors engine load, trigger contacts used to determine when and which pair of injectors will operate, three temperature sensors used to control air fuel mixture enrichment, a cold starting valve for additional cold starting fuel enrichment, a throttle valve switch used to cut off fuel during deceleration, and the brain box used to analyze information about engine operating characteristics and, after processing this information, to control the electrically operated injectors.

It is absolutely imperative that no adjustments other than those found in the following pages be performed. The controls for this fuel injection system are extremely sensitive and easily damaged when subject to abuse. Never attempt to test the brain box without proper training and the proper equipment. The dealer is the best place to have any needed work performed.

✱✱ CAUTION

Whenever a fuel injection component is to be removed or installed, the battery should be disconnected and the ignition turned OFF.

It is not recommended that the inexperienced mechanic work on any portion of the fuel injection system.

AIR FLOW CONTROLLED

1974 Type 4 models equipped with automatic transmission, as well as all 1975–81 Type 1 and Type 2 models, are equipped with an improved system known as the Air Flow Controlled Electronic Fuel Injection System. With this system, some of the electronic sensors and wiring are eliminated, and the control box is smaller. Instead fuel is metered according to intake air flow.

The system consists of the following components:
- Intake air sensor—measures intake air volume and temperature and sends voltage signals to the control unit (brain box). It also controls the electric fuel pump by shutting it off when intake air stops. It is located between the air cleaner and the intake air distributor.
- Ignition contact breaker points—these are the regular points inside the distributor. When the points open, all four injectors are triggered. The points also send engine speed signals to the control unit. No separate triggering contacts are used.

FUEL SYSTEM 5-17

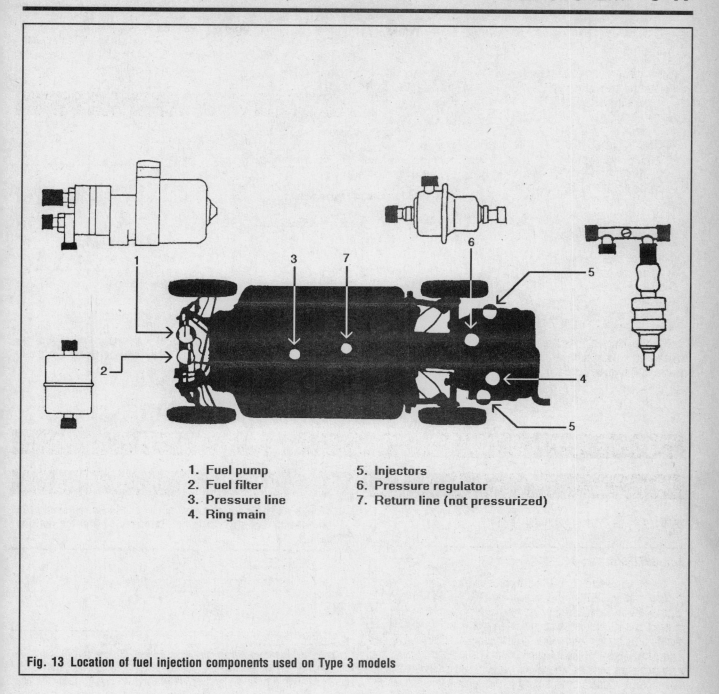

1. Fuel pump
2. Fuel filter
3. Pressure line
4. Ring main
5. Injectors
6. Pressure regulator
7. Return line (not pressurized)

Fig. 13 Location of fuel injection components used on Type 3 models

- Throttle valve switch—provides only for full load enrichment. This switch is not adjustable.
- Temperature sensor I—senses intake temperature as before. It is now located in the intake air sensor.
- Temperature sensor II—senses cylinder head temperature as before.
- Control unit (brain box—contains only 80 components compared to the old system's 300.
- Pressure regulator—is connected by a vacuum hose to the intake air distributor and is no longer adjustable. It adjusts fuel pressure according to manifold vacuum.
- Auxiliary air regulator—provides more air during cold warmup.

Electronic Control (Brain) Box

All work concerning the brain box is to be performed by the dealer. Do not remove the brain box and take it to a dealer because the dealer will not be able to test it without the vehicle. Do not disconnect the brain box unless the battery is disconnected and the ignition is **OFF**.

Electric Fuel Pump

All Type 3 and Type 4 models, as well as 1975 and later Type 1 and Type 2 models have an electric pump. The fuel pump is lo-

5-18 FUEL SYSTEM

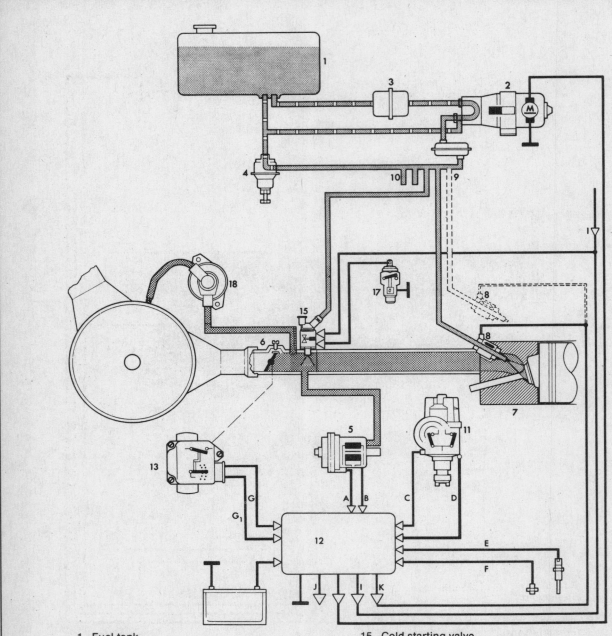

1. Fuel tank
2. Fuel pump
3. Fuel filter
4. Pressure regulator
5. Pressure sensor
6. Intake air distributor
7. Cylinder head
8. Injectors
9. Fuel distributor pipe
10. Fuel distributor pipe with connection for cold starting device
11. Distributor with trigger contacts (distributor contact I, distributor contact II)
12. Control unit
13. Throttle valve switch with acceleration enrichment
15. Cold starting valve
17. Thermostat for cold starting device
18. Auxiliary air regulator
A + B. from pressure sensor (load condition signal)
C + D. from distributor contacts (engine speed and releasing signal)
E + F. from temperature sensors (warmup signal)
G. from throttle valve switch (fuel supply cut-off when coasting)
G1. Acceleration enrichment
I. from starter, terminal 50 solenoid switch (signal for enrichment mixture when starting)
J. to the injectors, cylinders 1 and 4
K. to the injectors, cylinders 2 and 3

Fig. 14 Schematic of the electronic fuel injection system used on 1970 and later Federal models and 1970–71 California models

FUEL SYSTEM 5-19

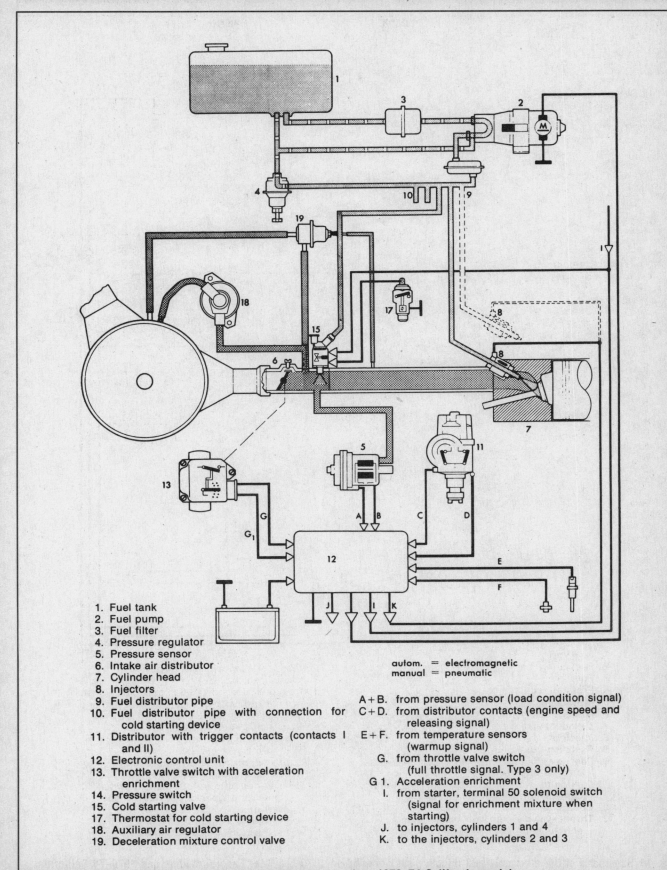

1. Fuel tank
2. Fuel pump
3. Fuel filter
4. Pressure regulator
5. Pressure sensor
6. Intake air distributor
7. Cylinder head
8. Injectors
9. Fuel distributor pipe
10. Fuel distributor pipe with connection for cold starting device
11. Distributor with trigger contacts (contacts I and II)
12. Electronic control unit
13. Throttle valve switch with acceleration enrichment
14. Pressure switch
15. Cold starting valve
17. Thermostat for cold starting device
18. Auxiliary air regulator
19. Deceleration mixture control valve

autom. = electromagnetic
manual = pneumatic

A+B. from pressure sensor (load condition signal)
C+D. from distributor contacts (engine speed and releasing signal)
E+F. from temperature sensors (warmup signal)
G. from throttle valve switch (full throttle signal. Type 3 only)
G 1. Acceleration enrichment
I. from starter, terminal 50 solenoid switch (signal for enrichment mixture when starting)
J. to injectors, cylinders 1 and 4
K. to the injectors, cylinders 2 and 3

Fig. 15 Schematic of the electronic fuel injection system used on 1972–74 California models

5-20 FUEL SYSTEM

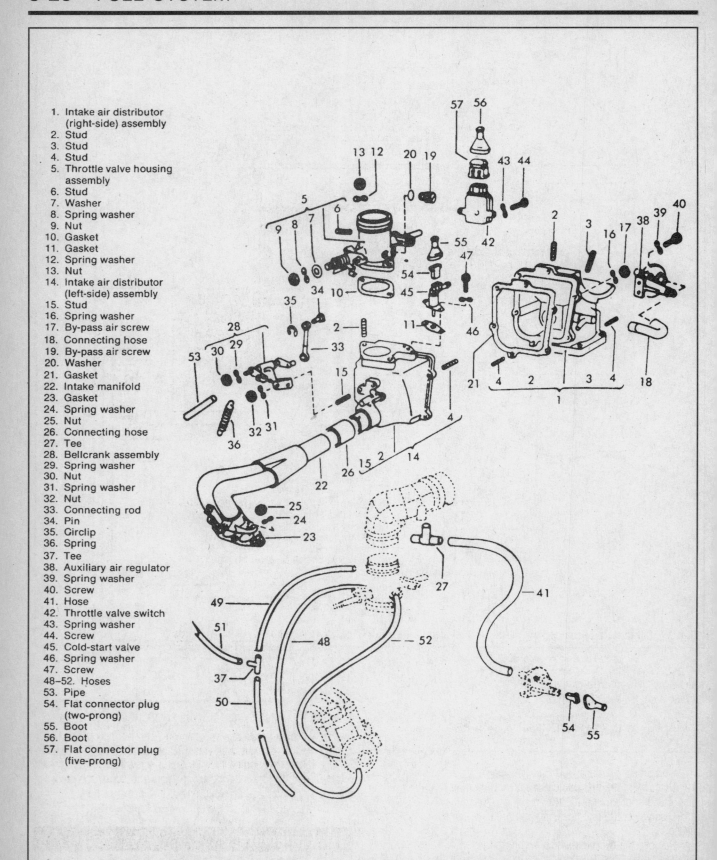

1. Intake air distributor (right-side) assembly
2. Stud
3. Stud
4. Stud
5. Throttle valve housing assembly
6. Stud
7. Washer
8. Spring washer
9. Nut
10. Gasket
11. Gasket
12. Spring washer
13. Nut
14. Intake air distributor (left-side) assembly
15. Stud
16. Spring washer
17. By-pass air screw
18. Connecting hose
19. By-pass air screw
20. Washer
21. Gasket
22. Intake manifold
23. Gasket
24. Spring washer
25. Nut
26. Connecting hose
27. Tee
28. Bellcrank assembly
29. Spring washer
30. Nut
31. Spring washer
32. Nut
33. Connecting rod
34. Pin
35. Circlip
36. Spring
37. Tee
38. Auxiliary air regulator
39. Spring washer
40. Screw
41. Hose
42. Throttle valve switch
43. Spring washer
44. Screw
45. Cold-start valve
46. Spring washer
47. Screw
48–52. Hoses
53. Pipe
54. Flat connector plug (two-prong)
55. Boot
56. Boot
57. Flat connector plug (five-prong)

Fig. 16 Exploded view of the airflow controlled fuel injection system used on 1975–81 Type 1 and Type 2 models (Type 1 shown, Type 2 similar)

FUEL SYSTEM 5-21

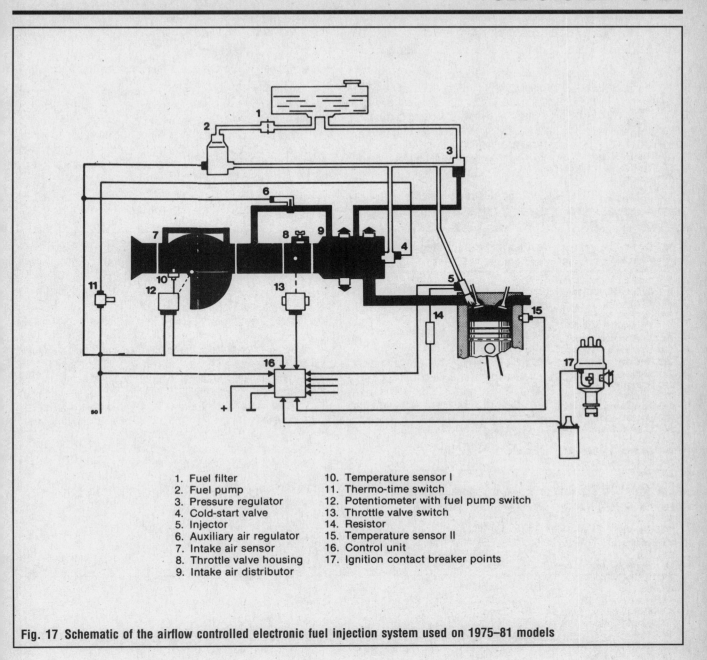

1. Fuel filter
2. Fuel pump
3. Pressure regulator
4. Cold-start valve
5. Injector
6. Auxiliary air regulator
7. Intake air sensor
8. Throttle valve housing
9. Intake air distributor
10. Temperature sensor I
11. Thermo-time switch
12. Potentiometer with fuel pump switch
13. Throttle valve switch
14. Resistor
15. Temperature sensor II
16. Control unit
17. Ignition contact breaker points

Fig. 17 Schematic of the airflow controlled electronic fuel injection system used on 1975–81 models

cated near the front axle on Types 1, 3, and 4, and near the fuel tank on Type 2.

REMOVAL & INSTALLATION

1. Disconnect the negative battery cable.
2. Disconnect the fuel pump wiring. Pull the plug from the pump but do not pull on the wiring.
3. Disconnect the fuel hoses and plug them to prevent any leakage.
4. Remove the two nuts which secure the pump and then remove the pump.
5. Reconnect the fuel pump hoses and wiring and install the pump on the vehicle.
6. Connect the battery cable.

ADJUSTMENTS

Electric fuel pump pressure is 28 psi. Fuel pump pressure is determined by a pressure regulator which diverts part of the fuel pump output to the gas tank when 28 psi is reached. The regulator, located on the engine firewall, has a screw and locknut on its end. Loosen the locknut and turn the screw to adjust the pressure. Do not force the screw in or out if it does not turn.

Fuel Injectors

There are two types of injectors. One type is secured in place by a ring that holds a single injector. The second type of injector is secured to the intake manifold in pairs by a common bracket.

5-22 FUEL SYSTEM

CHECKING FUEL INJECTOR OPERATION

When the engine fails to run or runs erratically, always check the ignition system before assuming that the problem lies under that inscrutable tangle known as the fuel injection system. If the ignition system is functioning correctly, remove the injectors from the engine, leaving the fuel lines and wiring connected to them. Disconnect the high tension cable from the ignition coil to prevent the engine from starting and have an assistant engage the starter briefly.

➡ See precautions in Section 2 for electronic ignition.

✱✱ CAUTION

Raw fuel will be sprayed from the injectors if they are working, so have a container ready and do not smoke.

If a cone-shaped mist of fuel sprays in pulses from all of the injectors, they are working properly. If one or more injectors fail to spray, check the fuel injection signal.

To check for injector leakage, wipe the tips of the injectors dry with a clean, lint-free rag, then turn the ignition to **ON** (not **START**). If the injector tips become wet with fuel but do not drip more than one or two drops per minute, it is a good possibility that the pressure regulator is set too high; if the injector loses more than one or two drops per minute, the injector is faulty and should be replaced.

CHECKING THE FUEL INJECTOR SIGNAL

➡ **You will need a mechanic's stethoscope to perform the second part of this test.**

Unplug the electrical connector on the fuel injector and connect the leads of a test light to the two terminals of the connector. Disconnect the ignition coil high tension cable to prevent the engine from starting and have an assistant engage the starter.

➡ **See Precautions in Section 2 for electronic ignition.**

The test light should blink, indicating the pulses of the injector. Reconnect all fuel injector connections and run the engine. Place the end of the stethoscope on the fuel injector body. You should hear a clicking sound indicating that the injector is working properly. Failure to hear a clicking sound or hearing a sound different from the rest of the injectors probably indicates a faulty injector.

REMOVAL & INSTALLATION

Single Injectors
▶ See Figure 18

1. Disconnect the negative battery cable.
2. Detach the fuel injector wiring.
3. Remove the nut which secures the injector bracket to the manifold.
4. If the injector is not going to be replaced, do not disconnect the fuel line. Disconnect the injector wiring.
5. Gently slide the injector bracket up the injector and pull the injector from the intake manifold. Be careful not to damage the inner and outer rubber sealing rings. These sealing rings are used to seal the injector to the manifold and must be replaced if they show any sign of deterioration.
6. Installation is the reverse of removal. Be careful not to damage the injector tip or contaminate the injector with dirt.

Paired Injectors
▶ See Figure 19

1. Disconnect the negative battery cable.
2. Disconnect the injector wiring.

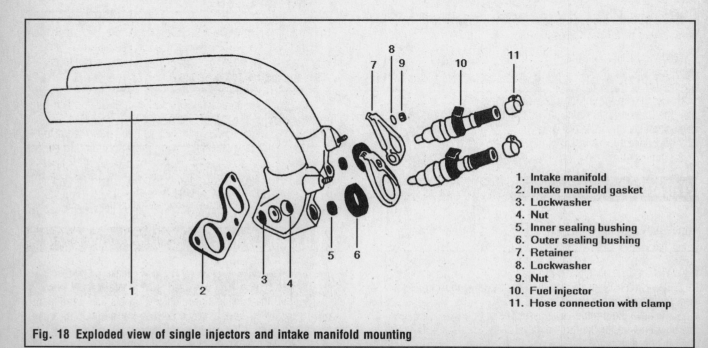

Fig. 18 Exploded view of single injectors and intake manifold mounting

1. Intake manifold
2. Intake manifold gasket
3. Lockwasher
4. Nut
5. Inner sealing bushing
6. Outer sealing bushing
7. Retainer
8. Lockwasher
9. Nut
10. Fuel injector
11. Hose connection with clamp

FUEL SYSTEM 5-23

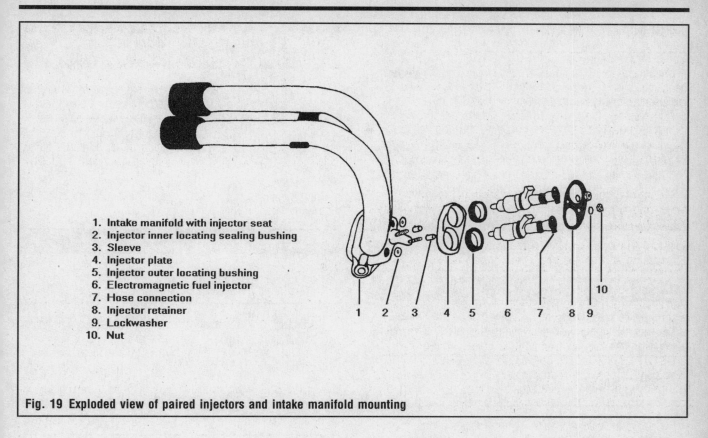

1. Intake manifold with injector seat
2. Injector inner locating sealing bushing
3. Sleeve
4. Injector plate
5. Injector outer locating bushing
6. Electromagnetic fuel injector
7. Hose connection
8. Injector retainer
9. Lockwasher
10. Nut

Fig. 19 Exploded view of paired injectors and intake manifold mounting

3. Remove the two nuts which secure the injector bracket to the manifold. Slide the bracket up the injector. Do not disconnect the fuel lines if the injector is not going to be replaced.
4. Gently slide the pair of injectors out of their bores along with the rubber sealing rings, injector plate, and the inner and outer injector locating bushings. It may be necessary to remove the inner bushings from the intake manifold after the injectors are removed since they sometimes lodge within the manifold.

➡ There are two sleeves that fit over the injector bracket studs. Be careful not to lose them.

5. Upon installation, place the injector bracket, the outer locating bushings, the injector plate, and the inner locating bushings on the pair of injectors in that order.
6. Gently slip the injector assembly into the manifold and install the bracket nuts. Be careful not to damage the injector tips or contaminate the injectors with dirt.
7. Reconnect the injector wiring.
8. Connect the battery.

Throttle Valve Switch

REMOVAL & INSTALLATION

1. Remove the air filter.
2. The switch is located on the throttle valve housing. Disconnect the throttle valve return spring.
3. Remove the throttle valve assembly but do not disconnect the bowden wire for the throttle valve or the connecting hoses to the ignition distributor.
4. Remove the throttle valve switch securing screws and remove the switch.
5. Reverse the above steps to install. It will be necessary to adjust the switch after installation.

ADJUSTMENT (NON-AIR FLOW CONTROLLED ONLY)

The throttle valve switch is used to shut off the fuel supply during deceleration. The switch is supposed to operate when the throttle valve is opened 2°. A degree scale is stamped into the attachment plate for adjustment purposes.

1. Completely close the throttle valve.
2. Loosen the switch attaching screws and turn the switch carefully to the right until it hits its stop.
3. Turn the switch slowly to the left until it can be heard to click and then note the position of the switch according to the degree scale.
4. Continue to turn the switch another 2°. The distance between any two marks on the degree scale is 2°.
5. Tighten the screws and recheck the adjustment.

Cold Start Valve

The cold start valve is located on the air intake distributor and is held by two screws. The cold start valve operates only when the engine and the outside air are cold. It injects fuel into the air intake distributor for several seconds when the starter is engaged and then shuts off. If the valve is not working, it will be very difficult to start the engine; if the valve is leaking, it could cause

5-24 FUEL SYSTEM

flooding while starting, especially if the engine is hot. The cold start valve is controlled by a thermo-time switch or, on some models, a thermostat. Power for the valve is fed from terminal 50 of the starter.

CHECKING VALVE LEAKAGE

Remove the cold start valve from the engine after detaching the electrical connector. Tape the connector's end to prevent sparks. Leave the fuel line attached. Disconnect the high tension ignition coil wire to prevent the engine from starting and have an assistant engage the starter while you hold the valve nozzle up in a rag.

➡ See precautions in Section 2 for electronic ignitions.

If fuel forms on the nozzle and begins to drip off, the valve is leaking and should be replaced, or fuel pressure is too high.

✱✱ CAUTION

Fuel might be expelled from the valve, so do not smoke and do not allow any electrical connections (coil wire, etc.) to cause sparks.

CHECKING VALVE OPERATION

1. Remove the cold start valve from the engine leaving the fuel line attached.
2. Remove the high tension wire from the middle of the ignition coil and put tape over the coil terminal to prevent sparks.

➡ See precautions in Section 2 for electronic ignition.

3. If you haven't done so already, detach the electrical connector from the cold start valve.
4. Have a container ready to catch gasoline and using a jumper wire, connect one of the terminals on the cold start valve to ground. Connect the other terminal of the cold start valve to terminal 15 of the ignition coil using another jumper cable.
5. Point the nozzle of the cold start valve into the container and have an assistant turn the ignition to the **ON** position.

✱✱ CAUTION

Be careful not to allow the jumper wires to touch each other and do not smoke, as fuel will be expelled from the valve if it is working.

6. A spray of fuel should come from the valve while the ignition is on. If not, the cold start valve is not working correctly.
7. Reconnect the cold start valve electrical connection. Turn the ignition to **ON**. If the cold start valve injects fuel with the ignition in the **ON** (not **START**) position, there is a good possibility that the cold start valve wire that belongs on terminal 50 of the starter has been mistakingly connected to terminal 30.

This test has been for the cold start valve itself, not the other components of the cold start system (thermo-time valve, relay, etc.). If your vehicle still starts hard on cold mornings, the problem could be with the thermo-time switch (thermo-switch on some models). It is best to let your dealer troubleshoot these components.

Trigger Contacts (Non-Air Flow Controlled Only)

REMOVAL & INSTALLATION

The trigger contacts are located in the base of the distributor and are secured by two screws. These contacts are supplied in pairs and are not adjustable. Do not attempt to replace just one set of contacts.

One set of contacts controls a pair of injectors and tells the injectors when to fire.

Fuel Pressure Regulator

REMOVAL & INSTALLATION

Disconnect the hoses from the regulator and remove the regulator from its bracket. The fuel pump pressure is adjustable; however, lack of fuel pressure is usually due to other defects in the system and the regulator should be adjusted only as a last resort.

ADJUSTMENT (NON-AIR FLOW CONTROLLED ONLY)

▶ See Figure 20

1. Remove the air cleaner.
2. Connect a fuel pressure gauge.
3. Start the engine and operate at idle.
4. Loosen the locknut and adjust the fuel pressure to 28 psi with the screw.

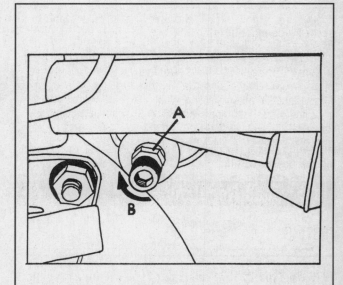

Fig. 20 Fuel pressure regulator adjustment for Type 3 models

FUEL SYSTEM   5-25

TESTING

It is possible to check the fuel pressure by inserting a fuel pressure gauge in the line after the pressure regulator. Insert the gauge using a T-fitting. Turn on the key and check the pressure. If the pressure is low, check for leaking injectors, restricted lines, clogged fuel filters, damaged pressure regulator, bad fuel pump, water in the gas and resultant corrosion of the injectors, or a leaking or jammed cold start valve.

Temperature Sensors I and II

REMOVAL & INSTALLATION

The air temperature sensor is located in the air distributor housing and may be unscrewed from the housing. The second temperature switch is located in the cylinder head on the left side and senses cylinder head temperature. It is removed with a special wrench. To test these switches, attach an ohmmeter and measure the resistance of the switch as the temperature is raised gradually to 212°. As the temperature rises, the resistance of the first switch should drop from about 200 ohms to 80 ohms. The cylinder head switch resistance should drop from about 1,700 ohms to 190 ohms at 212°.

The third switch is actually a thermoswitch and is an ON/OFF type switch. Below 41° it is ON to activate the cold starting valve. The switch is located next to the distributor and may be removed with a 24 mm wrench.

Pressure Sensor (Non-Air Flow Controlled Only)

REMOVAL & INSTALLATION

The sensor is secured to the firewall by two screws. Remove the screws and disconnect the wiring.

Remove the pressure connection and immediately plug the connection into the sensor. Always keep the connection plugged as the bellows inside the sensor is sensitive to the smallest pieces of dirt. Reverse the above steps to install.

Do not disassemble the sensor. There are no adjustments possible.

➡ Do not reverse the square electrical plug when reconnecting the sensor wiring.

Idle Speed Regulator

TESTING

1975–76 Type 2; 1971 and Later Type 4 With Automatic Transaxle

This vacuum operated assembly slightly opens the throttle when the transmission is put in drive to prevent the engine from stalling under the load.

To test, adjust the regulator as described in Section 2. When the plunger clearance is correct, pull the vacuum hose off the idle speed regulator: the plunger should extend and open the throttle slightly. When you reconnect the hose, the plunger should pull back. If not, check the hose for blockage or rips, or replace the regulator.

Auxiliary Air Regulator

♦ See Figure 21

TESTING

The auxiliary air regulator lets extra air into the intake manifold during engine warm-up which allows the engine to run smoothly while it is cold. As the engine warms up, a rotary valve inside the air regulator twists shut slowly, cutting off more and more air intake until almost none is passing through the unit.

To test the auxiliary air regulator, the engine must be cold. Unplug the fabric covered air intake hose (not the one that leads into the intake air distributor). There should be suction at the hose. Cover the hose: engine speed should drop. After a reasonable amount of time (time will be longer in cold weather), the suction should diminish until it is almost nonexsistant. If the regulator does not close at all, disconnect its electrical connector and connect a test light to the terminals of the connector (not the regulator). With the engine running the test light should light. If it does, the auxiliary air regulator is probably defective.

Fig. 21 Auxiliary air regulator used on Type 4 models— regulator used on Type 3 models equipped with automatic transaxles similar

FUEL SYSTEM

FUEL TANK

Tank Assembly

REMOVAL & INSTALLATION

Type 1, Type 3 and Type 4

1. Disconnect the battery ground cable.
2. Drain the fuel from the tank.
3. Detach the fuel line and the wires from the sending unit, after removing the luggage compartment liner.
4. Remove the ventilation hoses from the fuel tank.
5. Loosen the hose clamps on the filler neck hose and remove the hose.
6. Remove the bolts from the four fuel tank retaining plates. Lift off the retaining plates and remove the tank from the car.
7. Installation is the reverse of removal.

Type 2 Through 1979

1. On all models except the pick-up truck, remove the engine. See Section 3 for procedures. On 1972 and later models, it is possible to remove the rear panel without removing the engine, but it is a tight squeeze.
2. If not done already, disconnect the battery cable and drain the fuel from the tank.
3. Remove the screws from the bulkhead behind the fuel tank (enclosed models) or from the panels in the lower storage compartment (pick-up truck models).
4. Disconnect the fuel filler hose and all lines and hoses. Plug the fuel lines.

The fuel tank (arrow) is mounted in the forward luggage space—Super Beetle shown, regular Beetle similar

5. Remove the fuel tank hold-down straps and remove the tank toward the engine compartment (enclosed models) or through the lower storage compartment (pick-up trucks).
6. Installation is the reverse of removal.

Type 2 1980–81

The fuel tank is located behind the front axle assembly. It can be removed from beneath the vehicle after being drained by removing all hoses and wiring, then removing the two braces from beneath the tank and any other retaining bolts.

UNDERSTANDING AND
  TROUBLESHOOTING ELECTRICAL
  SYSTEMS 6-2
SAFETY PRECAUTIONS 6-2
UNDERSTANDING BASIC
  ELECTRICITY 6-2
    THE WATER ANALOGY 6-2
    CIRCUITS 6-2
    AUTOMOTIVE CIRCUITS 6-3
    SHORT CIRCUITS 6-3
TROUBLESHOOTING 6-3
    BASIC TROUBLESHOOTING
      THEORY 6-4
    TEST EQUIPMENT 6-4
    TESTING 6-6
WIRING HARNESSES 6-8
    WIRING REPAIR 6-8
ADD-ON ELECTRICAL EQUIPMENT 6-11
HEATER 6-12
HEATER BLOWER 6-13
    REMOVAL & INSTALLATION 6-13
HEATER CABLES 6-13
    REMOVAL & INSTALLATION 6-13
WINDSHIELD WIPERS 6-14
WIPER BLADE AND ARM 6-14
    REMOVAL & INSTALLATION 6-14
WIPER MOTOR 6-14
    REMOVAL & INSTALLATION 6-14
LINKAGE 6-15
    REMOVAL & INSTALLATION 6-15
INSTRUMENT CLUSTER 6-15
SPEEDOMETER 6-15
    REMOVAL & INSTALLATION 6-15
SPEEDOMETER CABLE 6-16
    REMOVAL & INSTALLATION 6-16
FUEL GAUGE AND CLOCK
  ASSEMBLY 6-16
    REMOVAL & INSTALLATION 6-16
IGNITION SWITCH 6-16
**SEAT BELT/STARTER
  INTERLOCK 6-16**
LIGHTING 6-16
HEADLIGHTS 6-16
    REMOVAL & INSTALLATION 6-16
REAR BRAKE/TURN SIGNAL/BACK-UP
  LIGHTS 6-17
    REMOVAL & INSTALLATION 6-17
FRONT TURN SIGNAL/SIDE MARKER
  LIGHTS 6-18
    REMOVAL & INSTALLATION 6-18
LICENSE PLATE LIGHT 6-19
    REMOVAL & INSTALLATION 6-19
FUSES 6-20
TRAILER WIRING 6-21
WIRING DIAGRAMS 6-22
SPECIFICATION CHARTS
  FUSES 6-20

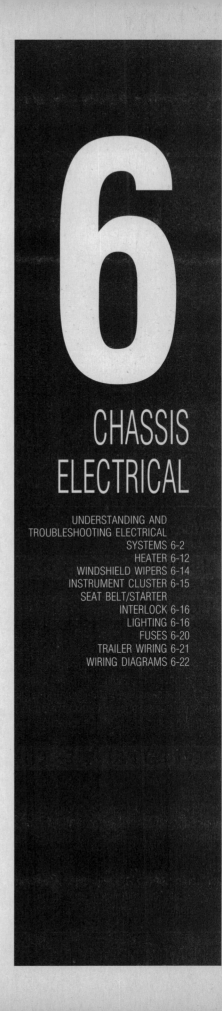

6

CHASSIS ELECTRICAL

UNDERSTANDING AND
TROUBLESHOOTING ELECTRICAL
SYSTEMS 6-2
HEATER 6-12
WINDSHIELD WIPERS 6-14
INSTRUMENT CLUSTER 6-15
SEAT BELT/STARTER
INTERLOCK 6-16
LIGHTING 6-16
FUSES 6-20
TRAILER WIRING 6-21
WIRING DIAGRAMS 6-22

6-2 CHASSIS ELECTRICAL

UNDERSTANDING AND TROUBLESHOOTING ELECTRICAL SYSTEMS

Over the years import and domestic manufacturers have incorporated electronic control systems into their production lines. In fact, electronic control systems are so prevalent that all new cars and trucks built today are equipped with at least one on-board computer. These electronic components (with no moving parts) should theoretically last the life of the vehicle, provided that nothing external happens to damage the circuits or memory chips.

While it is true that electronic components should never wear out, in the real world malfunctions do occur. It is also true that any computer-based system is extremely sensitive to electrical voltages and cannot tolerate careless or haphazard testing/service procedures. An inexperienced individual can literally cause major damage looking for a minor problem by using the wrong kind of test equipment or connecting test leads/connectors with the ignition switch **ON**. When selecting test equipment, make sure the manufacturer's instructions state that the tester is compatible with whatever type of system is being serviced. Read all instructions carefully and double check all test points before installing probes or making any test connections.

The following section outlines basic diagnosis techniques for dealing with automotive electrical systems. Along with a general explanation of the various types of test equipment available to aid in servicing modern automotive systems, basic repair techniques for wiring harnesses and connectors are also given. Read the basic information before attempting any repairs or testing. This will provide the background of information necessary to avoid the most common and obvious mistakes that can cost both time and money. Although the replacement and testing procedures are simple in themselves, the systems are not, and unless one has a thorough understanding of all components and their function within a particular system, the logical test sequence these systems demand cannot be followed. Minor malfunctions can make a big difference, so it is important to know how each component affects the operation of the overall system in order to find the ultimate cause of a problem without replacing good components unnecessarily. It is not enough to use the correct test equipment; the test equipment must be used correctly.

Safety Precautions

✱✱ CAUTION

Whenever working on or around any electrical or electronic systems, always observe these general precautions to prevent the possibility of personal injury or damage to electronic components.

- Never install or remove battery cables with the key **ON** or the engine running. Jumper cables should be connected with the key **OFF** to avoid power surges that can damage electronic control units. Engines equipped with computer controlled systems should avoid both giving and getting jump starts due to the possibility of serious damage to components from arcing in the engine compartment if connections are made with the ignition **ON**.
- Always remove the battery cables before charging the battery. Never use a high output charger on an installed battery or attempt to use any type of "hot shot" (24 volt) starting aid.
- Exercise care when inserting test probes into connectors to insure good contact without damaging the connector or spreading the pins. Always probe connectors from the rear (wire) side, NOT the pin side, to avoid accidental shorting of terminals during test procedures.
- Never remove or attach wiring harness connectors with the ignition switch **ON**, especially to an electronic control unit.
- Do not drop any components during service procedures and never apply 12 volts directly to any component (like a solenoid or relay) unless instructed specifically to do so. Some component electrical windings are designed to safely handle only 4 or 5 volts and can be destroyed in seconds if 12 volts are applied directly to the connector.
- Remove the electronic control unit if the vehicle is to be placed in an environment where temperatures exceed approximately 176°F (80°C), such as a paint spray booth or when arc/gas welding near the control unit location.

Understanding Basic Electricity

Understanding the basic theory of electricity makes electrical troubleshooting much easier. Several gauges are used in electrical troubleshooting to see inside the circuit being tested. Without a basic understanding, it will be difficult to understand testing procedures.

THE WATER ANALOGY

Electricity is the flow of electrons—hypothetical particles thought to constitute the basic stuff of electricity. Many people have been taught electrical theory using an analogy with water. In a comparison with water flowing in a pipe, the electrons would be the water. As the flow of water can be measured, the flow of electricity can be measured. The unit of measurement is amperes, frequently abbreviated amps. An ammeter will measure the actual amount of current flowing in the circuit.

Just as the water pressure is measured in units such as pounds per square inch, electrical pressure is measured in volts. When a voltmeter's two probes are placed on two live portions of an electrical circuit with different electrical pressures, current will flow through the voltmeter and produce a reading which indicates the difference in electrical pressure between the two parts of the circuit.

While increasing the voltage in a circuit will increase the flow of current, the actual flow depends not only on voltage, but on the resistance of the circuit. The standard unit for measuring circuit resistance is an ohm, measured by an ohmmeter. The ohmmeter is somewhat similar to an ammeter, but incorporates its own source of power so that a standard voltage is always present.

CIRCUITS

An actual electric circuit consists of four basic parts. These are: the power source, such as a generator or battery; a hot wire, which conducts the electricity under a relatively high voltage to the component supplied by the circuit; the load, such as a lamp, motor, resistor or relay coil; and the ground wire, which carries the current back to the source under very low voltage. In such a circuit the bulk of the resistance exists between the point where the hot wire is connected to the load, and the point where the load is grounded. In an automobile, the vehicle's frame or body, which is made of steel, is used as a part of the ground circuit for many of the electrical devices.

Remember that, in electrical testing, the voltmeter is connected in parallel with the circuit being tested (without disconnecting any wires) and measures the difference in voltage between the locations of the two probes; that the ammeter is connected in series with the load (the circuit is separated at one point and the ammeter inserted so it becomes a part of the circuit); and the ohmmeter is self-powered, so that all the power in the circuit should be off and the portion of the circuit to be measured contacted at either end by one of the probes of the meter.

For any electrical system to operate, it must make a complete circuit. This simply means that the power flow from the battery must make a complete circle. When an electrical component is operating, power flows from the battery to the component, passes through the component causing it to perform it to function (such as lighting a light bulb) and then returns to the battery through the ground of the circuit. This ground is usually (but not always) the metal part of the vehicle on which the electrical component is mounted.

Perhaps the easiest way to visualize this is to think of connecting a light bulb with two wires attached to it to your vehicle's battery. The battery in your vehicle has two posts (negative and positive). If one of the two wires attached to the light bulb was attached to the negative post of the battery and the other wire was attached to the positive post of the battery, you would have a complete circuit. Current from the battery would flow out one post, through the wire attached to it and then to the light bulb, where it would pass through causing it to light. It would then leave the light bulb, travel through the other wire, and return to the other post of the battery.

AUTOMOTIVE CIRCUITS

The normal automotive circuit differs from this simple example in two ways. First, instead of having a return wire from the bulb to the battery, the light bulb return the current to the battery through the chassis of the vehicle. Since the negative battery cable is attached to the chassis and the chassis is made of electrically conductive metal, the chassis of the vehicle can serve as a ground wire to complete the circuit. Secondly, most automotive circuits contain switches to turn components on and off.

Some electrical components which require a large amount of current to operate also have a relay in their circuit. Since these circuits carry a large amount of current, the thickness of the wire in the circuit (gauge size) is also greater. If this large wire were connected from the component to the control switch on the instrument panel, and then back to the component, a voltage drop would occur in the circuit. To prevent this potential drop in voltage, an electromagnetic switch (relay) is used. The large wires in the circuit are connected from the vehicle battery to one side of the relay, and from the opposite side of the relay to the component. The relay is normally open, preventing current from passing through the circuit. An additional, smaller wire is connected from the relay to the control switch for the circuit. When the control switch is turned on, it grounds the smaller wire from the relay and completes the circuit.

SHORT CIRCUITS

If you were to disconnect the light bulb (from the previous example of a light-bulb being connected to the battery by two wires) from the wires and touch the two wires together (please take our word for this; don't try it), the result will be a shower of sparks. A similar thing happens (on a smaller scale) when the power supply wire to a component or the electrical component itself becomes grounded before the normal ground connection for the circuit. To prevent damage to the system, the fuse for the circuit blows to interrupt the circuit—protecting the components from damage. Because grounding a wire from a power source makes a complete circuit—less the required component to use the power—the phenomenon is called a short circuit. The most common causes of short circuits are: the rubber insulation on a wire breaking or rubbing through to expose the current carrying core of the wire to a metal part of the car, or a shorted switch.

Some electrical systems on the vehicle are protected by a circuit breaker which is, basically, a self-repairing fuse. When either of the described events takes place in a system which is protected by a circuit breaker, the circuit breaker opens the circuit the same way a fuse does. However, when either the short is removed from the circuit or the surge subsides, the circuit breaker resets itself and does not have to be replaced as a fuse does.

Troubleshooting

When diagnosing a specific problem, organized troubleshooting is a must. The complexity of a modern automobile demands that you approach any problem in a logical, organized manner. There are certain troubleshooting techniques that are standard:

1. Establish when the problem occurs. Does the problem appear only under certain conditions? Were there any noises, odors, or other unusual symptoms?

2. Isolate the problem area. To do this, make some simple tests and observations; then eliminate the systems that are working properly. Check for obvious problems such as broken wires, dirty connections or split/disconnected vacuum hoses. Always check the obvious before assuming something complicated is the cause.

3. Test for problems systematically to determine the cause once the problem area is isolated. Are all the components functioning properly? Is there power going to electrical switches and motors? Is there vacuum at vacuum switches and/or actuators? Is there a mechanical problem such as bent linkage or loose mounting screws? Performing careful, systematic checks will often turn up most causes on the first inspection without wasting time checking components that have little or no relationship to the problem.

4. Test all repairs after the work is done to make sure that the problem is fixed. Some causes can be traced to more than one component, so a careful verification of repair work is important in order to pick up additional malfunctions that may cause a problem to reappear or a different problem to arise. A blown fuse, for example, is a simple problem that may require more than another fuse to repair. If you don't look for a problem that caused a fuse to blow, a shorted wire (for example) may go undetected.

Experience has shown that most problems tend to be the result

6-4 CHASSIS ELECTRICAL

of a fairly simple and obvious cause, such as loose or corroded connectors or air leaks in the intake system. This makes careful inspection of components during testing essential to quick and accurate troubleshooting.

BASIC TROUBLESHOOTING THEORY

Electrical problems generally fall into one of three areas:
- The component that is not functioning is not receiving current.
- The component itself is not functioning.
- The component is not properly grounded.

Problems that fall into the first category are by far the most complicated. It is the current supply system to the component which contains all the switches, relay, fuses, etc.

The electrical system can be checked with a test light and a jumper wire. A test light is a device that looks like a pointed screwdriver with a wire attached to it. It has a light bulb in its handle. A jumper wire is a piece of insulated wire with an alligator clip attached to each end.

If a light bulb is not working, you must follow a systematic plan to determine which of the three causes is the villain.

1. Turn on the switch that controls the inoperable bulb.
2. Disconnect the power supply wire from the bulb.
3. Attach the ground wire to the test light to a good metal ground.
4. Touch the probe end of the test light to the end of the power supply wire that was disconnected from the bulb. If the bulb is receiving current, the test light will go on.

➡ **If the bulb is one which works only when the ignition key is turned on (turn signal), make sure the key is turned on.**

If the test light does not go on, then the problem is in the circuit between the battery and the bulb. As mentioned before, this includes all the switches, fuses, and relays in the system. Turn to a wiring diagram and find the bulb on the diagram. Follow the wire that runs back to the battery. The problem is an open circuit between the battery and the bulb. If the fuse is blown and, when replaced, immediately blows again, there is a short circuit in the system which must be located and repaired. If there is a switch in the system, bypass it with a jumper wire. This is done by connecting one end of the jumper wire to the power supply wire into the switch and the other end of the jumper wire to the wire coming out of the switch. If the test light illuminates with the jumper wire installed, the switch or whatever was bypassed is defective.

➡ **Never substitute the jumper wire for the bulb, as the bulb is the component required to use the power from the power source.**

5. If the bulb in the test light goes on, then the current is getting to the bulb that is not working in the car. This eliminates the first of the three possible causes. Connect the power supply wire and connect a jumper wire from the bulb to a good metal ground. Do this with the switch which controls the bulb works with jumper wire installed, then it has a bad ground. This is usually caused by the metal area on which the bulb mounts to the vehicle being coated with some type of foreign matter.

6. If neither test located the source of the trouble, then the light bulb itself is defective.

The above test procedure can be applied to any of the components of the chassis electrical system by substituting the component that is not working for the light bulb. Remember that for any electrical system to work, all connections must be clean and tight.

TEST EQUIPMENT

➡ **Pinpointing the exact cause of trouble in an electrical system can sometimes only be accomplished by the use of special test equipment. The following describes different types of commonly used test equipment and explains how to use them in diagnosis. In addition to the information covered below, the tool manufacturer's instructions booklet (provided with the tester) should be read and clearly understood before attempting any test procedures.**

Jumper Wires

Jumper wires are simple, yet extremely valuable, pieces of test equipment. They are basically test wires which are used to bypass sections of a circuit. The simplest type of jumper wire is a length of multi-strand wire with an alligator clip at each end. Jumper wires are usually fabricated from lengths of standard automotive wire and whatever type of connector (alligator clip, spade connector or pin connector) that is required for the particular vehicle being tested. The well equipped tool box will have several different styles of jumper wires in several different lengths. Some jumper wires are made with three or more terminals coming from a common splice for special purpose testing. In cramped, hard-to-reach areas it is advisable to have insulated boots over the jumper wire terminals in order to prevent accidental grounding, sparks, and possible fire, especially when testing fuel system components.

Jumper wires are used primarily to locate open electrical circuits, on either the ground (−) side of the circuit or on the hot (+) side. If an electrical component fails to operate, connect the jumper wire between the component and a good ground. If the component operates only with the jumper installed, the ground circuit is open. If the ground circuit is good, but the component does not operate, the circuit between the power feed and component may be open. By moving the jumper wire successively back from the lamp toward the power source, you can isolate the area

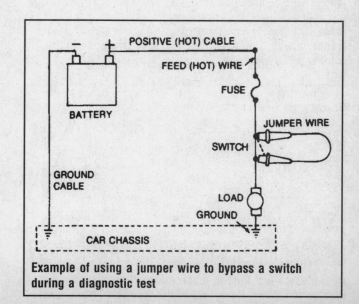

Example of using a jumper wire to bypass a switch during a diagnostic test

CHASSIS ELECTRICAL   6-5

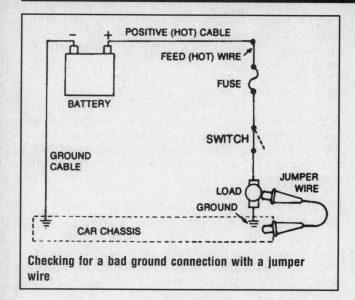

Checking for a bad ground connection with a jumper wire

of the circuit where the open is located. When the component stops functioning, or the power is cut off, the open is in the segment of wire between the jumper and the point previously tested.

You can sometimes connect the jumper wire directly from the battery to the hot terminal of the component, but first make sure the component uses 12 volts in operation. Some electrical components, such as fuel injectors, are designed to operate on about 4 volts and running 12 volts directly to the injector terminals can cause damage.

By inserting an in-line fuse holder between a set of test leads, a fused jumper wire can be used for bypassing open circuits. Use a 5 amp fuse to provide protection against voltage spikes. When in doubt, use a voltmeter to check the voltage input to the component and measure how much voltage is normally being applied.

✱✱ CAUTION

Never use jumpers made from wire that is of lighter gauge than that which is used in the circuit under test. If the jumper wire is of too small a gauge, it may overheat and possibly melt. Never use jumpers to bypass high resistance loads in a circuit. Bypassing resistances, in effect, creates a short circuit. This may, in turn, cause damage and fire. Jumper wires should only be used to bypass lengths of wire.

Unpowered Test Lights

The 12 volt test light is used to check circuits and components while electrical current is flowing through them. It is used for voltage and ground tests. Twelve volt test lights come in different styles but all have three main parts; a ground clip, a probe, and a light. The most commonly used 12 volt test lights have pick-type probes. To use a 12 volt test light, connect the ground clip to a good ground and probe wherever necessary with the pick. The pick should be sharp so that it can be probed into tight spaces.

✱✱ CAUTION

Do not use a test light to probe electronic ignition spark plug or coil wires. Never use a pick-type test light to probe

wiring on computer controlled systems unless specifically instructed to do so. Any wire insulation that is pierced by the test light probe should be taped and sealed with silicone after testing.

Like the jumper wire, the 12 volt test light is used to isolate opens in circuits. But, whereas the jumper wire is used to bypass the open to operate the load, the 12 volt test light is used to locate the presence of voltage in a circuit. If the test light glows, you know that there is power up to that point; if the 12 volt test light does not glow when its probe is inserted into the wire or connector, you know that there is an open circuit (no power). Move the test light in successive steps back toward the power source until the light in the handle does glow. When it glows, the open is between the probe and point which was probed previously.

➡ **The test light does not detect that 12 volts (or any particular amount of voltage) is present; it only detects that some voltage is present. It is advisable before using the test light to touch its terminals across the battery posts to make sure the light is operating properly.**

Self-Powered Test Lights

The self-powered test light usually contains a 1.5 volt penlight battery. One type of self-powered test light is similar in design to the 12 volt unit. This type has both the battery and the light in the handle, along with a pick-type probe tip. The second type has the light toward the open tip, so that the light illuminates the contact point. The self-powered test light is a dual purpose piece of test equipment. It can be used to test for either open or short circuits when power is isolated from the circuit (continuity test). A powered test light should not be used on any computer controlled system or component unless specifically instructed to do so. Many engine sensors can be destroyed by even this small amount of voltage applied directly to the terminals.

Voltmeters

A voltmeter is used to measure voltage at any point in a circuit, or to measure the voltage drop across any part of a circuit. It can also be used to check continuity in a wire or circuit by indicating current flow from one end to the other. Analog voltmeters usually have various scales on the meter dial and a selector switch to allow the selection of different voltages. The voltmeter has a positive and a negative lead. To avoid damage to the meter, always connect the negative lead to the negative (−) side of the circuit (to ground or nearest the ground side of the circuit) and connect the positive lead to the positive (+) side of the circuit (to the power source or the nearest power source). Note that the negative voltmeter lead will always be black and that the positive voltmeter will always be some color other than black (usually red).

Depending on how the voltmeter is connected into the circuit, it has several uses. A voltmeter can be connected either in parallel or in series with a circuit and it has a very high resistance to current flow. When connected in parallel, only a small amount of current will flow through the voltmeter current path; the rest will flow through the normal circuit current path and the circuit will work normally. When the voltmeter is connected in series with a circuit, only a small amount of current can flow through the circuit. The circuit will not work properly, but the voltmeter reading will show if the circuit is complete or not.

Ohmmeters

The ohmmeter is designed to read resistance (which is measured in ohms or Ω) in a circuit or component. Although there are several different styles of ohmmeters, all analog meters will usually have a selector switch which permits the measurement of different ranges of resistance (usually the selector switch allows the multiplication of the meter reading by 10, 100, 1000, and 10,000). A calibration knob allows the meter to be set at zero for accurate measurement. Since all ohmmeters are powered by an internal battery, the ohmmeter can be used as a self-powered test light. When the ohmmeter is connected, current from the ohmmeter flows through the circuit or component being tested. Since the ohmmeter's internal resistance and voltage are known values, the amount of current flow through the meter depends on the resistance of the circuit or component being tested.

The ohmmeter can be used to perform a continuity test for opens or shorts (either by observation of the meter needle or as a self-powered test light), and to read actual resistance in a circuit. It should be noted that the ohmmeter is used to check the resistance of a component or wire while there is no voltage applied to the circuit. Current flow from an outside voltage source (such as the vehicle battery) can damage the ohmmeter, so the circuit or component should be isolated from the vehicle electrical system before any testing is done. Since the ohmmeter uses its own voltage source, either lead can be connected to any test point.

➡**When checking diodes or other solid state components, the ohmmeter leads can only be connected one way in order to measure current flow in a single direction. Make sure the positive (+) and negative (−) terminal connections are as described in the test procedures to verify the one-way diode operation.**

In using the meter for making continuity checks, do not be concerned with the actual resistance readings. Zero resistance, or any ohm reading, indicates continuity in the circuit. Infinite resistance indicates an open in the circuit. A high resistance reading where there should be none indicates a problem in the circuit. Checks for short circuits are made in the same manner as checks for open circuits except that the circuit must be isolated from both power and normal ground. Infinite resistance indicates no continuity to ground, while zero resistance indicates a dead short to ground.

Ammeters

An ammeter measures the amount of current flowing through a circuit in units called amperes or amps. Amperes are units of electron flow which indicate how fast the electrons are flowing through the circuit. Since Ohms Law dictates that current flow in a circuit is equal to the circuit voltage divided by the total circuit resistance, increasing voltage also increases the current level (amps). Likewise, any decrease in resistance will increase the amount of amps in a circuit. At normal operating voltage, most circuits have a characteristic amount of amperes, called "current draw" which can be measured using an ammeter. By referring to a specified current draw rating, measuring the amperes, and comparing the two values, one can determine what is happening within the circuit to aid in diagnosis. An open circuit, for example, will not allow any current to flow so the ammeter reading will be zero. More current flows through a heavily loaded circuit or when the charging system is operating.

An ammeter is always connected in series with the circuit being tested. All of the current that normally flows through the circuit must also flow through the ammeter; if there is any other path for the current to follow, the ammeter reading will not be accurate. The ammeter itself has very little resistance to current flow and therefore will not affect the circuit, but it will measure current draw only when the circuit is closed and electricity is flowing. Excessive current draw can blow fuses and drain the battery, while a reduced current draw can cause motors to run slowly, lights to dim and other components to not operate properly. The ammeter can help diagnose these conditions by locating the cause of the high or low reading.

Multimeters

Different combinations of test meters can be built into a single unit designed for specific tests. Some of the more common combination test devices are known as Volt/Amp testers, Tach/Dwell meters, or Digital Multimeters. The Volt/Amp tester is used for charging system, starting system or battery tests and consists of a voltmeter, an ammeter and a variable resistance carbon pile. The voltmeter will usually have at least two ranges for use with 6, 12 and/or 24 volt systems. The ammeter also has more than one range for testing various levels of battery loads and starter current draw. The carbon pile can be adjusted to offer different amounts of resistance. The Volt/Amp tester has heavy leads to carry large amounts of current and many later models have an inductive ammeter pickup that clamps around the wire to simplify test connections. On some models, the ammeter also has a zero-center scale to allow testing of charging and starting systems without switching leads or polarity. A digital multimeter is a voltmeter, ammeter and ohmmeter combined in an instrument which gives a digital readout. These are often used when testing solid state circuits because of their high input impedance (usually 10 megohms or more).

The tach/dwell meter that combines a tachometer and a dwell (cam angle) meter is a specialized kind of voltmeter. The tachometer scale is marked to show engine speed in rpm and the dwell scale is marked to show degrees of distributor shaft rotation. In most electronic ignition systems, dwell is determined by the control unit, but the dwell meter can also be used to check the duty cycle (operation) of some electronic engine control systems. Some tach/dwell meters are powered by an internal battery, while others take their power from the vehicle battery in use. The battery powered testers usually require calibration (much like an ohmmeter) before testing.

TESTING

Open Circuits

To use the self-powered test light or a multimeter to check for open circuits, first isolate the circuit from the vehicle's 12 volt power source by disconnecting the battery or wiring harness connector. Connect the test light or ohmmeter ground clip to a good ground and probe sections of the circuit sequentially with the test light. (start from either end of the circuit). If the light is out/or there is infinite resistance, the open is between the probe and the circuit ground. If the light is on/or the meter shows continuity, the open is between the probe and end of the circuit toward the power source.

CHASSIS ELECTRICAL   6-7

Short Circuits

By isolating the circuit both from power and from ground, and using a self-powered test light or multimeter, you can check for shorts to ground in the circuit. Isolate the circuit from power and ground. Connect the test light or ohmmeter ground clip to a good ground and probe any easy-to-reach test point in the circuit. If the light comes on or there is continuity, there is a short somewhere in the circuit. To isolate the short, probe a test point at either end of the isolated circuit (the light should be on/there should be continuity). Leave the test light probe engaged and open connectors, switches, remove parts, etc., sequentially, until the light goes out/continuity is broken. When the light goes out, the short is between the last circuit component opened and the previous circuit opened.

➡ The battery in the test light and does not provide much current. A weak battery may not provide enough power to illuminate the test light even when a complete circuit is made (especially if there are high resistances in the circuit). Always make sure that the test battery is strong. To check the battery, briefly touch the ground clip to the probe; if the light glows brightly the battery is strong enough for testing. Never use a self-powered test light to perform checks for opens or shorts when power is applied to the electrical system under test. The 12 volt vehicle power will quickly burn out the light bulb in the test light.

Available Voltage Measurement

Set the voltmeter selector switch to the 20V position and connect the meter negative lead to the negative post of the battery. Connect the positive meter lead to the positive post of the battery and turn the ignition switch **ON** to provide a load. Read the voltage on the meter or digital display. A well charged battery should register over 12 volts. If the meter reads below 11.5 volts, the battery power may be insufficient to operate the electrical system properly. This test determines voltage available from the battery and should be the first step in any electrical trouble diagnosis procedure. Many electrical problems, especially on computer controlled systems, can be caused by a low state of charge in the battery. Excessive corrosion at the battery cable terminals can cause a poor contact that will prevent proper charging and full battery current flow.

Normal battery voltage is 12 volts when fully charged. When the battery is supplying current to one or more circuits it is said to be "under load." When everything is off the electrical system is under a "no-load" condition. A fully charged battery may show about 12.5 volts at no load; will drop to 12 volts under medium load; and will drop even lower under heavy load. If the battery is partially discharged the voltage decrease under heavy load may be excessive, even though the battery shows 12 volts or more at no load. When allowed to discharge further, the battery's available voltage under load will decrease more severely. For this reason, it is important that the battery be fully charged during all testing procedures to avoid errors in diagnosis and incorrect test results.

Voltage Drop

When current flows through a resistance, the voltage beyond the resistance is reduced (the larger the current, the greater the reduction in voltage). When no current is flowing, there is no voltage drop because there is no current flow. All points in the circuit which are connected to the power source are at the same voltage as the power source. The total voltage drop always equals the total source voltage. In a long circuit with many connectors, a series of small, unwanted voltage drops due to corrosion at the connectors can add up to a total loss of voltage which impairs the operation of the normal loads in the circuit. The maximum allowable voltage drop under load is critical, especially if there is more than one high resistance problem in a circuit because all voltage drops are cumulative. A small drop is normal due to the resistance of the conductors.

INDIRECT COMPUTATION OF VOLTAGE DROPS

1. Set the voltmeter selector switch to the 20 volt position.
2. Connect the meter negative lead to a good ground.
3. While operating the circuit, probe all loads in the circuit with the positive meter lead and observe the voltage readings. A drop should be noticed after the first load. But, there should be little or no voltage drop before the first load.

DIRECT MEASUREMENT OF VOLTAGE DROPS

1. Set the voltmeter switch to the 20 volt position.
2. Connect the voltmeter negative lead to the ground side of the load to be measured.
3. Connect the positive lead to the positive side of the resistance or load to be measured.
4. Read the voltage drop directly on the 20 volt scale.

Too high a voltage indicates too high a resistance. If, for example, a blower motor runs too slowly, you can determine if perhaps there is too high a resistance in the resistor pack. By taking voltage drop readings in all parts of the circuit, you can isolate the problem. Too low a voltage drop indicates too low a resistance. Take the blower motor for example again. If a blower motor runs too fast in the MED and/or LOW position, the problem might be isolated in the resistor pack by taking voltage drop readings in all parts of the circuit to locate a possibly shorted resistor.

HIGH RESISTANCE TESTING

1. Set the voltmeter selector switch to the 4 volt position.
2. Connect the voltmeter positive lead to the positive post of the battery.
3. Turn on the headlights and heater blower to provide a load.
4. Probe various points in the circuit with the negative voltmeter lead.
5. Read the voltage drop on the 4 volt scale. Some average maximum allowable voltage drops are:
- FUSE PANEL: 0.7 volts
- IGNITION SWITCH: 0.5 volts
- HEADLIGHT SWITCH: 0.7 volts
- IGNITION COIL (+): 0.5 volts
- ANY OTHER LOAD: 1.3 volts

➡ Voltage drops are all measured while a load is operating; without current flow, there will be no voltage drop.

Resistance Measurement

The batteries in an ohmmeter will weaken with age and temperature, so the ohmmeter must be calibrated or "zeroed" before taking measurements. To zero the meter, place the selector switch in its lowest range and touch the two ohmmeter leads together. Turn the calibration knob until the meter needle is exactly on zero.

➡All analog (needle) type ohmmeters must be zeroed before use, but some digital ohmmeter models are automatically calibrated when the switch is turned on. Self-calibrating digital ohmmeters do not have an adjusting knob, but its a good idea to check for a zero readout before use by touching the leads together. All computer controlled systems require the use of a digital ohmmeter with at least 10 megohms impedance for testing. Before any test procedures are attempted, make sure the ohmmeter used is compatible with the electrical system or damage to the on-board computer could result.

To measure resistance, first isolate the circuit from the vehicle power source by disconnecting the battery cables or the harness connector. Make sure the key is **OFF** when disconnecting any components or the battery. Where necessary, also isolate at least one side of the circuit to be checked in order to avoid reading parallel resistances. Parallel circuit resistances will always give a lower reading than the actual resistance of either of the branches. When measuring the resistance of parallel circuits, the total resistance will always be lower than the smallest resistance in the circuit. Connect the meter leads to both sides of the circuit (wire or component) and read the actual measured ohms on the meter scale. Make sure the selector switch is set to the proper ohm scale for the circuit being tested to avoid misreading the ohmmeter test value.

✷✷ WARNING

Never use an ohmmeter with power applied to the circuit. Like the self-powered test light, the ohmmeter is designed to operate on its own power supply. The normal 12 volt automotive electrical system current could damage the meter!

Wiring Harnesses

The average automobile contains about ½ mile of wiring, with hundreds of individual connections. To protect the many wires from damage and to keep them from becoming a confusing tangle, they are organized into bundles, enclosed in plastic or taped together and called wiring harnesses. Different harnesses serve different parts of the vehicle. Individual wires are color coded to help trace them through a harness where sections are hidden from view.

Automotive wiring or circuit conductors can be in any one of three forms:

1. Single strand wire
2. Multi-strand wire
3. Printed circuitry

Single strand wire has a solid metal core and is usually used inside such components as alternators, motors, relays and other devices. Multi-strand wire has a core made of many small strands of wire twisted together into a single conductor. Most of the wiring in an automotive electrical system is made up of multi-strand wire, either as a single conductor or grouped together in a harness. All wiring is color coded on the insulator, either as a solid color or as a colored wire with an identification stripe. A printed circuit is a thin film of copper or other conductor that is printed on an insulator backing. Occasionally, a printed circuit is sandwiched between two sheets of plastic for more protection and flexibility. A complete printed circuit, consisting of conductors, insulating material and connectors for lamps or other components is called a printed circuit board. Printed circuitry is used in place of individual wires or harnesses in places where space is limited, such as behind instrument panels.

Since automotive electrical systems are very sensitive to changes in resistance, the selection of properly sized wires is critical when systems are repaired. A loose or corroded connection or a replacement wire that is too small for the circuit will add extra resistance and an additional voltage drop to the circuit. A ten percent voltage drop can result in slow or erratic motor operation, for example, even though the circuit is complete. The wire gauge number is an expression of the cross-section area of the conductor. The most common system for expressing wire size is the American Wire Gauge (AWG) system.

Gauge numbers are assigned to conductors of various cross-section areas. As gauge number increases, area decreases and the conductor becomes smaller. A 5 gauge conductor is smaller than a 1 gauge conductor and a 10 gauge is smaller than a 5 gauge. As the cross-section area of a conductor decreases, resistance increases and so does the gauge number. A conductor with a higher gauge number will carry less current than a conductor with a lower gauge number.

➡**Gauge wire size refers to the size of the conductor, not the size of the complete wire. It is possible to have two wires of the same gauge with different diameters because one may have thicker insulation than the other.**

12 volt automotive electrical systems generally use 10, 12, 14, 16 and 18 gauge wire. Main power distribution circuits and larger accessories usually use 10 and 12 gauge wire. Battery cables are usually 4 or 6 gauge, although 1 and 2 gauge wires are occasionally used. Wire length must also be considered when making repairs to a circuit. As conductor length increases, so does resistance. An 18 gauge wire, for example, can carry a 10 amp load for 10 feet without excessive voltage drop; however if a 15 foot wire is required for the same 10 amp load, it must be a 16 gauge wire.

An electrical schematic shows the electrical current paths when a circuit is operating properly. It is essential to understand how a circuit works before trying to figure out why it doesn't. Schematics break the entire electrical system down into individual circuits and show only one particular circuit. In a schematic, no attempt is made to represent wiring and components as they physically appear on the vehicle; switches and other components are shown as simply as possible. Face views of harness connectors show the cavity or terminal locations in all multi-pin connectors to help locate test points.

If you need to backprobe a connector while it is on the component, the order of the terminals must be mentally reversed. The wire color code can help in this situation, as well as a keyway, lock tab or other reference mark.

WIRING REPAIR

Soldering is a quick, efficient method of joining metals permanently. Everyone who has the occasion to make wiring repairs should know how to solder. Electrical connections that are soldered are far less likely to come apart and will conduct electricity much better than connections that are only "pig-tailed" together. The most popular (and preferred) method of soldering is with an electrical soldering gun. Soldering irons are available in many

CHASSIS ELECTRICAL 6-9

sizes and wattage ratings. Irons with higher wattage ratings deliver higher temperatures and recover lost heat faster. A small soldering iron rated for no more than 50 watts is recommended, especially on electrical systems where excess heat can damage the components being soldered.

There are three ingredients necessary for successful soldering; proper flux, good solder and sufficient heat. A soldering flux is necessary to clean the metal of tarnish, prepare it for soldering and to enable the solder to spread into tiny crevices. When soldering, always use a rosin core solder which is non-corrosive and will not attract moisture once the job is finished. Other types of flux (acid core) will leave a residue that will attract moisture and cause the wires to corrode. Tin is a unique metal with a low melting point. In a molten state, it dissolves and alloys easily with many metals. Solder is made by mixing tin with lead. The most common proportions are 40/60, 50/50 and 60/40, with the percentage of tin listed first. Low priced solders usually contain less tin, making them very difficult for a beginner to use because more heat is required to melt the solder. A common solder is 40/60 which is well suited for all-around general use, but 60/40 melts easier and is preferred for electrical work.

Soldering Techniques

Successful soldering requires that the metals to be joined be heated to a temperature that will melt the solder, usually 360–460°F (182–238°C). Contrary to popular belief, the purpose of the soldering iron is not to melt the solder itself, but to heat the parts being soldered to a temperature high enough to melt the solder when it is touched to the work. Melting flux-cored solder on the soldering iron will usually destroy the effectiveness of the flux.

➡**Soldering tips are made of copper for good heat conductivity, but must be "tinned" regularly for quick transference of heat to the project and to prevent the solder from sticking to the iron. To "tin" the iron, simply heat it and touch the flux-cored solder to the tip; the solder will flow over the hot tip. Wipe the excess off with a clean rag, but be careful as the iron will be hot.**

After some use, the tip may become pitted. If so, simply dress the tip smooth with a smooth file and "tin" the tip again. Flux-cored solder will remove oxides but rust, bits of insulation and oil or grease must be removed with a wire brush or emery cloth. For maximum strength in soldered parts, the joint must start off clean and tight. Weak joints will result in gaps too wide for the solder to bridge.

If a separate soldering flux is used, it should be brushed or swabbed on only those areas that are to be soldered. Most solders contain a core of flux and separate fluxing is unnecessary. Hold the work to be soldered firmly. It is best to solder on a wooden board, because a metal vise will only rob the piece to be soldered of heat and make it difficult to melt the solder. Hold the soldering tip with the broadest face against the work to be soldered. Apply solder under the tip close to the work, using enough solder to give a heavy film between the iron and the piece being soldered, while moving slowly and making sure the solder melts properly. Keep the work level or the solder will run to the lowest part and favor the thicker parts, because these require more heat to melt the solder. If the soldering tip overheats (the solder coating on the face of the tip burns up), it should be retinned. Once the soldering is completed, let the soldered joint stand until cool. Tape and seal all soldered wire splices after the repair has cooled.

Wire Harness Connectors

Most connectors in the engine compartment or that are otherwise exposed to the elements are protected against moisture and dirt which could create oxidation and deposits on the terminals.

These special connectors are weather-proof. All repairs require the use of a special terminal and the tool required to service it. This tool is used to remove the pin and sleeve terminals. If removal is attempted with an ordinary pick, there is a good chance that the terminal will be bent or deformed. Unlike standard blade type terminals, these weather-proof terminals cannot be straightened once they are bent. Make certain that the connectors are properly seated and all of the sealing rings are in place when connecting leads. On some models, a hinge-type flap provides a backup or secondary locking feature for the terminals. Most secondary locks are used to improve connector reliability by retaining the terminals if the small terminal lock tangs are not positioned properly.

Molded-on connectors require complete replacement of the connection. This means splicing a new connector assembly into the harness. All splices should be soldered to insure proper contact. Use care when probing the connections or replacing terminals in them as it is possible to short between opposite terminals. If this happens to the wrong terminal pair, it is possible to damage certain components. Always use jumper wires between connectors for circuit checking and never probe through weatherproof seals.

Open circuits are often difficult to locate by sight because corrosion or terminal misalignment are hidden by the connectors. Merely wiggling a connector on a sensor or in the wiring harness may correct the open circuit condition. This should always be considered when an open circuit or a failed sensor is indicated. Intermittent problems may also be caused by oxidized or loose connections. When using a circuit tester for diagnosis, always probe connections from the wire side. Be careful not to damage sealed connectors with test probes.

All wiring harnesses should be replaced with identical parts, using the same gauge wire and connectors. When signal wires are spliced into a harness, use wire with high temperature insulation only. It is seldom necessary to replace a complete harness. If replacement is necessary, pay close attention to insure proper harness routing. Secure the harness with suitable plastic wire clamps to prevent vibrations from causing the harness to wear in spots or contact any hot components.

➡**Weatherproof connectors cannot be replaced with standard connectors. Instructions are provided with replacement connector and terminal packages. Some wire harnesses have mounting indicators (usually pieces of colored tape) to mark where the harness is to be secured.**

In making wiring repairs, its important that you always replace damaged wires with wiring of the same gauge as the wire being replaced. The heavier the wire, the smaller the gauge number. Wires are color-coded to aid in identification and whenever possible the same color coded wire should be used for replacement. A wire stripping and crimping tool is necessary to install solderless terminal connectors. Test all crimps by pulling on the wires; it should not be possible to pull the wires out of a good crimp.

Wires which are open, exposed or otherwise damaged are repaired by simple splicing. Where possible, if the wiring harness is accessible and the damaged place in the wire can be located, it is best to open the harness and check for all possible damage. In an inaccessible harness, the wire must be bypassed with a new insert, usually taped to the outside of the old harness.

6-10 CHASSIS ELECTRICAL

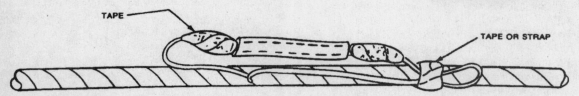

TYPICAL REPAIR USING THE SPECIAL #17 GA. (9.00" LONG-YELLOW) FUSE LINK REQUIRED FOR THE AIR/COND. CIRCUITS (2) #687E and #261A LOCATED IN THE ENGINE COMPARTMENT

REMOVE EXISTING VINYL TUBE SHIELDING REINSTALL OVER FUSE LINK BEFORE CRIMPING FUSE LINK TO WIRE ENDS

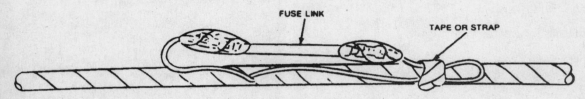

TYPICAL REPAIR FOR ANY IN-LINE FUSE LINK USING THE SPECIFIED GAUGE FUSE LINK FOR THE SPECIFIC CIRCUIT

TYPICAL REPAIR USING THE EYELET TERMINAL FUSE LINK OF THE SPECIFIED GAUGE FOR ATTACHMENT TO A CIRCUIT WIRE END

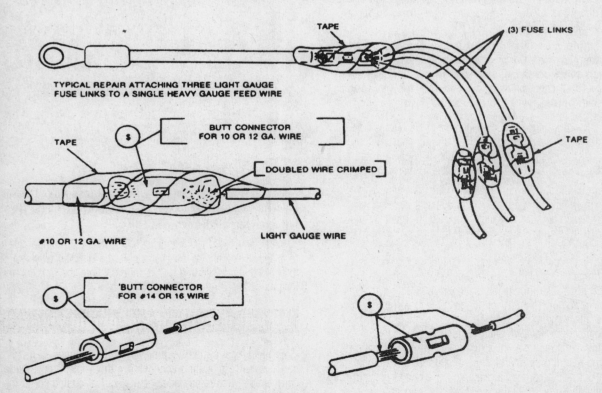

FUSIBLE LINK REPAIR PROCEDURE

General fusible link repair—never replace a fusible link with regular wire or a fusible link rated at a higher amperage than the one being replaced

CHASSIS ELECTRICAL 6-11

When replacing fusible links, be sure to use fusible link wire, NOT ordinary automotive wire. Make sure the fusible segment is of the same gauge and construction as the one being replaced and double the stripped end when crimping the terminal connector for a good contact. The melted (open) fusible link segment of the wiring harness should be cut off as close to the harness as possible, then a new segment spliced in as described. In the case of a damaged fusible link that feeds two harness wires, the harness connections should be replaced with two fusible link wires so that each circuit will have its own separate protection.

➡ **Most of the problems caused in the wiring harness are due to bad ground connections. Always check all vehicle ground connections for corrosion or looseness before performing any power feed checks to eliminate the chance of a bad ground affecting the circuit.**

Hard-Shell Connectors

Unlike molded connectors, the terminal contacts in hard-shell connectors can be replaced. Weatherproof hard-shell connectors with the leads molded into the shell have non-replaceable terminal ends. Replacement usually involves the use of a special terminal removal tool that depresses the locking tangs (barbs) on the connector terminal and allows the connector to be removed from the rear of the shell. The connector shell should be replaced if it shows any evidence of burning, melting, cracks, or breaks. Replace individual terminals that are burnt, corroded, distorted or loose.

➡ **The insulation crimp must be tight to prevent the insulation from sliding back on the wire when the wire is pulled. The insulation must be visibly compressed under the crimp tabs, and the ends of the crimp should be turned in for a firm grip on the insulation.**

The wire crimp must be made with all wire strands inside the crimp. The terminal must be fully compressed on the wire strands with the ends of the crimp tabs turned in to make a firm grip on the wire. Check all connections with an ohmmeter to insure a good contact. There should be no measurable resistance between the wire and the terminal when connected.

Fusible Links

The fuse link is a short length of special, Hypalon (high temperature) insulated wire, integral with the engine compartment wiring harness and should not be confused with standard wire. It is several wire gauges smaller than the circuit which it protects. Under no circumstances should a fuse link replacement repair be made using a length of standard wire cut from bulk stock or from another wiring harness.

To repair any blown fuse link use the following procedure:
1. Determine which circuit is damaged, its location and the cause of the open fuse link. If the damaged fuse link is one of three fed by a common No. 10 or 12 gauge feed wire, determine the specific affected circuit.
2. Disconnect the negative battery cable.
3. Cut the damaged fuse link from the wiring harness and discard it. If the fuse link is one of three circuits fed by a single feed wire, cut it out of the harness at each splice end and discard it.
4. Identify and procure the proper fuse link with butt connectors for attaching the fuse link to the harness.

➡ **Heat shrink tubing must be slipped over the wire before crimping and soldering the connection.**

5. To repair any fuse link in a 3-link group with one feed:
   a. After cutting the open link out of the harness, cut each of the remaining undamaged fuse links close to the feed wire weld.
   b. Strip approximately ½ in. (13mm) of insulation from the detached ends of the two good fuse links. Insert two wire ends into one end of a butt connector, then carefully push one stripped end of the replacement fuse link into the same end of the butt connector and crimp all three firmly together.

➡ **Care must be taken when fitting the three fuse links into the butt connector as the internal diameter is a snug fit for three wires. Make sure to use a proper crimping tool. Pliers, side cutters, etc. will not apply the proper crimp to retain the wires and withstand a pull test.**

   c. After crimping the butt connector to the three fuse links, cut the weld portion from the feed wire and strip approximately ½ in. (13mm) of insulation from the cut end. Insert the stripped end into the open end of the butt connector and crimp very firmly.
   d. To attach the remaining end of the replacement fuse link, strip approximately ½ in. (13mm) of insulation from the wire end of the circuit from which the blown fuse link was removed, and firmly crimp a butt connector or equivalent to the stripped wire. Then, insert the end of the replacement link into the other end of the butt connector and crimp firmly.
   e. Using rosin core solder with a consistency of 60 percent tin and 40 percent lead, solder the connectors and the wires at the repairs then insulate with electrical tape or heat shrink tubing.
6. To replace any fuse link on a single circuit in a harness, cut out the damaged portion, strip approximately ½ in. (13mm) of insulation from the two wire ends and attach the appropriate replacement fuse link to the stripped wire ends with two proper size butt connectors. Solder the connectors and wires, then insulate.
7. To repair any fuse link which has an eyelet terminal on one end such as the charging circuit, cut off the open fuse link behind the weld, strip approximately ½ in. (13mm) of insulation from the cut end and attach the appropriate new eyelet fuse link to the cut stripped wire with an appropriate size butt connector. Solder the connectors and wires at the repair, then insulate.
8. Connect the negative battery cable to the battery and test the system for proper operation.

➡ **Do not mistake a resistor wire for a fuse link. The resistor wire is generally longer and has print stating, "Resistor-don't cut or splice."**

When attaching a single No. 16, 17, 18 or 20 gauge fuse link to a heavy gauge wire, always double the stripped wire end of the fuse link before inserting and crimping it into the butt connector for positive wire retention.

Add-On Electrical Equipment

The electrical system in your vehicle is designed to perform under reasonable operating conditions without interference between components. Before any additional electrical equipment is installed, it is recommended that you consult your dealer or a reputable repair facility that is familiar with the vehicle and its systems.

If the vehicle is equipped with mobile radio equipment and/or

6-12 CHASSIS ELECTRICAL

mobile telephone, it may have an effect upon the operation of any on-board computer control modules. Radio Frequency Interference (RFI) from the communications system can be picked up by the vehicle's wiring harnesses and conducted into the control module, giving it the wrong messages at the wrong time. Although well shielded against RFI, the computer should be further protected by taking the following measures:

• Install the antenna as far as possible from the control module. For instance, if the module is located behind the center console area, then the antenna should be mounted at the rear of the vehicle.

• Keep the antenna wiring a minimum of eight inches away from any wiring running to control modules and from the module itself. NEVER wind the antenna wire around any other wiring.

• Mount the equipment as far from the control module as possible. Be very careful during installation not to drill through any wires or short a wire harness with a mounting screw.

• Insure that the electrical feed wire(s) to the equipment are properly and tightly connected. Loose connectors can cause interference.

• Make certain that the equipment is properly grounded to the vehicle. Poor grounding can damage expensive equipment.

HEATER

▶ **See Figures 1 and 2**

The only Volkswagens with an electric heater blower fan are the Type 2/1700/1800/2000 through 1979 and the Type 4, which also comes standard with an auxilliary gas heating system. The 1980 Type 2 heater fan is driven by the alternator. Do not confuse the fresh air circulation fan used on many Type 3 models, all 1971 and later Super Beetles and Type 1 Convertibles, and all 1976 and later standard Type 1 models with a heater blower fan, as the fresh air fan circulates only fresh, unheated air through the vehicle.

The auxiliary gas heating system on the Type 4 switches on when the engine is not producing enough heat for the passenger compartment. The gas heater runs on gasoline from the fuel tank and is available as an option on Type 2 and Type 3 models.

On all models, the primary heat circulator is the engine cooling fan, which directs air flow through the heat exchangers in the exhaust system and into the passenger compartment when the heating ducts are closed.

Procedures for removing the heat exchangers and heater duct assemblies are given in Section 3.

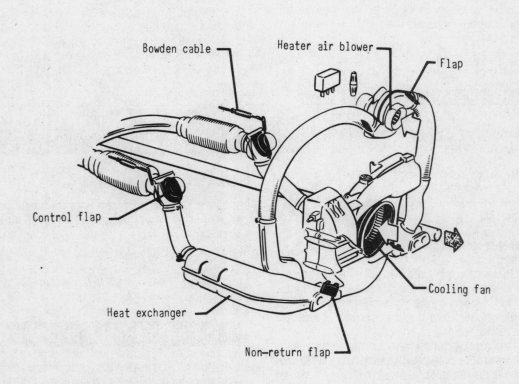

Fig. 1 Heating system components used on Type 2 models equipped with suitcase engines

CHASSIS ELECTRICAL 6-13

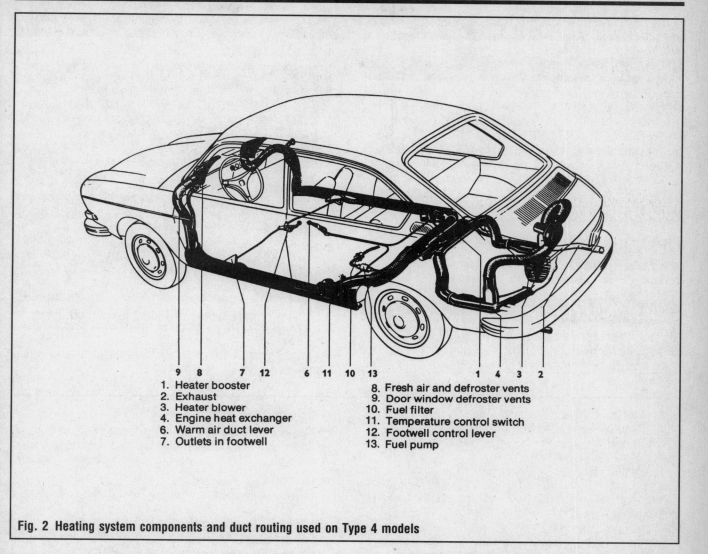

1. Heater booster
2. Exhaust
3. Heater blower
4. Engine heat exchanger
6. Warm air duct lever
7. Outlets in footwell
8. Fresh air and defroster vents
9. Door window defroster vents
10. Fuel filter
11. Temperature control switch
12. Footwell control lever
13. Fuel pump

Fig. 2 Heating system components and duct routing used on Type 4 models

Heater Blower

REMOVAL & INSTALLATION

Type 2 Through 1979, and Type 4

The heater blower is located above the engine.
1. Disconnect the negative battery cable.
2. Loosen the two clamps and disconnect the warm air hoses from the blower.
3. Disconnect and mark the blower motor electrical connections and remove the blower housing retaining screws or nuts.
4. Remove the blower housing.
5. To remove the blower motor and fan from the housing, remove the three screws in the blower end plate and remove the assembly. Installation is the reverse of removal.

1980 Type 2

The heater fan assembly unbolts from the back of the alternator.

Heater Cables

REMOVAL & INSTALLATION

1. Disconnect the cables at the heat exchangers. The cables are usually held in the flap levers at the heat exchangers with sleeve bolts with 10 mm heads and locked with 9 mm head nuts.
2. Remove the cable ends from the sleeve bolts and pull the rubber boots off the cable guide tubes and off the disconnected cables.
3. Disconnect the cables at the heater controls. They are usually held by pins with cotter pin retainers. You will have to unbolt and pull out the temperature control lever to unfasten the cables on Types 1, 3, and 4. On the Type 2, disconnect the cables from the temperature control lever on the dash.
4. Remove the cables by pulling them out from the heater control side.
5. Feed new cables into the guide tubes at the heater controls after you have sprayed them completely with silicone to prevent rust and wear.

CHASSIS ELECTRICAL

WINDSHIELD WIPERS

Wiper Blade and Arm

REMOVAL & INSTALLATION

To remove the wiper blade, turn the blade at an angle to the arm, then move the blade down so its spring clip moves off the hook in the arm and remove the blade. Reverse to install.

To remove the arm, on models before 1973, remove the cap nut where the arm attaches to the windshield wiper drive shaft and lift off the arm. On 1973 and later models, the retaining nut is covered by a plastic cap. Remove the cap, remove the nut and remove the arm. When installing wiper arms, make sure they park at the bottom of the swept area and that they cover the entire swept area when in operation.

Wiper Motor

REMOVAL & INSTALLATION

Type 1

1. Disconnect the battery ground cable.
2. Remove the wiper arms.
3. Remove the wiper bearing nuts as well as the washers. Take off the outer bearing seals.
4. On 1973–80 Super Beetle and convertible models, remove the screws holding the cover of the fresh air box in the luggage compartment, remove the cover and remove the plastic rain cover, if necessary. On all other models, remove the back of the instrument panel from the luggage compartment.
5. On all models except 1973–80 Super Beetles and Convertibles, remove the fresh air box/rain drain by removing the three screws at the top and the nut at the bottom; next remove the glove box and the right side air vent.
6. Disconnect the wiper motor wiring harness and remove the screw which secures the wiper frame to the body.
7. Remove the frame and motor with the linkage.

➡ The ball joints at the ends of the linkage may be slipped apart by gently popping the ball and socket apart with a small prytool. Always lubricate the joints upon reassembly.

8. Remove the lock and spring washers from the motor driveshaft and remove the connecting rod. Matchmark the motor and frame to ensure proper realignment when the motor is reinstalled.
9. Remove the nut located at the base of the motor driveshaft, and remove the motor from the frame.
10. To install, reverse the above steps and heed the following reminders.
11. The pressed lug on the wiper frame must engage the groove in the wiper bearing. Make sure that the wiper spindles are perpendicular to the plane of the windshield.
12. Check the linkage bushings for wear.
13. The hollow side of the links must face toward the frame with the angled end of the driving link toward the right bearing.
14. The inner bearing seal should be placed so that the shoulder of the rubber molding faces the wiper arm.

Type 2

1. Disconnect the ground wire from the battery.
2. Remove both wiper arms.
3. Remove the bearing cover and nut.
4. Remove the heater branch connections under the instrument panel.
5. Disconnect the wiper motor wiring.
6. Remove the wiper motor securing screw and remove the motor.
7. Reverse the above steps to install.

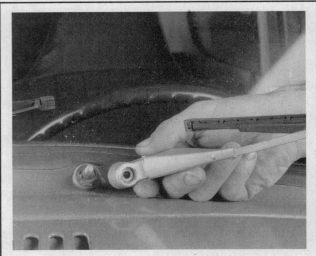

. . . then lift the arm up and off of the arm driveshaft

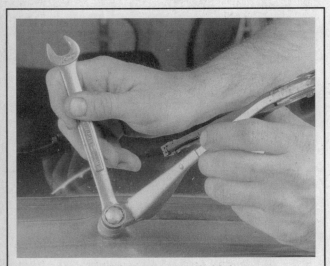

Remove the windshield wiper arm hold-down nut . . .

CHASSIS ELECTRICAL   6-15

Type 3

1. Disconnect the negative battery cable.
2. Remove the ashtray and glove compartment.
3. Remove the fresh air controls.
4. Remove the cover for the heater and water drainage hoses.
5. Disconnect the motor wiring.
6. Remove the wiper arms.
7. Remove the bearing covers and nuts, washers, and outer bearing seals.
8. Remove the wiper motor securing screws and remove the motor.
9. Reverse the above steps to install.

Type 4

1. Disconnect the negative battery cable.
2. Remove the wiper arms.
3. Remove the bearing cover and remove the nut under it.
4. Remove the steering column cover and the hoses running between the fresh air control box and the vents.
5. Remove the clock but do not disconnect the wiring.
6. Remove the left fresh air and defroster vent. Disconnect the air hose from the vent.
7. Disconnect the wiring for the motor at the windshield wiper switch. Remove the ground wire from the motor gear cover.
8. Remove the motor securing screw and remove the motor frame and motor assembly downward and to the right.
9. Installation is the reverse of the removal procedure.

Linkage

REMOVAL & INSTALLATION

The windshield wiper linkage is secured at the ends by a ball and socket type joint. The ball and joint may be gently pryed apart with the aid of a small prytool. Always lubricate the joints with grease before assembly.

Wiper Arm Shaft

1. Remove the wiper arm.
2. Remove the bearing cover or the shaft seal, depending on type.
3. On Type 4 models, remove the shaft circlip.
4. Remove the large wiper shaft bearing securing nut and remove the accompanying washer and rubber seal.
5. Disconnect the wiper linkage from the wiper arm shaft.
6. Working from inside the car, slide the shaft out of its bearing.

➡ It may be necessary to lightly tap the shaft out of its bearing. Use a soft-faced hammer.

7. Installation is the reverse of the removal procedure.

INSTRUMENT CLUSTER

Speedometer

REMOVAL & INSTALLATION

Except Type 2

1. Disconnect the negative battery cable.

➡ On the Type 3, it is necessary to remove the fuse panel to gain access to the speedometer.

2. Disconnect the speedometer light bulb wires.
3. Unscrew the knurled nut which secures the speedometer cable to the back of the speedometer. Pull the cable from the back of the speedometer.
4. Using a 4 mm allen wrench, remove the two knurled nuts which secure the speedometer brackets. Remove the brackets.
5. Remove the speedometer from the dashboard by sliding it out toward the steering wheel.
6. Reverse the above steps to install. Before fully tightening the nuts for the speedometer brackets, make sure the speedometer is correctly positioned in the dash.

Type 2

1970–79 MODELS

1. Disconnect the negative battery cable.
2. Remove the fresh air control lever knobs.

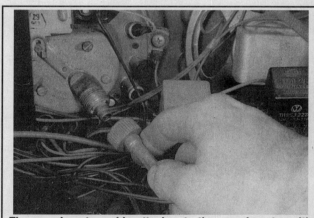

The speedometer cable attaches to the speedometer with a threaded collar—access to the back of the speedometer must be gained through the luggage compartment

3. Remove the Phillips head screws at the four corners of the instrument panel, being careful not to lose the spring clips on 1973 and later models.
4. Disconnect the speedometer cable.
5. Lift the instrument panel out of the dash far enough to disconnect the wiring, then remove the instrument panel. Remove the two long screws, separate the panel halves, then remove the

6-16 CHASSIS ELECTRICAL

screws and remove the speedometer. Installation is the reverse of removal.

1980–81 MODELS

1. Remove the upper dashboard cover. The cover simply pulls off. Disconnect the negative battery cable.
2. Remove retaining screws and disconnect all switches from the instrument panel. Squeeze the tabs on the switches to remove.
3. Disconnect the speedometer cable, all electrical connections, then remove the instrument panel. The speedometer can now be removed from the panel.

Installation is the reverse of removal.

Speedometer Cable

REMOVAL & INSTALLATION

1. Unscrew the cable from the back of the speedometer.
2. At the left front wheel, pry off the circlip retaining the square end of the speedometer shaft to the wheel bearing dust cap.
3. Pull the cable out of the back of the steering knuckle from under the car. With the circlip removed, the cable should pull right out. Pull the cable out of its holding grommets and remove.

Installation is the reverse of removal.

Fuel Gauge and Clock Assembly

REMOVAL & INSTALLATION

1. Disconnect the negative battery cable.
2. Disconnect the wiring from the back of the assembly.

➡ On the Type 3 it is necessary to remove the fuse panel to gain access to this assembly.

3. Remove the knurled nuts and brackets which secure the assembly in the dash. Use a 4 mm allen wrench.
4. Remove the assembly by gently sliding it toward the steering wheel and out of the dash.
5. The fuel gauge is secured into the base of the clock by two screws. Remove the screws and slip the fuel gauge out of the clock.
6. Installation is the reverse of the removal procedure. Make sure the clock and fuel gauge assembly is properly centered in the dash before fully tightening the nuts.

Ignition Switch

Ignition switch removal and installation is covered in Section 8.

SEAT BELT/STARTER INTERLOCK

All 1974 and some early production 1975 models are equipped with a seat belt/starter interlock system to prevent the driver and front seat passenger (if applicable) from starting the engine without first buckling his/her seat belts. The proper sequence is: sit in the seat(s), buckle up, start the engine. If this sequence is not followed, a warning system is activated which includes a "fasten belts" visual display and a buzzer. If the engine should stall with the ignition switch turned **ON**, the car may be restarted within three minutes. Late production 1975 models are not equipped with this system.

LIGHTING

Headlights

REMOVAL & INSTALLATION

1. Remove the screw(s) which secure(s) the headlight ring and remove the ring.
2. The sealed beam is held in place by a ring secured by three screws. Remove the screws and the ring. Do not confuse the headlight aiming screws with the screws for the ring. There are only two screws used for aiming.
3. Pull the wiring off the back of the sealed beam and remove the beam.
4. Installation is the reverse of the removal procedure.

To remove the headlight bulb, first remove the trim ring retaining screw(s) . . .

CHASSIS ELECTRICAL 6-17

. . . then pull the trim ring away from the headlight assembly

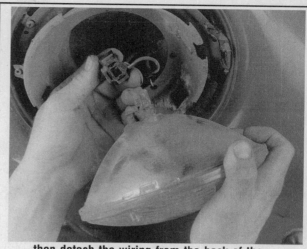

. . . then detach the wiring from the back of the headlight

Rear Brake/Turn Signal/Back-up Lights

REMOVAL & INSTALLATION

➡ This is a general procedure and applies to all models and types covered by this manual. A few steps may need to be slightly altered to comply with your particular vehicle.

1. Disconnect the negative battery cable.
2. Using a Phillips screwdriver, remove the rear light assembly lens retaining screws.
3. Pull the lens and gasket off of the rear fender.
4. Inspect the lens gasket for tearing or crumbling. If any such damage is evident, replace the old gasket with a new one upon installation.
5. Once the lens is removed from the rear light assembly, the individual light bulbs can be removed. To remove the light bulbs:

Remove the 3 retaining ring screws—do not confuse the retaining ring screws with the headlight aiming screws (arrow)

Once all 3 retaining ring screws are removed, pull the ring and headlight out of the mounting bracket . . .

To remove the rear brake/turn signal/back-up light bulbs, first remove the lens retaining screws . . .

6-18 CHASSIS ELECTRICAL

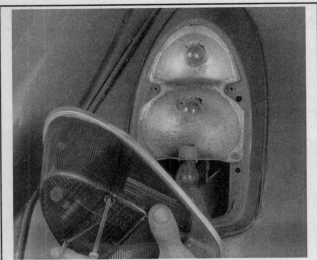
. . . then pull the lens off of the light assembly

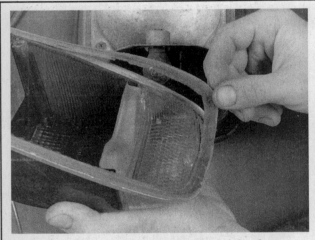
Make certain to inspect the sealing gasket for damage, such as dry rot or excessive cracking

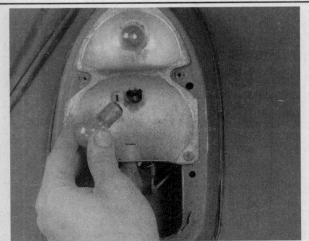
Once the lens is removed, the light bulbs may be inspected and replaced

　　a. Grasp the light bulb to be removed and gently press it in toward its socket.
　　b. While pressing the bulb in, turn the bulb counterclockwise until the two little pins on the bulb base clear the channels in the socket.
　　c. Pull the bulb up and out of its socket.
6. Installation is the reverse of the removal procedure. During installation, make sure that the light bulb contacts are clean and free of corrosion.

Front Turn Signal/Side Marker Lights

REMOVAL & INSTALLATION

➡This is a general procedure and applies to all models and types covered by this manual. A few steps may need to be slightly altered to comply with your particular vehicle.

1. Disconnect the negative battery cable.
2. Using a Phillips screwdriver, remove the front light assembly lens retaining screws.
3. Pull the lens and gasket off of the front fender.
4. Inspect the lens gasket for tearing or crumbling. If any such damage is evident, replace the old gasket with a new one upon installation.
5. Once the lens is removed from the light assembly, the individual light bulbs can be removed. To remove the light bulbs:
　　a. Grasp the light bulb to be removed and gently press it in toward its socket.
　　b. While pressing the bulb in, turn the bulb counterclockwise until the two little pins on the bulb base clear the channels in the socket.
　　c. Pull the bulb up and out of its socket.
6. Installation is the reverse of the removal procedure. During installation, make sure that the light bulb contacts are clean and free of corrosion.

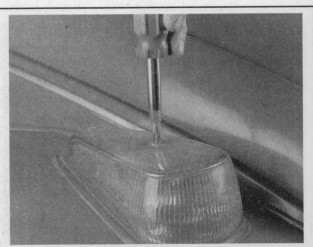
Remove the front turn signal/side marker light assembly lens retaining screws . . .

CHASSIS ELECTRICAL 6-19

... then lift the lens off of the light assembly

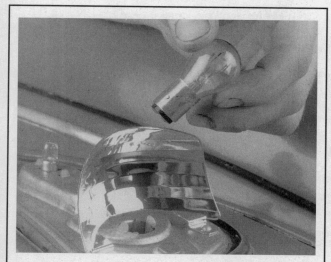

Once the lens is removed, either the front turn signal bulb . . .

... or the side marker light bulb can be removed from the assembly—note one of the two little pins found on all bulbs (arrow)

License Plate Light

REMOVAL & INSTALLATION

➡This is a general procedure and applies to all models and types covered by this manual. A few steps may need to be slightly altered to comply with your particular vehicle.

1. Disconnect the negative battery cable.
2. If necessary for easier access to the light assembly lens retaining screws, open the rear hood.
3. Using a Phillips screwdriver, remove the light assembly lens retaining screws.
4. Pull the lens and gasket, if equipped, off of the license plate light assembly.
5. If applicable, inspect the lens gasket for tearing or crumbling. If any such damage is evident, replace the old gasket with a new one upon installation.
6. Once the lens is removed from the light assembly, the light bulb can be removed. To remove the light bulb:
    a. Grasp the light bulb and gently press it in toward its socket.
    b. While pressing the bulb in, turn the bulb counterclockwise until the two little pins on the bulb base clear the channels in the socket.
    c. Pull the bulb up and out of its socket.
7. Installation is the reverse of the removal procedure. During installation, make sure that the light bulb contacts are clean and free of corrosion.

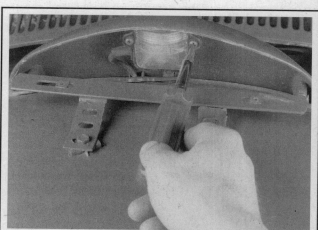

Remove the 2 license plate lens retaining screws to gain access to the bulb—opening the rear hood may provide easier access to this assembly

6-20 CHASSIS ELECTRICAL

FUSES

▶ See Figure 3

All major circuits are protected from overloading or short circuiting by fuses. A 12 position fusebox is located beneath the dashboard near the steering column, or located in the luggage compartment on some air conditioned models.

When a fuse blows, the cause should be investigated. Never install a fuse of a larger capacity than specified (see the Fuse Specifications chart), and never use foil, a bolt or a nail in place of a fuse. However, always carry a few spares in case of emergency. There are ten 8 amp (white) fuses and two 16 amp (red) fuses in the VW fusebox. Circuits No. 9 and 10 use the 16 amp fuses. To replace a fuse, pry off the clear plastic cover for the fusebox and depress a contact at either end of the subject fuse.

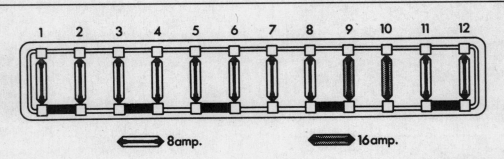

Fig. 3 Illustration of the common fuse box layout for all 1970–81 Volkswagen models covered by this manual

Fuses

Type 1

Circuit	Fuse
Left parking, side marker, and tail lights	8 amps
Right parking, side marker, and tail lights	8 amps
Left low beam	8 amps
Right low beam	8 amps
Left high beam	8 amps
Right high beam, high beam indicator	8 amps
License plate light	8 amps
Emergency flasher system	8 amps
Interior lights	16 amps
Windshield wiper, rear window defogger, fresh air fan	16 amps
Horn, stop lights, ATF warning light	8 amps
Fuel gauge, turn signals, brake warning light, oil pressure, turn signal and generator warning lights	8 amps

Type 2

Circuit	Fuse
Left tail and side marker lights	8 amps
Right tail and marker lights, license light, parking lights	8 amps
Left low beam	8 amps
Right low beam	8 amps
Left high beam, high beam indicator	8 amps
Right high beam	8 amps
Accessories	8 amps
Emergency flasher, front interior light	8 amps ①
Rear interior light, buzzer alarm, auxiliary heater	16 amps
Windshield wipers, rear window defogger	16 amps
Turn signals, warning lamps for alternator, oil pressure, fuel gauge, kickdown, and back-up lights	8 amps
Horn, stop lights, brake warning light	8 amps

Type 3

Circuit	Fuse
Right tail light, license plate light, parking and side marker light, luggage compartment light	8 amps
Left tail light	8 amps
Left low beam	8 amps
Right low beam	8 amps
Left high beam, high beam indicator	8 amps
Right high beam	8 amps
Electric fuel pump	8 amps
Emergency flasher, interior light	8 amps
Buzzer	16 amps
Windshield wipers, fresh air fan, rear window defogger	16 amps
Stop lights, turn signals, horn, brake warning light, back-up lights	8 amps
Accessories	8 amps

Type 4

Circuit	Fuse
Parking lights, left tail and left rear side marker lights	8 amps
Right tail light, right rear side marker light, license plate light, selector lever console light	8 amps
Left low beam	8 amps
Right low beam	8 amps
Left high beam	8 amps
Right high beam, high beam indicator	8 amps
Fuel pump	
Interior light, emergency flasher, buzzer	
Cigarette lighter, heater	16 amps
Window wiper, fresh air fan, heater, rear window defogger	16 amps
Turn signals, back-up lights, warning lights for alternator, oil pressure, fuel gauge	8 amps
Horn, brake warning light, stop lights	8 amps

① 1980–81 — 16 amp

CHASSIS ELECTRICAL 6-21

TRAILER WIRING

Wiring the vehicle for towing is fairly easy. There are a number of good wiring kits available and these should be used, rather than trying to design your own.

All trailers will need brake lights and turn signals as well as tail lights and side marker lights. Most areas require extra marker lights for overwide trailers. Also, most areas have recently required back-up lights for trailers, and most trailer manufacturers have been building trailers with back-up lights for several years.

Additionally, some Class I, most Class II and just about all Class III trailers will have electric brakes. Add to this number an accessories wire, to operate trailer internal equipment or to charge the trailer's battery, and you can have as many as seven wires in the harness.

Determine the equipment on your trailer and buy the wiring kit necessary. The kit will contain all the wires needed, plus a plug adapter set which includes the female plug, mounted on the bumper or hitch, and the male plug, wired into, or plugged into the trailer harness.

When installing the kit, follow the manufacturer's instructions.

The color coding of the wires is usually standard throughout the industry. One point to note: some domestic vehicles, and most imported vehicles, have separate turn signals. On most domestic vehicles, the brake lights and rear turn signals operate with the same bulb. For those vehicles with separate turn signals, you can purchase an isolation unit so that the brake lights won't blink whenever the turn signals are operated, or, you can go to your local electronics supply house and buy four diodes to wire in series with the brake and turn signal bulbs. Diodes will isolate the brake and turn signals. The choice is yours. The isolation units are simple and quick to install, but far more expensive than the diodes. The diodes, however, require more work to install properly, since they require the cutting of each bulb's wire and soldering in place of the diode.

One, final point, the best kits are those with a spring loaded cover on the vehicle mounted socket. This cover prevents dirt and moisture from corroding the terminals. Never let the vehicle socket hang loosely; always mount it securely to the bumper or hitch.

6-22 CHASSIS ELECTRICAL

WIRING DIAGRAMS

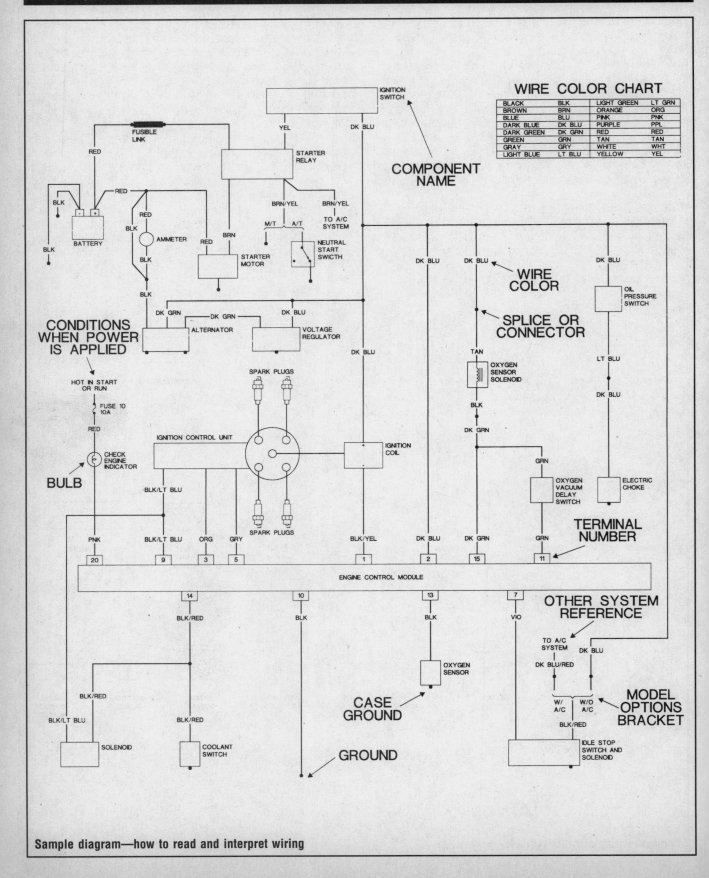

Sample diagram—how to read and interpret wiring

CHASSIS ELECTRICAL 6-23

WIRING DIAGRAM SYMBOLS

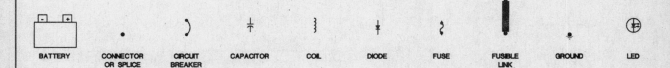

BATTERY — CONNECTOR OR SPLICE — CIRCUIT BREAKER — CAPACITOR — COIL — DIODE — FUSE — FUSIBLE LINK — GROUND — LED

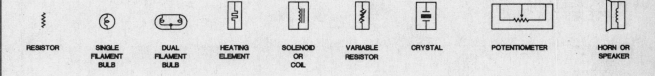

RESISTOR — SINGLE FILAMENT BULB — DUAL FILAMENT BULB — HEATING ELEMENT — SOLENOID OR COIL — VARIABLE RESISTOR — CRYSTAL — POTENTIOMETER — HORN OR SPEAKER

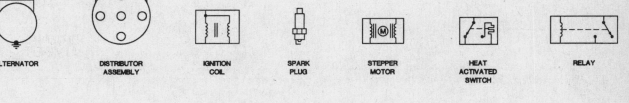

ALTERNATOR — DISTRIBUTOR ASSEMBLY — IGNITION COIL — SPARK PLUG — STEPPER MOTOR — HEAT ACTIVATED SWITCH — RELAY

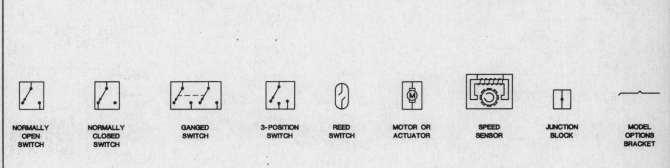

NORMALLY OPEN SWITCH — NORMALLY CLOSED SWITCH — GANGED SWITCH — 3-POSITION SWITCH — REED SWITCH — MOTOR OR ACTUATOR — SPEED SENSOR — JUNCTION BLOCK — MODEL OPTIONS BRACKET

Common wiring diagram symbols

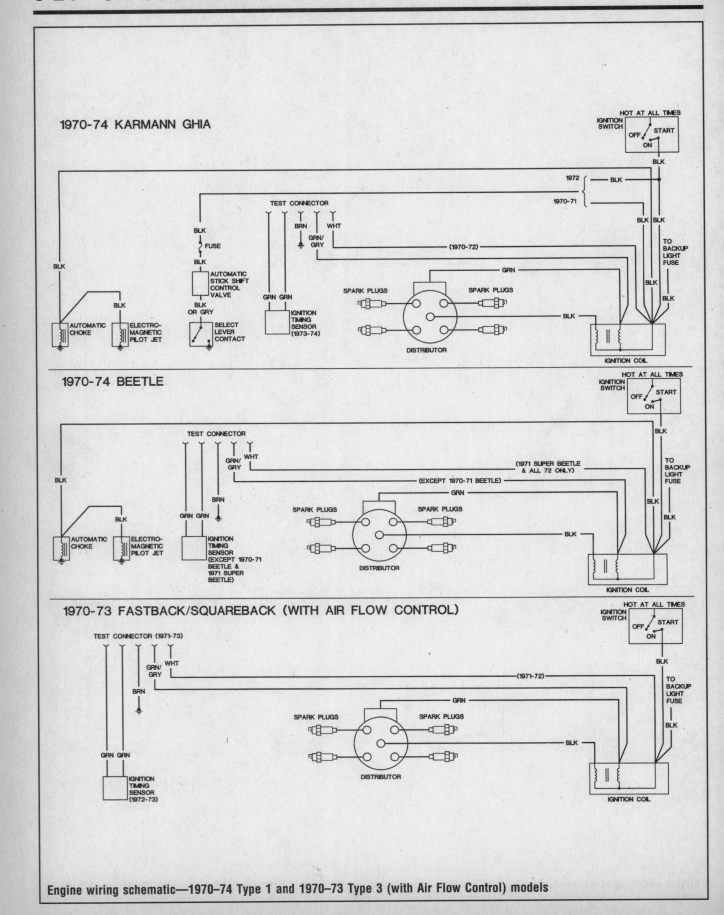

Engine wiring schematic—1970–74 Type 1 and 1970–73 Type 3 (with Air Flow Control) models

CHASSIS ELECTRICAL 6-25

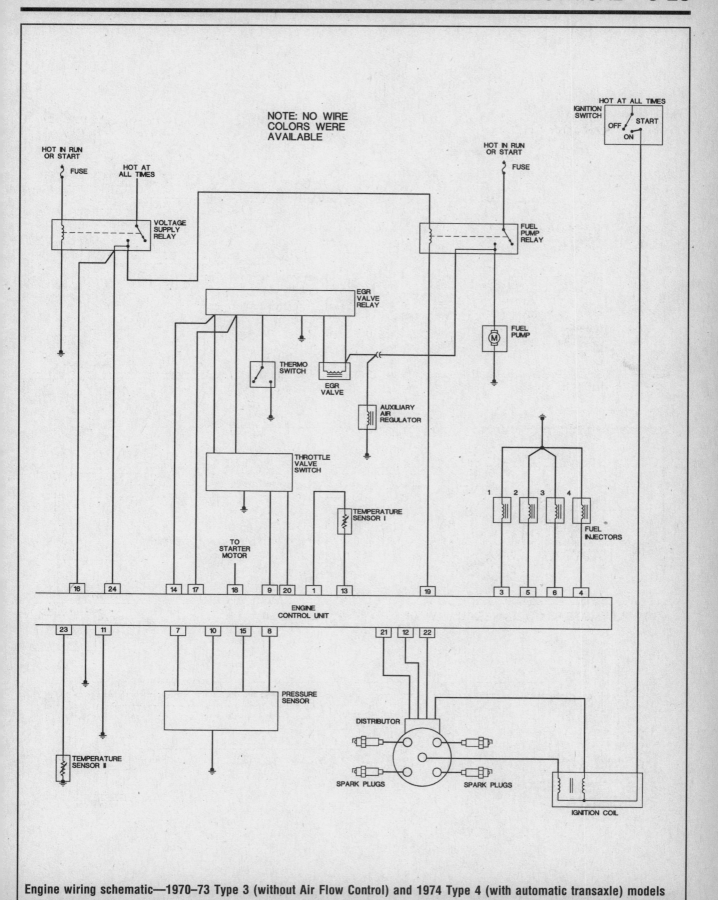

Engine wiring schematic—1970–73 Type 3 (without Air Flow Control) and 1974 Type 4 (with automatic transaxle) models

6-26 CHASSIS ELECTRICAL

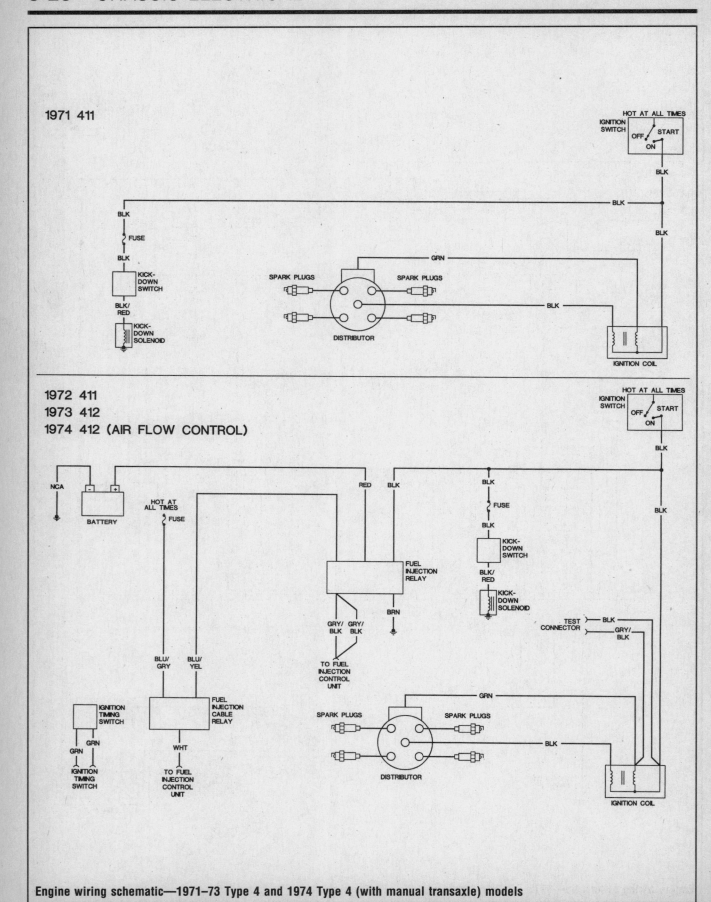

Engine wiring schematic—1971–73 Type 4 and 1974 Type 4 (with manual transaxle) models

CHASSIS ELECTRICAL 6-27

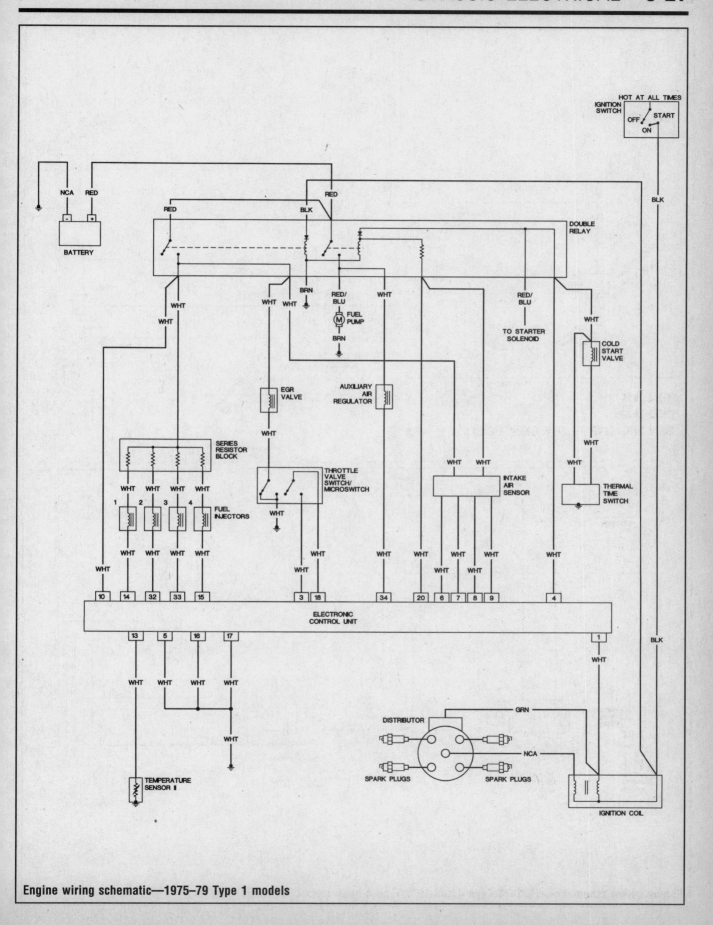

Engine wiring schematic—1975–79 Type 1 models

6-28 CHASSIS ELECTRICAL

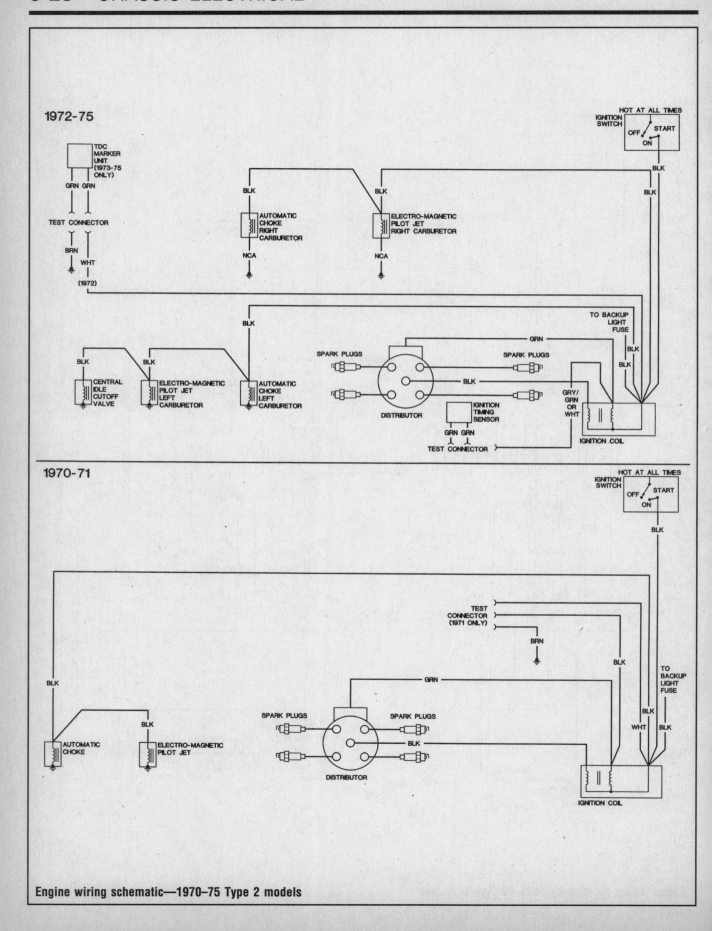

Engine wiring schematic—1970–75 Type 2 models

CHASSIS ELECTRICAL 6-29

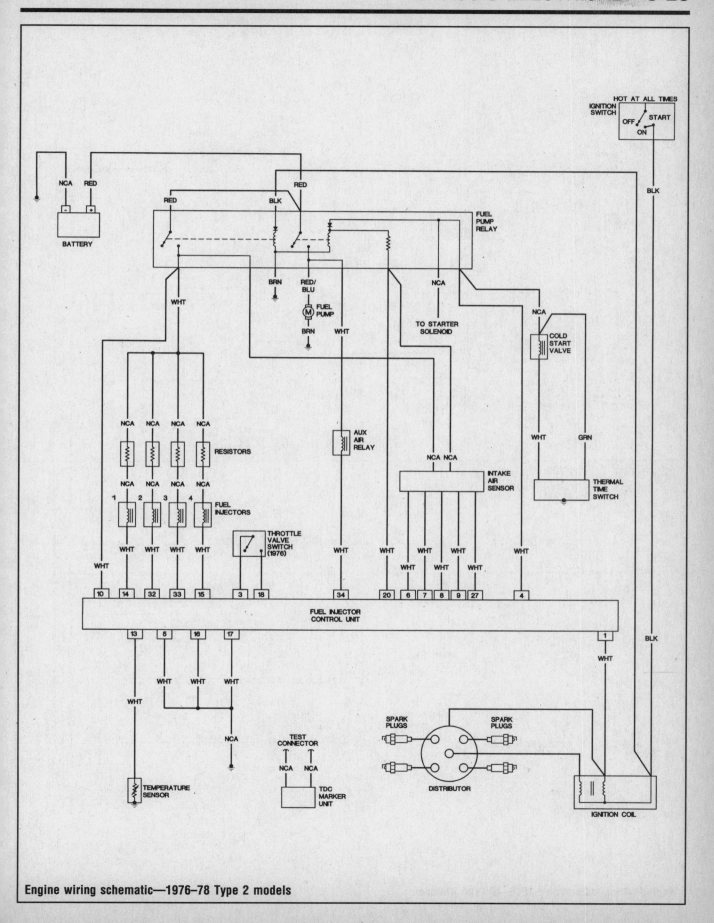

Engine wiring schematic—1976–78 Type 2 models

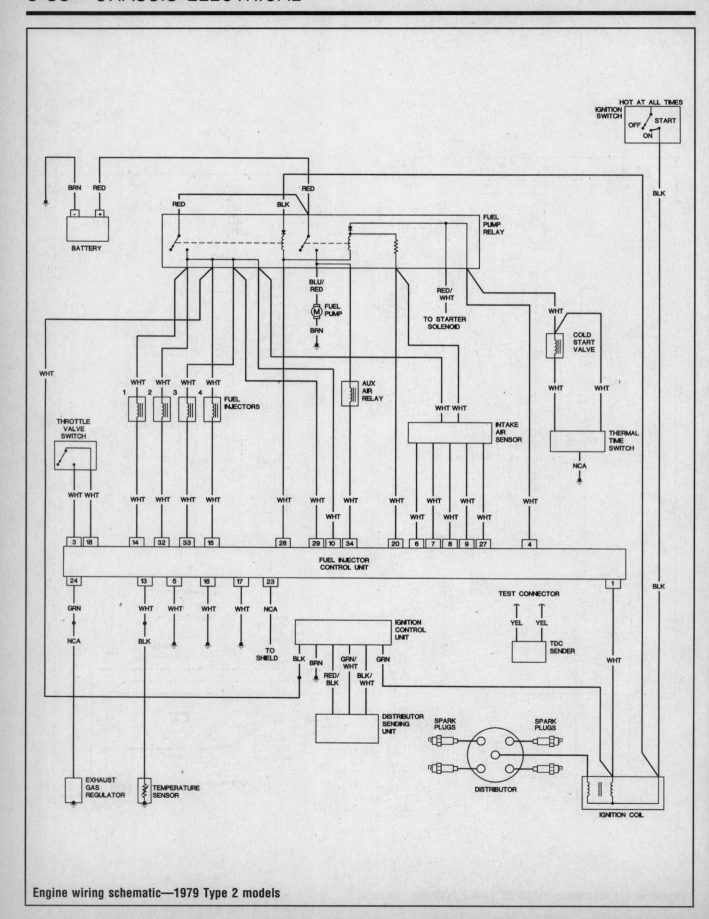

Engine wiring schematic—1979 Type 2 models

CHASSIS ELECTRICAL 6-31

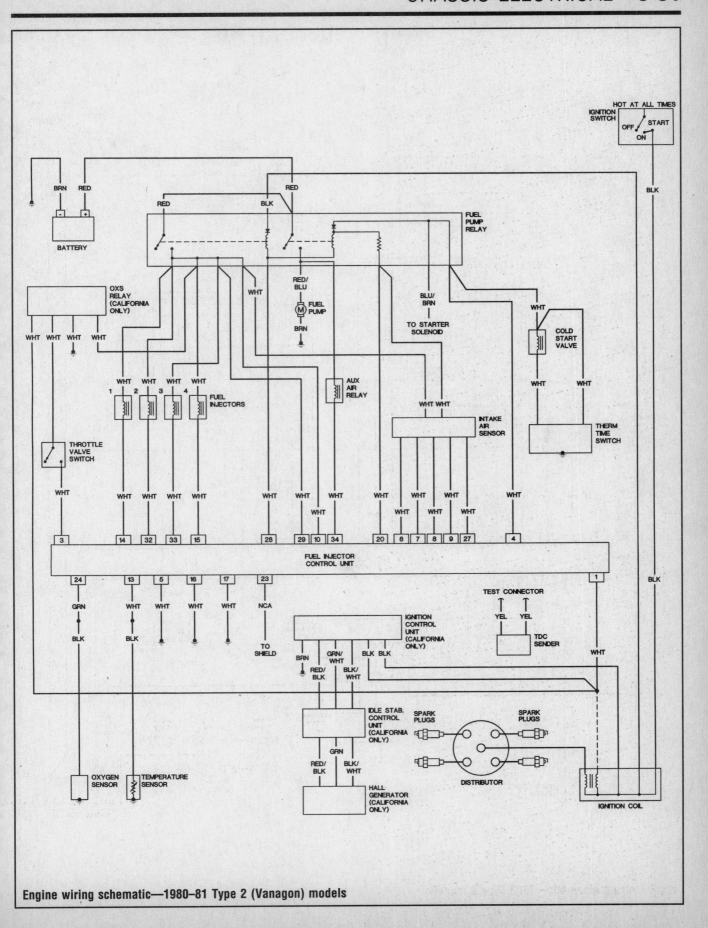

Engine wiring schematic—1980–81 Type 2 (Vanagon) models

6-32 CHASSIS ELECTRICAL

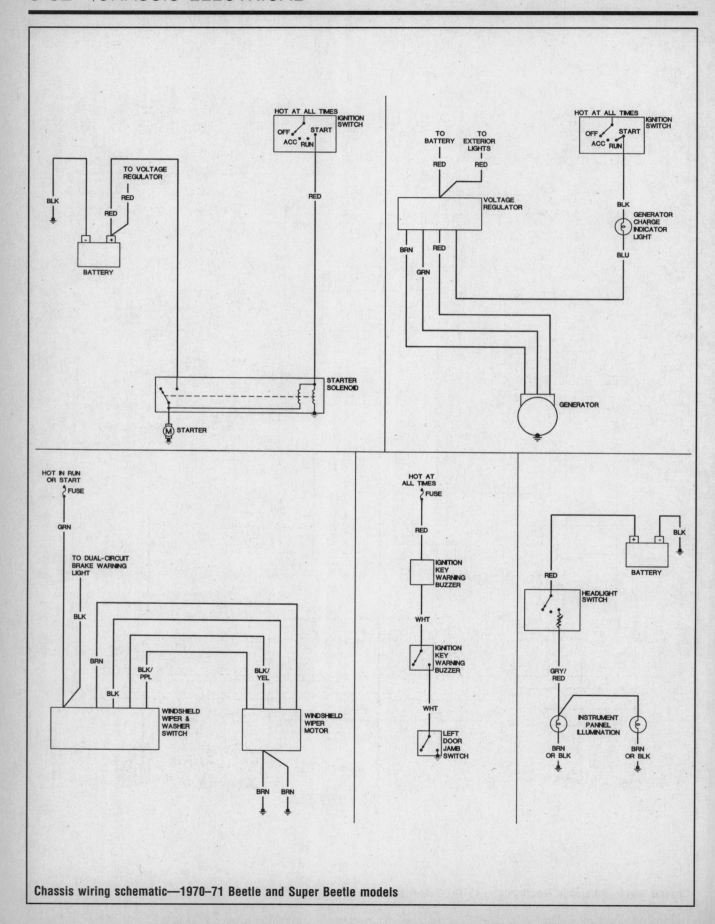

Chassis wiring schematic—1970–71 Beetle and Super Beetle models

CHASSIS ELECTRICAL   6-33

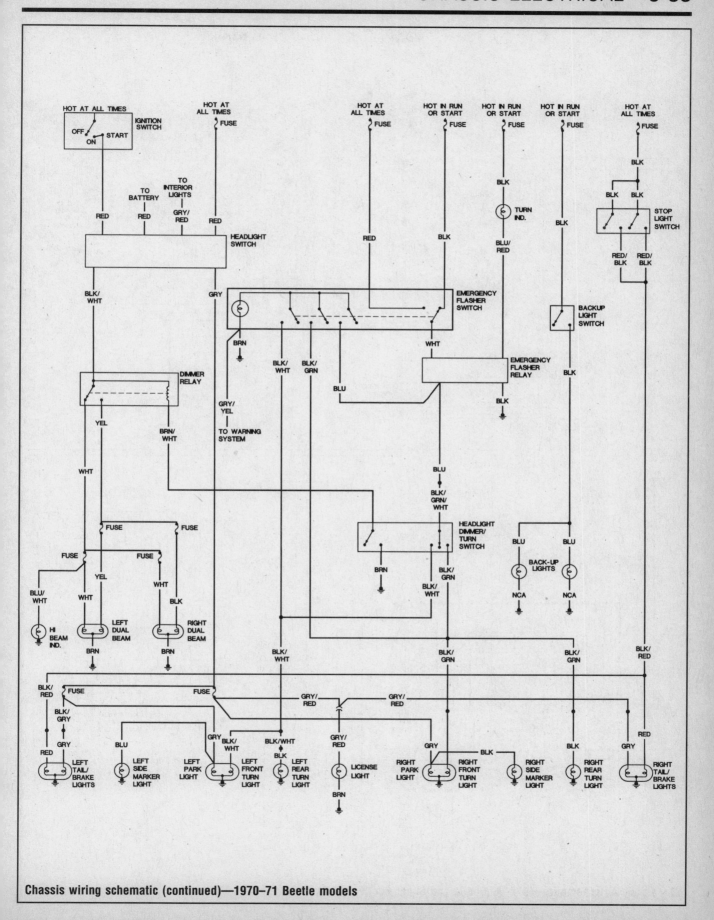

Chassis wiring schematic (continued)—1970–71 Beetle models

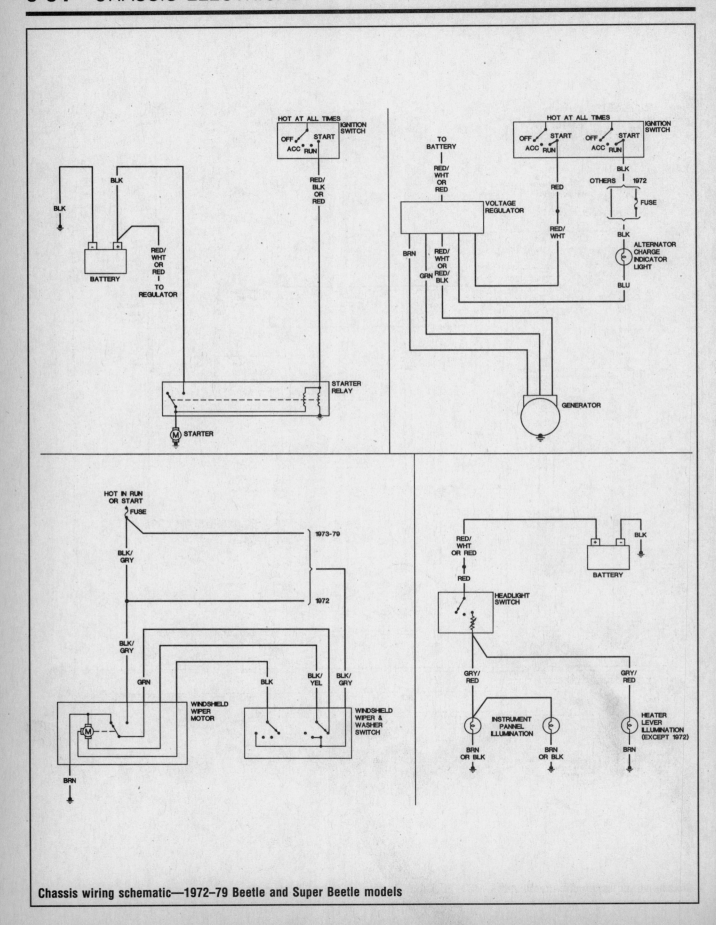

Chassis wiring schematic—1972-79 Beetle and Super Beetle models

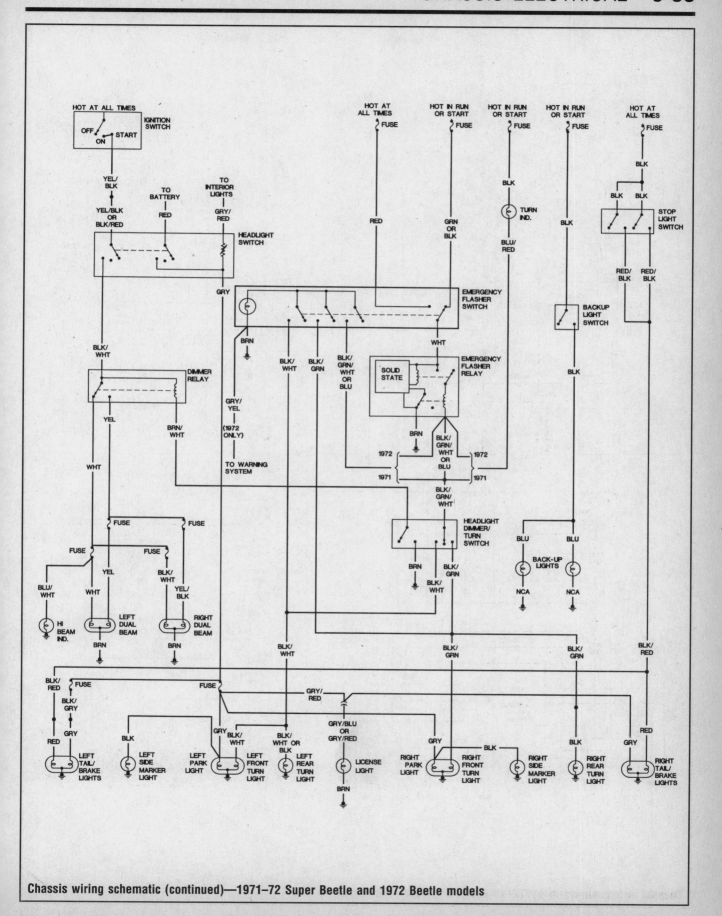

Chassis wiring schematic (continued)—1971–72 Super Beetle and 1972 Beetle models

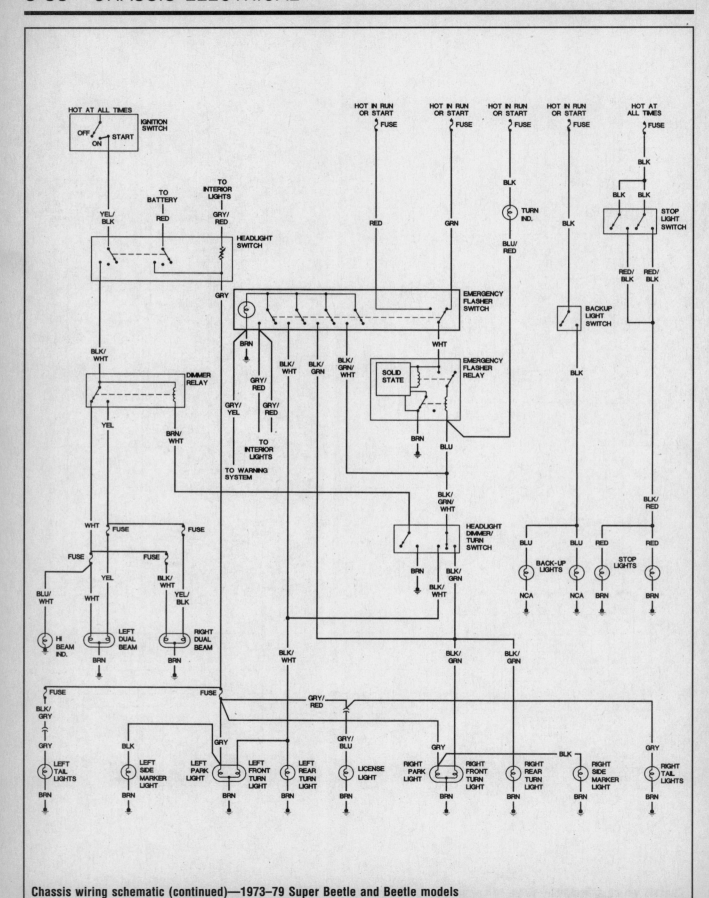

Chassis wiring schematic (continued)—1973–79 Super Beetle and Beetle models

CHASSIS ELECTRICAL 6-37

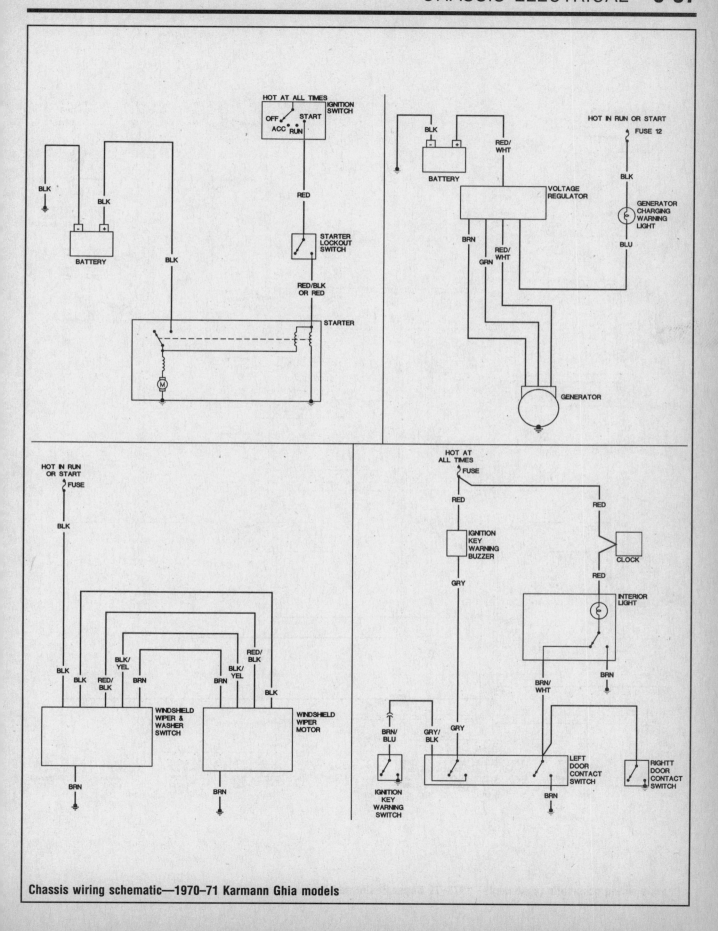

Chassis wiring schematic—1970–71 Karmann Ghia models

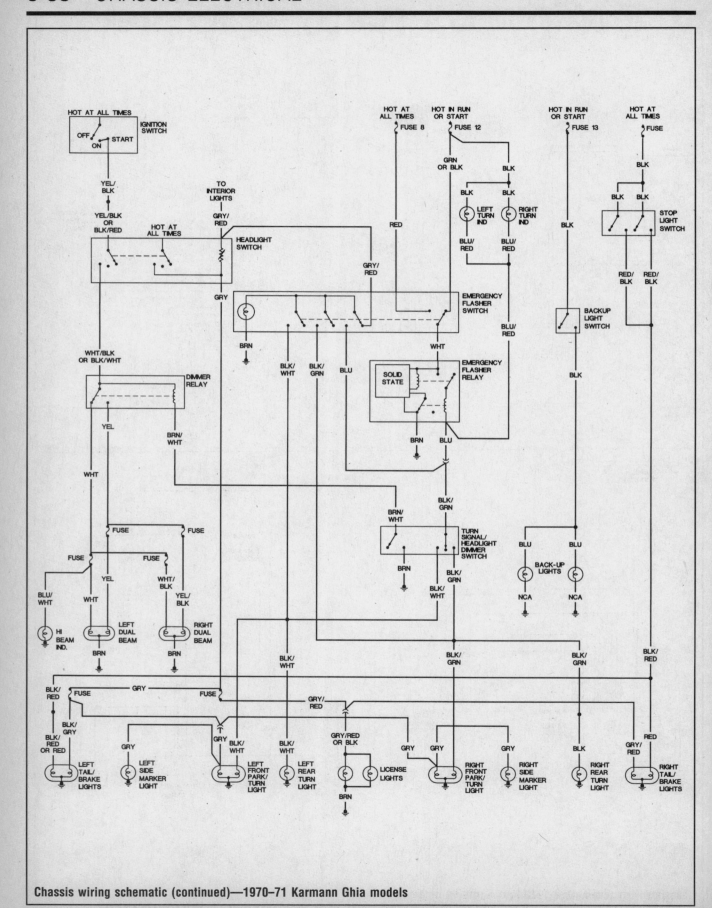

Chassis wiring schematic (continued)—1970–71 Karmann Ghia models

CHASSIS ELECTRICAL 6-39

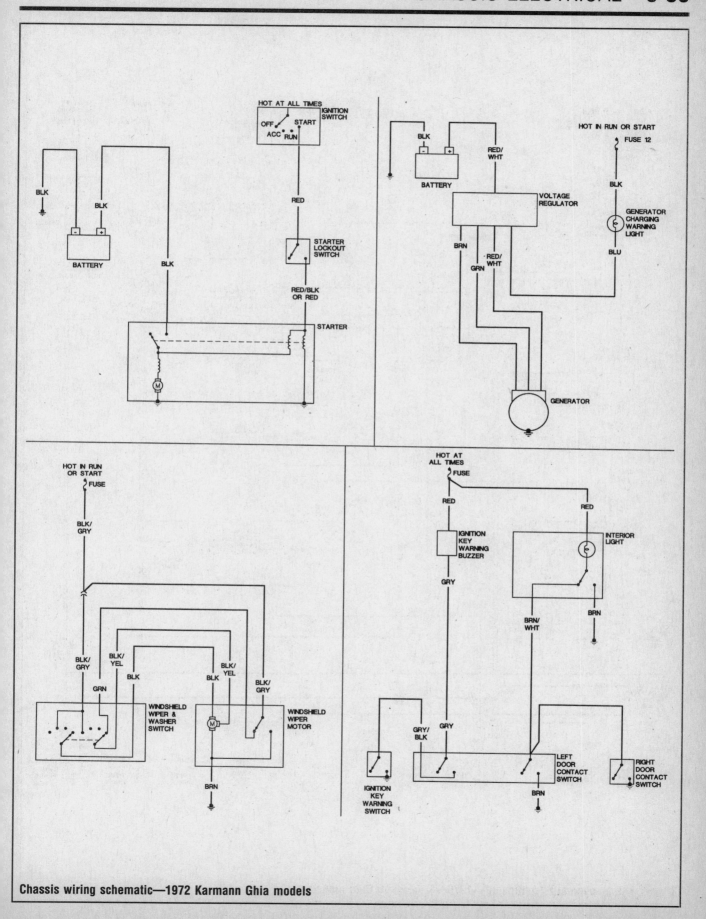

Chassis wiring schematic—1972 Karmann Ghia models

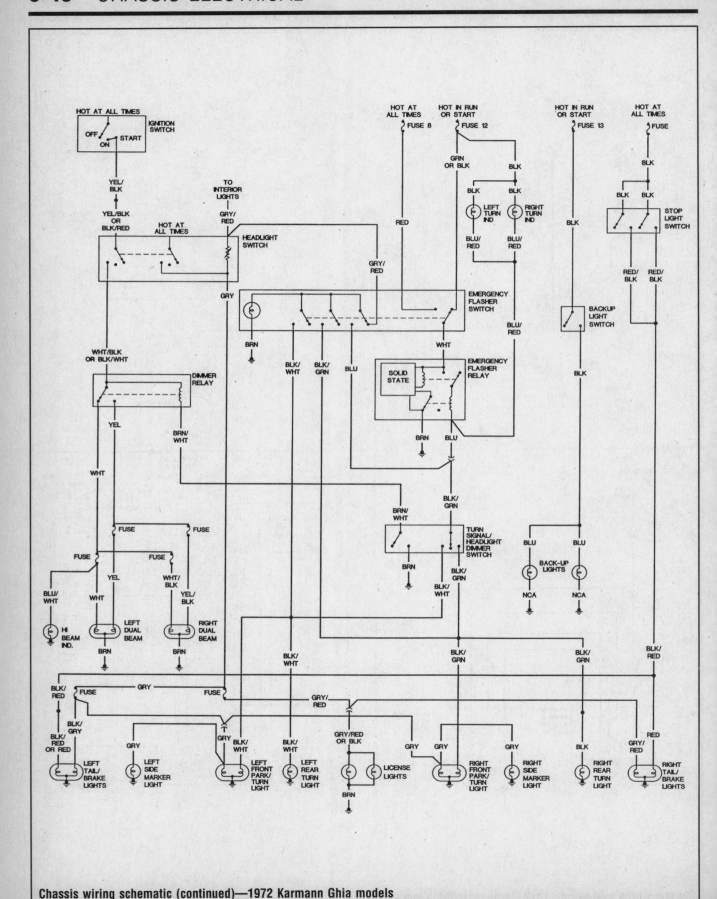

Chassis wiring schematic (continued)—1972 Karmann Ghia models

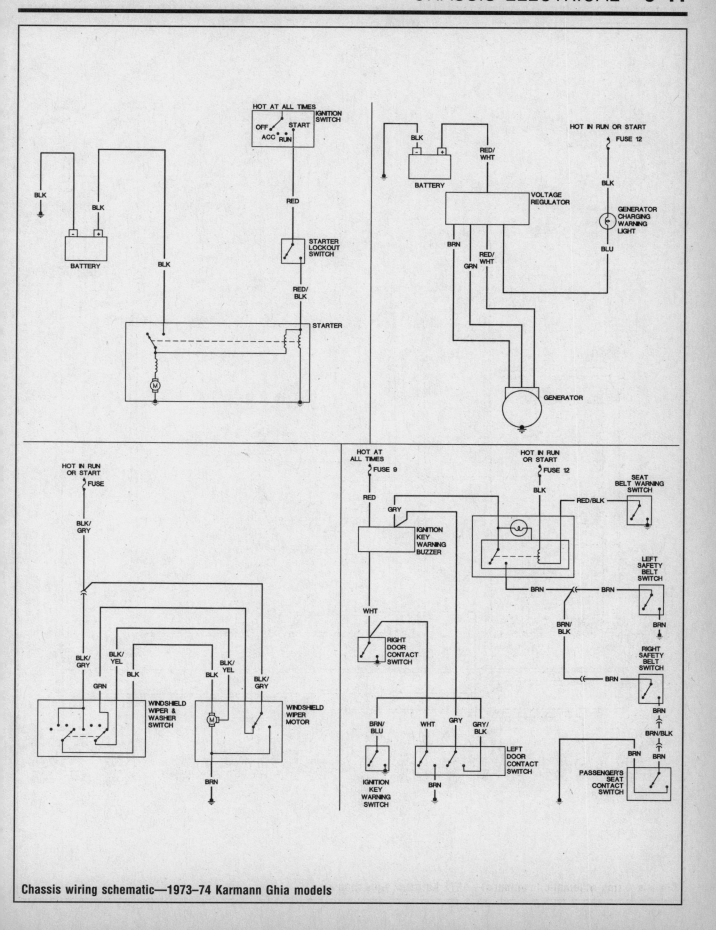

Chassis wiring schematic—1973-74 Karmann Ghia models

6-42 CHASSIS ELECTRICAL

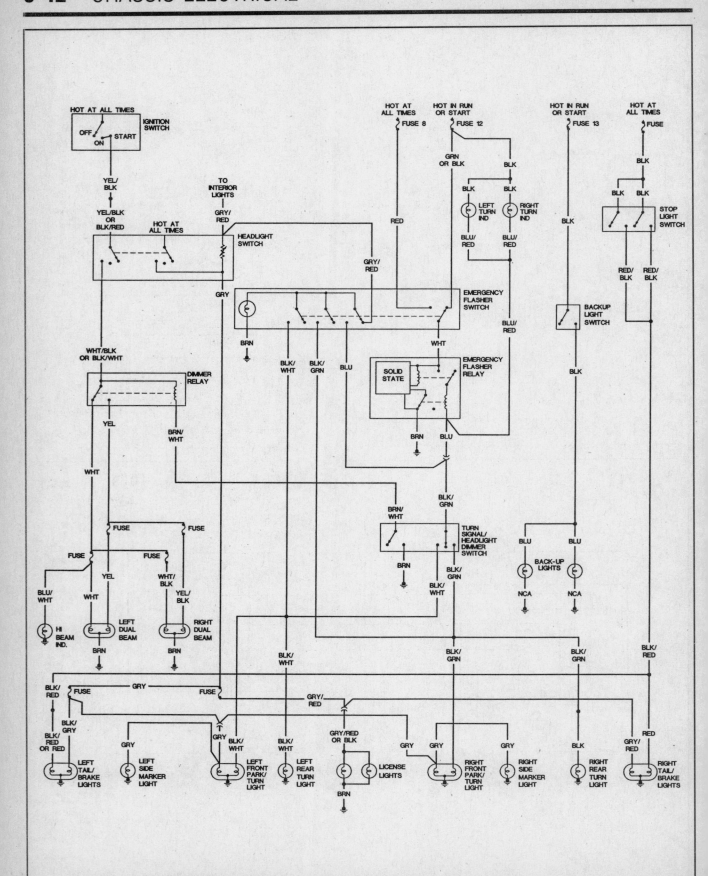

Chassis wiring schematic (continued)—1973–74 Karmann Ghia models

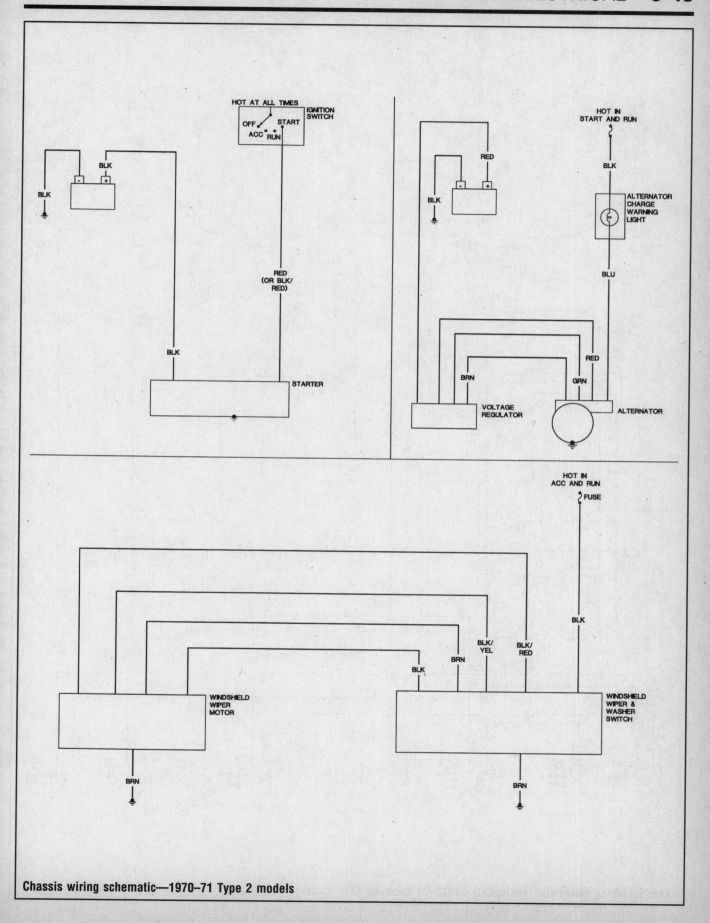

Chassis wiring schematic—1970–71 Type 2 models

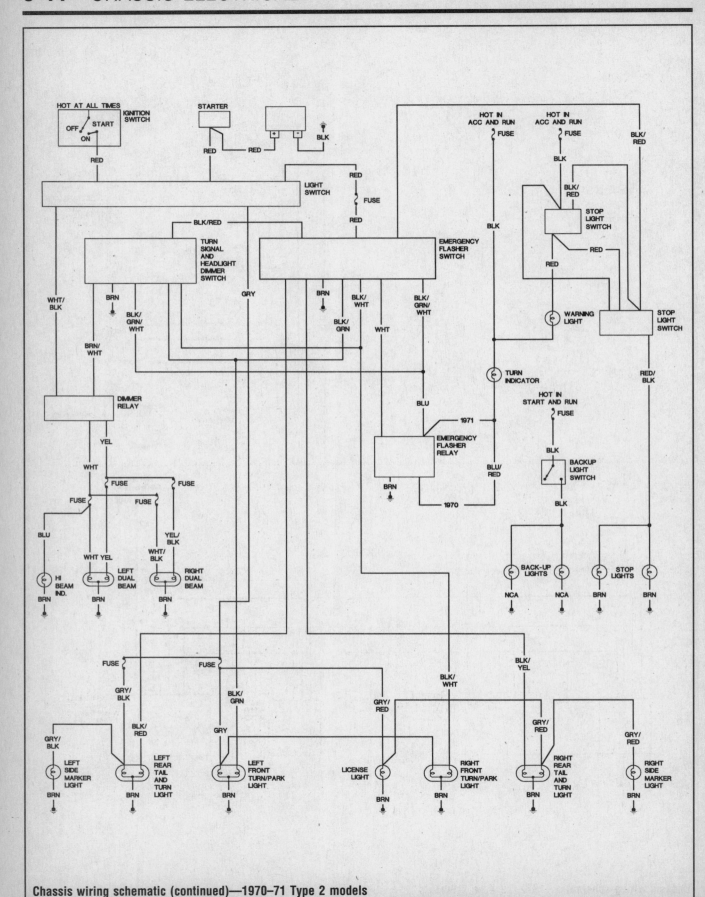

Chassis wiring schematic (continued)—1970–71 Type 2 models

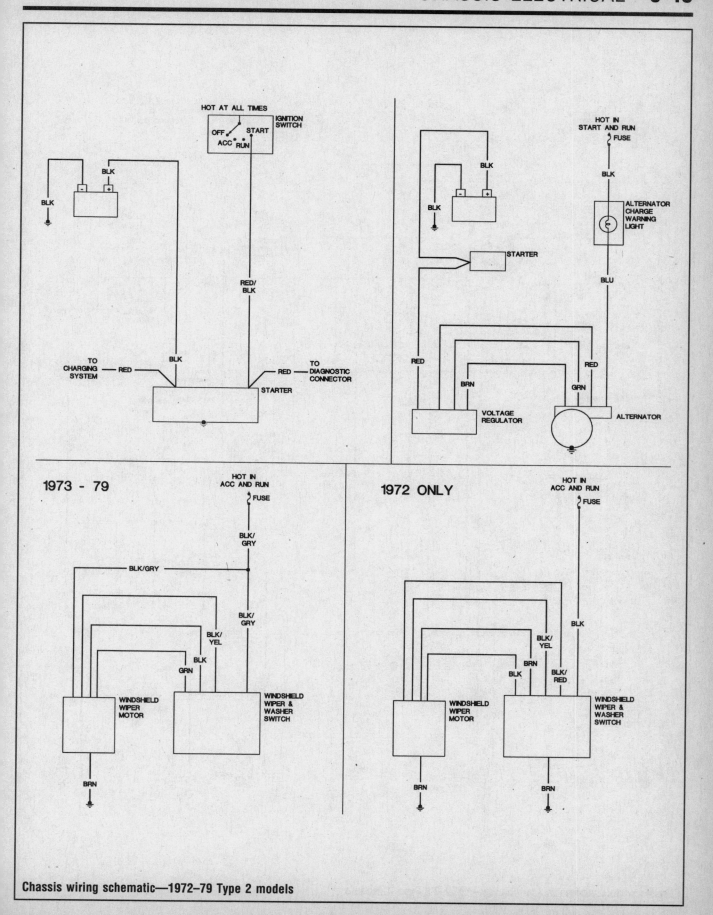

6-46 CHASSIS ELECTRICAL

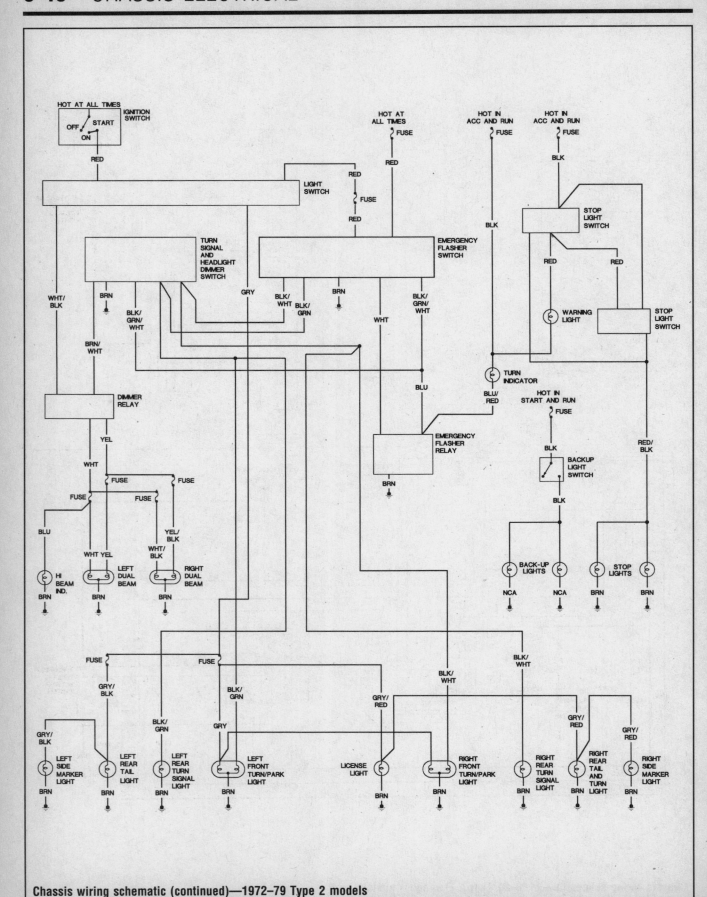

Chassis wiring schematic (continued)—1972–79 Type 2 models

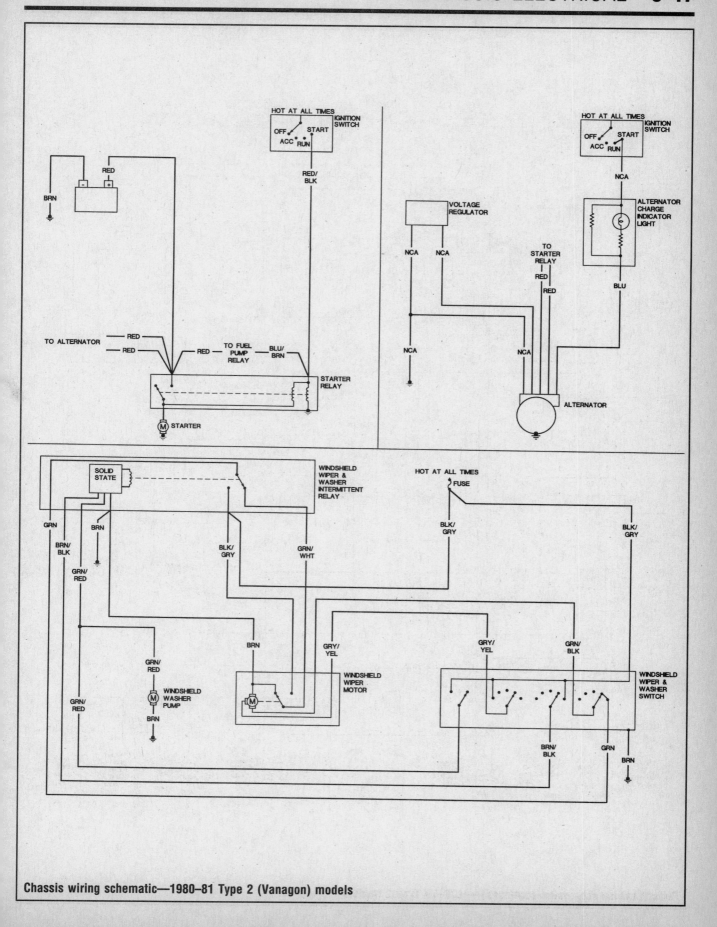

Chassis wiring schematic—1980–81 Type 2 (Vanagon) models

6-48 CHASSIS ELECTRICAL

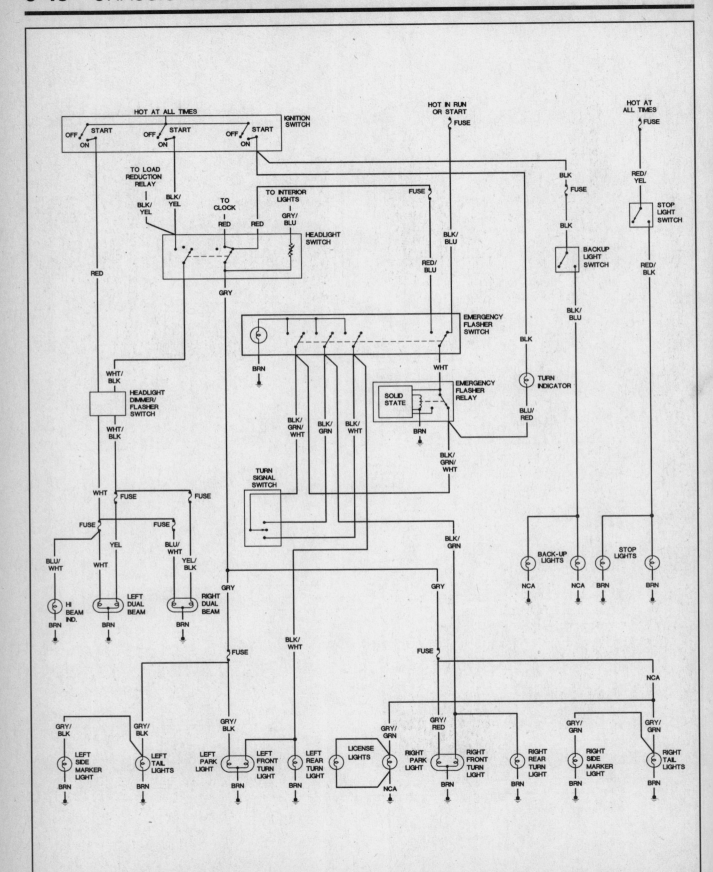

Chassis wiring schematic (continued)—1980–81 Type 2 (Vanagon) models

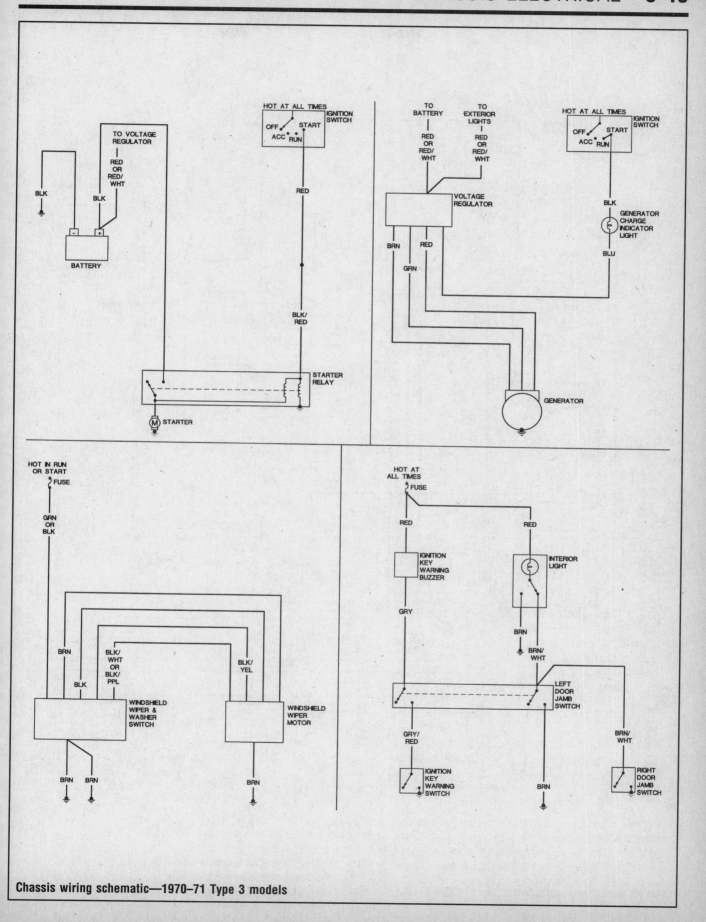

Chassis wiring schematic—1970–71 Type 3 models

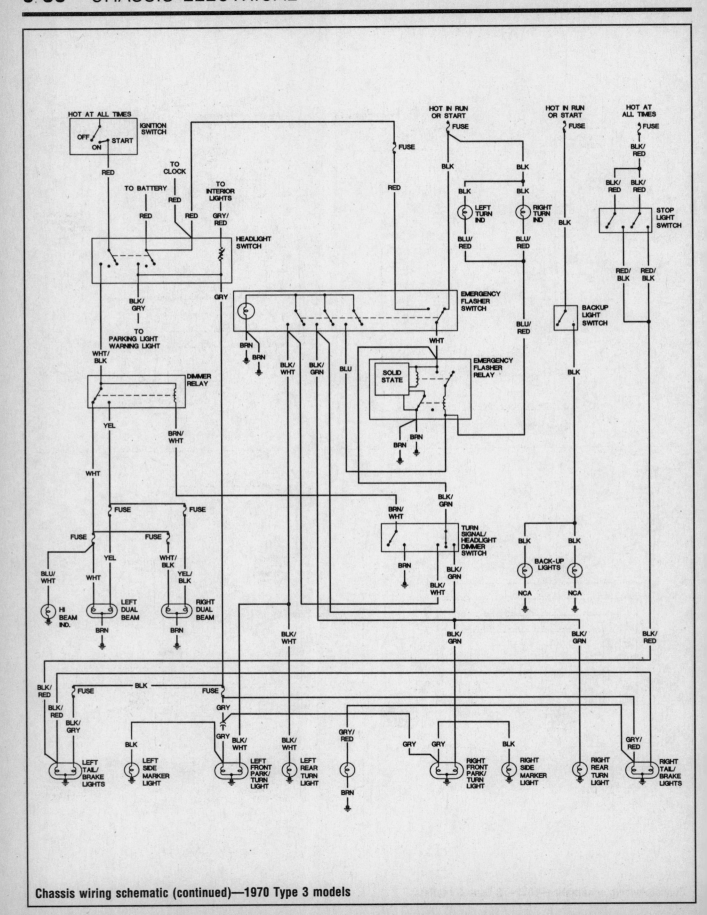

Chassis wiring schematic (continued)—1970 Type 3 models

CHASSIS ELECTRICAL 6-51

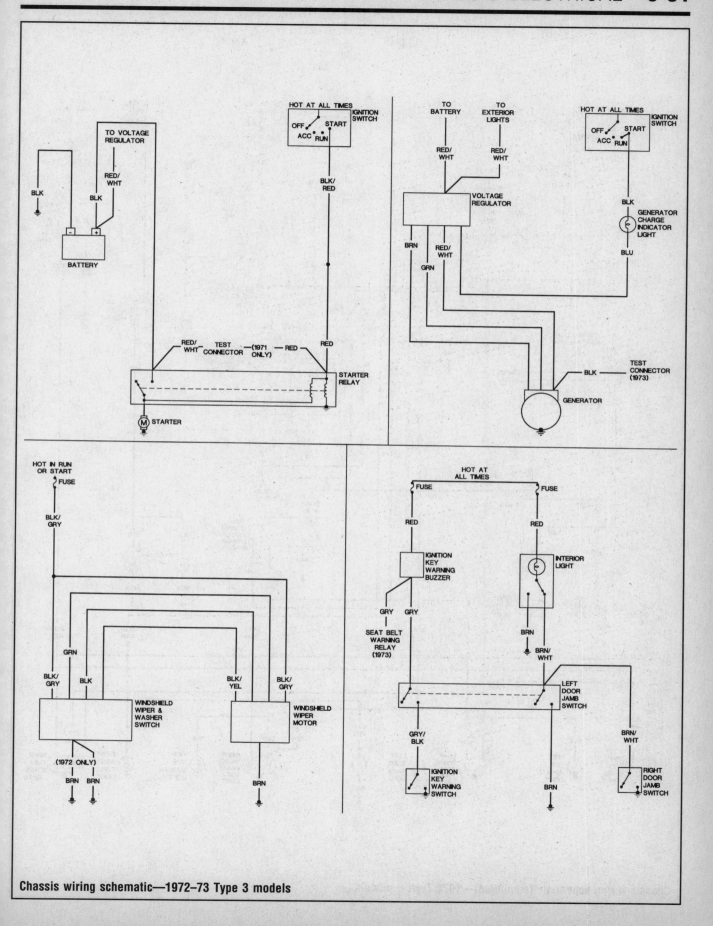

Chassis wiring schematic—1972–73 Type 3 models

6-52 CHASSIS ELECTRICAL

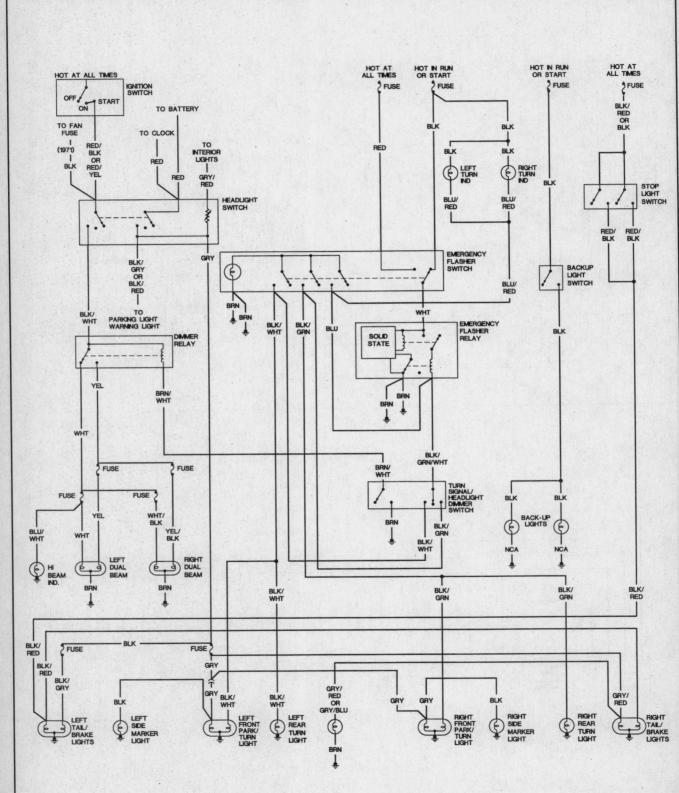

Chassis wiring schematic (continued)—1971–73 Type 3 models

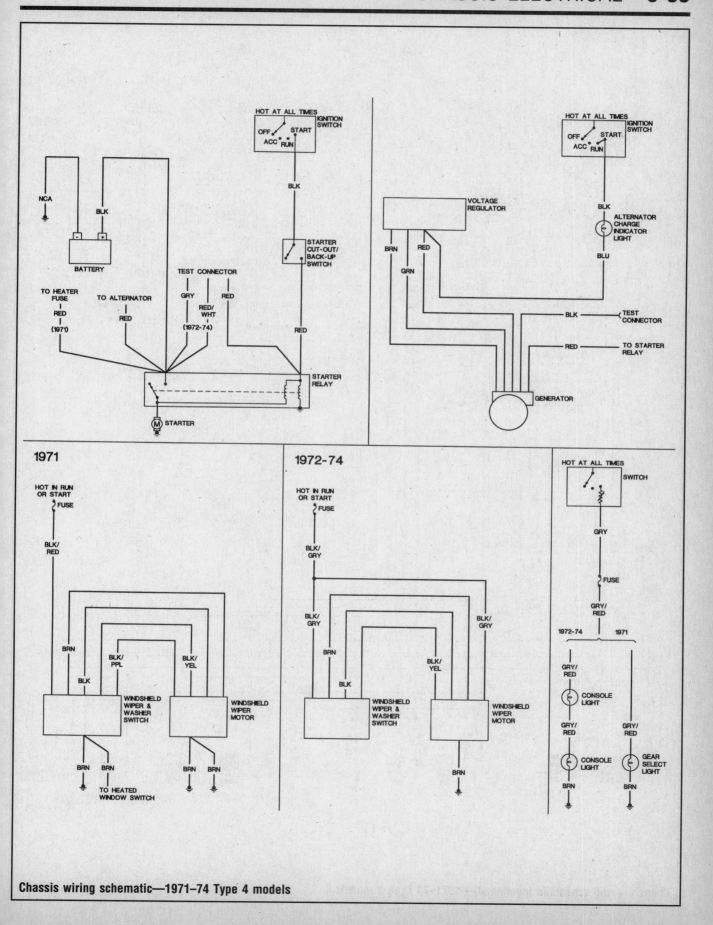

Chassis wiring schematic—1971-74 Type 4 models

6-54 CHASSIS ELECTRICAL

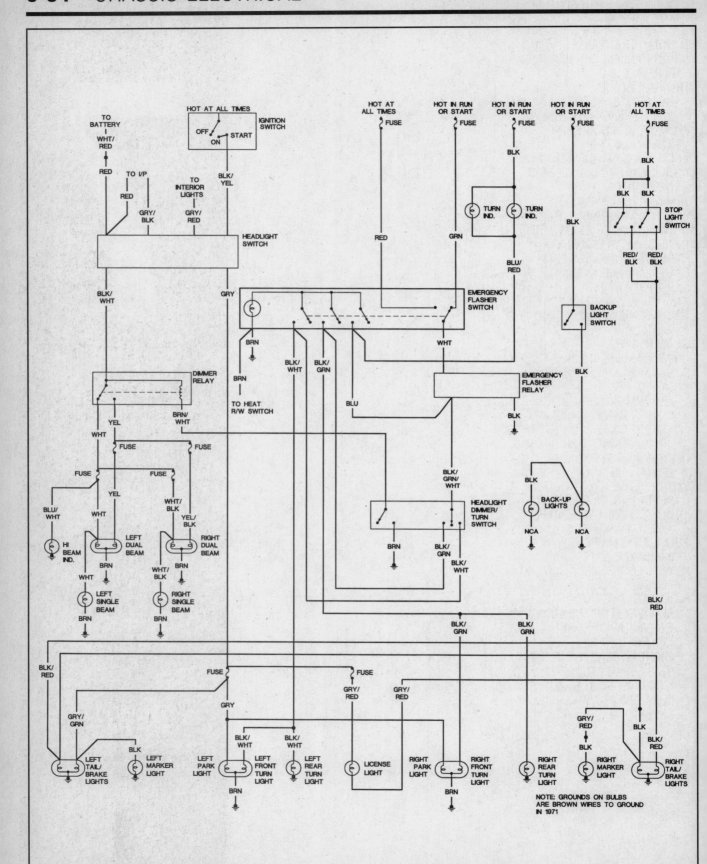

Chassis wiring schematic (continued)—1971-74 Type 4 models

MANUAL TRANSAXLE 7-2
UNDERSTANDING THE MANUAL
  TRANSAXLE 7-2
TRANSAXLE ASSEMBLY 7-2
  REMOVAL & INSTALLATION 7-2
  SHIFT LINKAGE ADJUSTMENT 7-4
AUTOMATIC STICK SHIFT
TRANSAXLE 7-5
UNDERSTANDING THE AUTOMATIC
  STICK SHIFT TRANSAXLE 7-5
  OPERATION 7-5
TRANSAXLE ASSEMBLY 7-6
  REMOVAL & INSTALLATION 7-6
  SHIFT LINKAGE ADJUSTMENT 7-7
DRIVESHAFT AND CONSTANT
VELOCITY (CV) JOINT 7-8
DRIVESHAFT AND CV-JOINT
  ASSEMBLY 7-8
  REMOVAL & INSTALLATION 7-8
  CONSTANT VELOCITY JOINT
    OVERHAUL 7-9
CLUTCH 7-10
UNDERSTANDING THE CLUTCH 7-10
DRIVEN DISC AND PRESSURE
  PLATE 7-10
  REMOVAL & INSTALLATION 7-10
  CLUTCH CABLE ADJUSTMENT 7-14
  CLUTCH CABLE
    REPLACEMENT 7-14
CLUTCH MASTER CYLINDER 7-15
  REMOVAL & INSTALLATION 7-15
CLUTCH SLAVE CYLINDER 7-15
  REMOVAL & INSTALLATION 7-15
  CLUTCH SYSTEM BLEEDING &
    ADJUSTMENT 7-15
FULLY AUTOMATIC
TRANSAXLE 7-17
TRANSAXLE ASSEMBLY 7-17
  REMOVAL & INSTALLATION 7-17
FLUID PAN 7-18
  REMOVAL & INSTALLATION 7-18
  FILTER SERVICE 7-20
ADJUSTMENTS 7-20
  FRONT (SECOND) BAND 7-20
  REAR (FIRST) BAND 7-20
  KICKDOWN SWITCH 7-21
  SHIFT LINKAGE 7-22

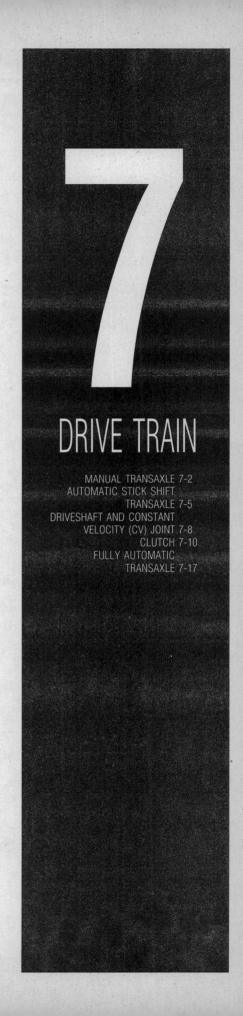

7
DRIVE TRAIN

MANUAL TRANSAXLE 7-2
AUTOMATIC STICK SHIFT
TRANSAXLE 7-5
DRIVESHAFT AND CONSTANT
VELOCITY (CV) JOINT 7-8
CLUTCH 7-10
FULLY AUTOMATIC
TRANSAXLE 7-17

7-2 DRIVE TRAIN

MANUAL TRANSAXLE

All of the Volkswagens covered in this manual are equipped with transaxles, so named because the transmission gears and the axle gears are contained in the same housing. On manual and automatic stick shift VWs, the transmission part of the assembly shares the same hypoid gear oil as the rear axle part. Automatic transaxles use ATF Dexron® in the transmission part, and hypoid gear oil in the rear axle.

All transaxles are mounted in a yoke at the rear of the car and bolt up to the front of the engine.

The transaxle case is constructed of aluminum alloy.

Understanding the Manual Transaxle

Because of the way an internal combustion engine breathes, it can produce torque, or twisting force, only within a narrow speed range. Most modern, overhead valve pushrod engines must turn at about 2500 rpm to produce their peak torque. By 4500 rpm they are producing so little torque that continued increases in engine speed produce no power increases. The torque peak on overhead camshaft engines is generally much higher, but within a much narrower speed range.

The manual transaxle and clutch are employed to vary the relationship between engine speed and the speed of the wheels so that adequate engine power can be produced under all circumstances. The clutch allows engine torque to be applied to the transaxle input shaft gradually, due to mechanical slippage. Consequently, the vehicle may be started smoothly from a full stop. The transaxle changes the ratio between the rotating speeds of the engine and wheels by the use of gears. The gear ratios allow full engine power to be applied to the wheels during acceleration at low speeds and at highway/passing speeds.

The power is usually transmitted from the input shaft to a mainshaft or output shaft. The gears of the mainshaft mesh with gears on the input shaft, allowing power to be carried from one to the other. All forward gears are in constant mesh and are free from rotating with the shaft unless the synchronizer and clutch are engaged. Shifting from one gear to the next causes one of the gears to be freed from rotating with the shaft and locks another to it. Gears are locked and unlocked by internal dog clutches which slide between the center of the gear and shaft. The forward gears employ synchronizers; friction members which smoothly bring gear and shaft to the same speed before the toothed dog clutches are engaged.

All manual transaxles (covered by this manual) employ four forward speeds and a reverse. All forward speeds have synchromesh engagement. The gears are helical and in constant mesh. Gear selection is accomplished by a floor-mounted lever working through a shift rod contained in the frame tunnel. The final drive pinion and ring gear are also helical cut.

Transaxle Assembly

REMOVAL & INSTALLATION

♦ See Figure 1

1. Disconnect the negative battery cable.
2. Remove the engine, as described in Section 3.
3. Remove the special 12-point socket head screws which secure the driveshafts to the transaxle. Remove the bolts from the transaxle end first and then remove the shafts.

➡ **It is not necessary to remove the driveshafts entirely from the car if the car does not have to be moved while the transaxle is out.**

4. Disconnect the clutch cable from the clutch lever and remove the clutch cable and its guide tube from the transaxle. Loosen the square head bolt at the shift linkage coupling located near the rear of the transaxle. Slide the coupling off the inner shift lever. There is an access plate under the rear seat to reach the coupling on Type 1 and 3 models. It is necessary to work under the car to reach the coupling on Type 2 models.
5. Label and disconnect the starter wiring.
6. Label and disconnect the back-up light switch wiring.
7. Remove the front transaxle mounting bolts.

To gain access to the shifter rod-to-transaxle coupling, remove the rear seat and remove the access plate hold-down screws

Lift the access plate up and off of the center frame tube

DRIVE TRAIN 7-3

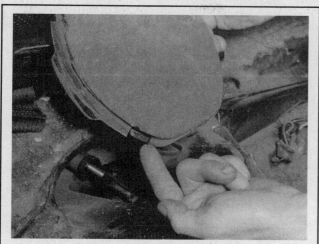

Note that the access plate is secured in place not only by the hold-down screws, but also by these holding flaps

. . . then remove the 8mm retaining bolt with a wrench or socket

Once the access plate is removed, the shift rod-to-transaxle coupling can be seen

Move the shift rod forward, by moving the shifter handle rearward, to disengage the coupling from the transaxle shift input shaft

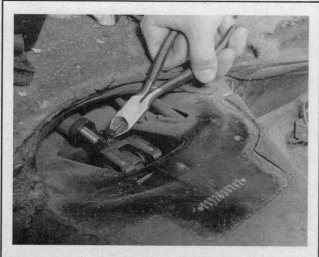

Snip the safety wire off of the retaining bolt . . .

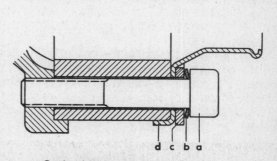

a. Socket head screws
b. Lockwasher
c. Spacer
d. Protective cap

Fig. 1 When installing the driveshafts, make certain to position the attaching bolt spacers and lockwashers in the correct order

DRIVE TRAIN

8. Support the transaxle with a floor jack and remove the transaxle carrier bolts.

➡ 1972 and later Type 2's have two upper carrier bolts which also join the engine to the transaxle. The transaxle must be supported when these bolts are removed to prevent damage which could be caused by letting the unit hang.

9. Carefully lower the jack and remove the transaxle from the car.

To install:

10. Jack the transaxle into position and loosely install the bolts.
11. Tighten the transaxle carrier bolts first, then tighten the front mounting nuts.
12. Install the driveshaft bolts with new lockwashers. The lockwashers should be positioned on the bolt with the convex side toward the screw head.
13. Reconnect the wiring, the clutch cable, and the shift linkage.

➡ It may be necessary to align the transaxle so that the driveshaft joints do not rub the frame.

14. Install the engine.

SHIFT LINKAGE ADJUSTMENT

◆ See Figures 2 and 3

The Volkswagen shift linkage is not adjustable. When shifting becomes difficult or there is an excessive amount of play in the linkage, check the shifting mechanism for worn parts. Make sure the shift linkage coupling is tightly connected to the inner shift lever located at the rear of the transaxle under a plate below the rear seat on Types 1, 3 and 4. On Type 2 models, Check the set-

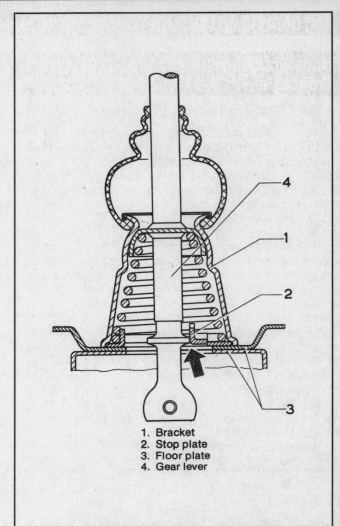

1. Bracket
2. Stop plate
3. Floor plate
4. Gear lever

Fig. 3 Cross-sectional view of the manual transaxle gear shifter assembly—the shifter is held to the floor by only two 13mm bolts

screw where the front shift rod connects to the center section below the passenger compartment, and the rear setscrew where the linkage goes into the transaxle. Worn parts may be found in the shift lever mechanism and the supports for the linkage rod sometimes wear out.

The gear shift lever can be removed after the front floor mat has been lifted. After the two retaining screws have been removed from the gear shift lever ball housing, the gear shift lever, ball housing, rubber boot, and spring are removed as a unit.

✱✱ CAUTION

Carefully mark the position of the stop plate and note the position of the turned up ramp at the side of the stop plate. Normally the ramp is turned up and on the right-hand side of the hole.

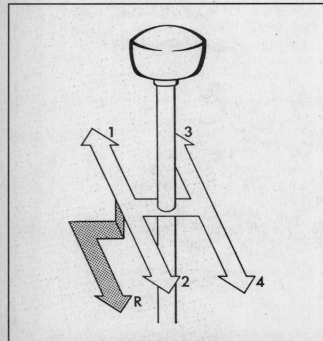

Fig. 2 Manual transaxle gearshift pattern—the shifter must be depressed to position the transaxle in Reverse

Installation is the reverse of removal. Lubricate all moving parts with grease. Test the gear shift pattern. If there is difficulty in shifting, adjust the stop plate back and forth in its slotted holes.

DRIVE TRAIN

AUTOMATIC STICK SHIFT TRANSAXLE

Understanding the Automatic Stick Shift Transaxle

▶ See Figures 4 and 5

An automatic clutch control three speed transaxle has been available on the Type 1. It is known as the Automatic Stick Shift.

It consists of a three speed gear box connected to the engine through a hydrodynamic torque converter. Between the converter and gearbox is a vacuum-operated clutch, which automatically separates the power flow from the torque converter while in the process of changing gear ratios.

While the torque converter components are illustrated here, the picture is for familiarization purposes only. The unit cannot be serviced. It is a welded unit, and must be replaced as a complete assembly.

The power flow passes from the engine via converter, clutch and gearbox to the final drive, which, as with the conventional gearbox, is located in the center of the transaxle housing.

The converter functions as a conventional clutch for starting and stopping. The shift clutch serves only for engaging and changing the speed ranges. Frictionwise, it is very lightly loaded.

There is an independent oil supply for the converter provided by an engine driven pump and a reservoir. The converter oil pump, driven off the engine oil pump, draws fluid from the reservoir and drives it around a circuit leading through the converter and back to the reservoir.

This circuit also furnishes cooling for the converter fluid.

OPERATION

The control valve is activated by a very light touch to the top of the shift selector knob which, in turn, is connected to an electromagnet. It has two functions.

At the beginning of the selection process, it has to conduct the vacuum promptly from the intake manifold to the clutch servo, so that the shift clutch disengages at once, and thus interrupts the power flow between converter and transmission. At the end of the selection process, it must, according to driving conditions, automatically ensure that the shift clutch engages at the proper speed. It may neither slip nor engage too harshly. The control valve can be adjusted for this purpose.

As soon as the selector lever is moved to the engaged position, the two contacts in the lever close the circuit. The electromagnet is then under voltage and operates the main valve. By this means the clutch servo is connected to the engine intake manifold, and at the same time the connection to the atmosphere is closed. In the vacuum space of the servo system, a vacuum is built up, the diaphragm of the clutch servo is moved by the difference with atmospheric pressure and the shift clutch is disengaged via its linkage. The power flow to the gearbox is interrupted and the required speed range can be engaged. The process of declutching, from movement of the selector lever up to full separation of the clutch, lasts about 1/10 seconds. The automatic can, therefore, declutch faster than would be possible by means of a foot-operated clutch pedal.

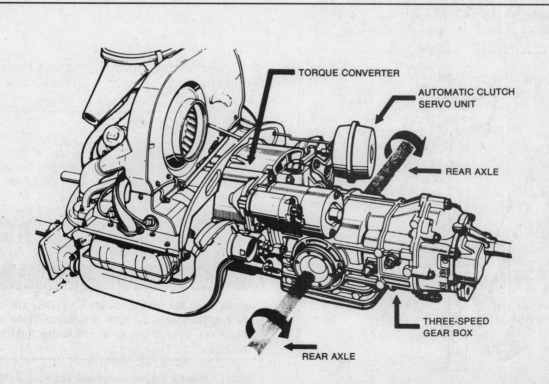

Fig. 4 Basic automatic stick shift transaxle components—note that the transaxle holds the engine in the vehicle

7-6 DRIVE TRAIN

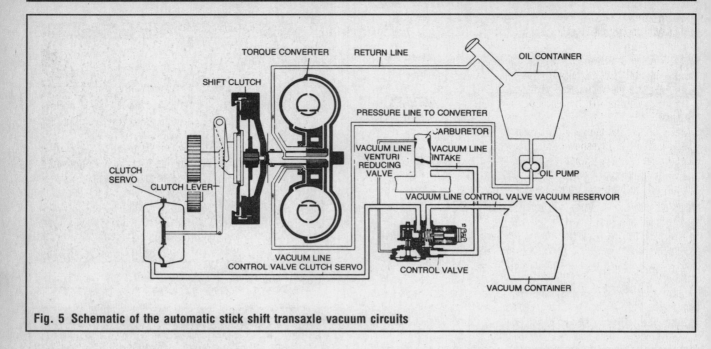

Fig. 5 Schematic of the automatic stick shift transaxle vacuum circuits

When the selector lever is released after changing the speed range, the switch interrupts the current flow to the electromagnet, which then returns to its rest position and closes the main valve. The vacuum is reduced by the reducing valve and the shift clutch re-engages.

Clutch engagement takes place, quickly or slowly, according to engine loading. The clutch will engage suddenly, for example, at full throttle, and can transform the full drive moment into acceleration of the car. Or, this can be effected slowly and gently if the braking force of the engine is to be used on overrun. In the part-load range, too, the duration of clutch re-engagement depends on the throttle opening, and thus the vacuum in the carburetor venturi.

Vanes on the outside of the converter housing aid in cooling. In the case of abnormal prolonged loading, however (lugging a trailer over mountain roads in second or third speed), converter heat may exceed maximum permissible temperature. This condition will cause a red warning light to function in the speedometer.

There is also a starter locking switch. This, combined with a bridging switch, is operated by the inner transmission shift lever. It performs two functions:

1. With a speed range engaged, the electrical connection to the starter is interrupted. The engine, therefore, can only be started in neutral.
2. The contacts in the selector lever are not closed in the neutral position. Instead, the bridging switch transmits a voltage to the electromagnets of the control valve. This ensures that the spearator clutch is also disengaged in the neutral shifter position.

Transaxle Assembly

REMOVAL & INSTALLATION

1. Disconnect the negative battery cable.
2. Remove the engine.

3. Make a bracket to hold the torque converter in place. If a bracket is not used, the converter will slide off the transmission input shaft.
4. Detach the gearshift rod coupling.
5. Disconnect the drive shafts at the transmission end. If the driveshafts are not going to be repaired, it is not necessary to detach the wheel end.
6. Disconnect the drive ATF hoses from the transmission. Seal the open ends. Disconnect the temperature switch, neutral safety switch, and the back-up light switch.
7. Pull off the vacuum servo hose.
8. Disconnect the starter wiring.
9. Remove the front transaxle mounting nuts.
10. Loosen the rear transaxle mounting bolts. Support the transaxle and remove the bolts.
11. Lower the axle and remove it from the car.

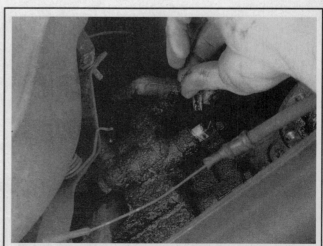

When removing the engine or transaxle, make certain to label and disconnect ALL wires, hoses and cables

DRIVE TRAIN

To install:

12. With the torque converter bracket still in place, raise the axle into the car.

13. Tighten the nuts for the front transmission mounting. Insert the rear mounting bolts, but do not tighten them at this time.

14. Replace the vacuum servo hose.

15. Connect the ATF hoses, using new washers. The washers are seals.

16. Connect the temperature switch and starter cables.

17. Install the driveshafts, using new washers. Turn the convex sides of the washers toward the screw head.

18. Align the transaxle so that the inner drive shaft joints do not rub on the frame fork and then tighten the rear mounting bolts.

19. Insert the shift rod coupling, tighten the screw, and secure it with wire.

20. Remove the torque converter bracket, and install the engine.

21. After installing the engine, bleed the ATF lines if return flow has not started after 2–3 minutes.

SHIFT LINKAGE ADJUSTMENT

The Volkswagen shift linkage is not adjustable. When shifting becomes difficult or there is an excessive amount of play in the linkage, check the shifting mechanism for worn parts. Make sure the shift linkage coupling is tightly connected to the inner shift lever located at the rear of the transaxle under a plate below the rear seat on Types 1, 3 and 4. On Type 2 models, Check the setscrew where the front shift rod connects to the center section below the passenger compartment, and the rear setscrew where the linkage goes into the transaxle. Worn parts may be found in the shift lever mechanism and the supports for the linkage rod sometimes wear out.

The gear shift lever can be removed after the front floor mat has been lifted. After the two retaining screws have been removed from the gear shift lever ball housing, the gear shift lever, ball housing, rubber boot, and spring are removed as a unit.

. . . remove the two 13mm hold-down bolts . . .

. . . then lift the shifter up and out of the center frame tube

After removing the rubber shifter boot . . .

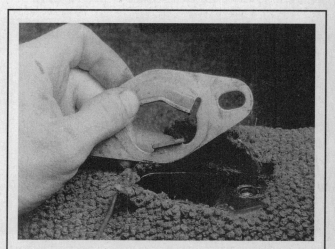

Note the orientation of the shifter plate, then remove it from the center tube

7-8 DRIVE TRAIN

✶✶ CAUTION

Carefully mark the position of the stop plate and note the position of the turned up ramp at the side of the stop plate. Normally the ramp is turned up and on the right-hand side of the hole.

Installation is the reverse of removal. Lubricate all moving parts with grease. Test the gear shift pattern. If there is difficulty in shifting, adjust the stop plate back and forth in its slotted holes.

DRIVESHAFT AND CONSTANT VELOCITY (CV) JOINT

Driveshaft and CV-Joint Assembly

REMOVAL & INSTALLATION

♦ See Figure 6

1. Remove the bolts which secure the joints at each end of the shaft, tilt the shaft down and remove the shaft.

2. Loosen the clamps which secure the rubber boot to the axle and slide the boot back on the axle.
3. Drive the stamped steel cover off of the joint with a drift.

➡ After the cover is removed, do not tilt the ball hub as the balls will fall out of the hub.

4. Remove the circlip from the end of the axle and press the axle out of the joint.
5. Installation is the reverse of the removal procedure. The po-

After raising the vehicle and supporting it safely on jackstands, remove the 6 outer driveshaft mounting bolts—these bolts require a special 12-point tool

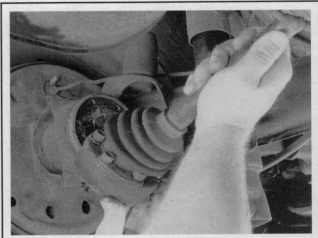

. . . lower the outer end of the driveshaft from its mounting flange . . .

Remove the inner 6 driveshaft-to-transaxle mounting bolts . . .

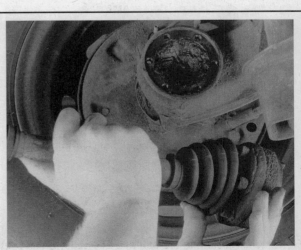

. . . then lower the driveshaft assembly out of the vehicle

DRIVE TRAIN 7-9

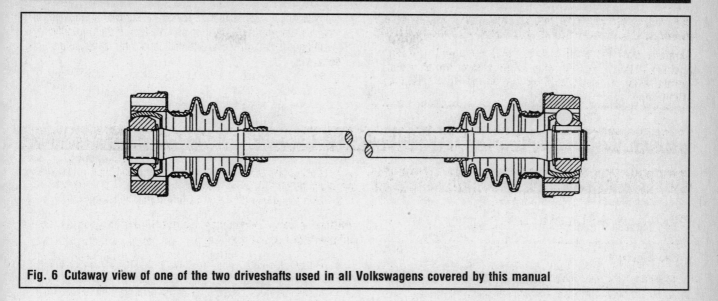

Fig. 6 Cutaway view of one of the two driveshafts used in all Volkswagens covered by this manual

sition of the dished washer is dependent on the type of transmission. On automatic transmissions, it is placed between the ball hub and the circlip. On manual transmissions, it is placed between the ball hub and the shoulder on the shaft. Be sure to pack the joint with grease.

➡ The chamfer on the splined inside diameter of the ball hub faces the shoulder on the driveshaft.

CONSTANT VELOCITY JOINT OVERHAUL

▸ See Figures 7 and 8

The Constant Velocity joint (CV-joint) must be disassembled to remove and replace old grease and to inspect. The individual pieces of the CV joint are machined matched; in the event that some part of the joint is bad, the entire joint (not including the axle shaft) must be replaced.

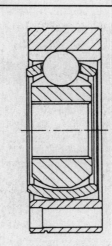

Fig. 8 Cross-sectional view of a CV-joint—essentially all of the CV-joints used on the models covered by this manual are disassembled and assembled in the same manner

1. Remove the CV-joint from the axle shaft, as described earlier in this section.
2. Pivot the ball hub and ball cage out of the joint until they are at a ninety degree angle to the case, and pull the ball hub and cage out as an assembly.
3. Press the balls out of the cage.
4. Align the two grooves and take the ball hub out of the cage.
5. Check the outer race, ball hub, ball cage and balls for wear and pitting. Check the ball cage for hairline cracks. Signs of polishing indicating the tracks of the ball bearings is no reason for replacement.

To assemble:

6. Fit the ball hub into the cage. It doesn't matter which way the hub is installed. Press in the ball bearings.

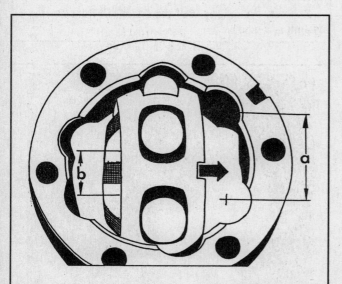

Fig. 7 When installing the ball hub and cage, insert it at a 90 degree angle and rotate it into the CV-joint in the direction of the arrow

7-10 DRIVE TRAIN

7. Fit the hub assembly into the outer race at a ninety degree angle to the outer race. When inserting, make sure that a wide separation between ball grooves on the outer race ("a" in the illustration) and a narrow separation between ball grooves on the hub ("b" in the illustration) are together when the hub and cage assembly is swung into the outer race (direction of arrow in illustration).

8. When pivoting the ball hub and cage assembly into the outer race, the hub should be pivoted out of the cage so that the balls spread apart enough to fit into the ball grooves.

9. Press the cage firmly down until the hub swings fully into position.

10. Pack the unit completely with axle grease. The joint is properly assembled when the ball hub can be moved over the full range of axial movement by hand.

CLUTCH

Understanding the Clutch

▸ See Figures 9 and 10

The clutch used in all models is a single dry disc mounted on the flywheel with a diaphragm spring type pressure plate. The release bearing is the ball bearing type and does not require lubrication. On Types 1, 2 (except Vanagon), and 3, the clutch is engaged mechanically via a cable which attaches to the clutch pedal. On the Type 4 and the Vanagon, the clutch is engaged hydraulically, using a clutch pedal operated master cylinder and a bell housing mounted slave cylinder.

❋❋ CAUTION

The clutch driven disc may contain asbestos, which has been determined to be a cancer causing agent. Never clean clutch surfaces with compressed air! Avoid inhaling any dust from any clutch surface! When cleaning clutch surfaces, use a commercially available brake cleaning fluid.

The purpose of the clutch is to disconnect and connect engine power at the transaxle. A vehicle at rest requires a lot of engine torque to get all that weight moving. An internal combustion engine does not develop a high starting torque (unlike steam engines) so it must be allowed to operate without any load until it builds up enough torque to move the vehicle. Torque increases with engine rpm. The clutch allows the engine to build up torque by physically disconnecting the engine from the transaxle, relieving the engine of any load or resistance.

The transfer of engine power to the transaxle (the load) must be smooth and gradual; if it weren't, drive line components would wear out or break quickly. This gradual power transfer is made possible by gradually releasing the clutch pedal. The clutch disc and pressure plate are the connecting link between the engine and transaxle. When the clutch pedal is released, the disc and plate contact each other (the clutch is engaged) physically joining the engine and transaxle. When the pedal is pushed inward, the disc and plate separate (the clutch is disengaged) disconnecting the engine from the transaxle.

Most clutches utilize a single plate, dry friction disc with a diaphragm-style spring pressure plate. The clutch disc has a splined hub which attaches the disc to the input shaft. The disc has friction material where it contacts the flywheel and pressure plate. Torsion springs on the disc help absorb engine torque pulses. The pressure plate applies pressure to the clutch disc, holding it tight against the surface of the flywheel. The clutch operating mechanism consists of a release bearing, fork and cylinder assembly.

The release fork and actuating linkage transfer pedal motion to the release bearing. In the engaged position (pedal released) the diaphragm spring holds the pressure plate against the clutch disc, so engine torque is transmitted to the input shaft. When the clutch pedal is depressed, the release bearing pushes the diaphragm spring center toward the flywheel. The diaphragm spring pivots the fulcrum, relieving the load on the pressure plate. Steel spring straps riveted to the clutch cover lift the pressure plate from the clutch disc, disengaging the engine drive from the transaxle and enabling the gears to be changed.

The clutch is operating properly if:

1. It will stall the engine when released with the vehicle held stationary.
2. The shift lever can be moved freely between 1st and reverse gears when the vehicle is stationary and the clutch disengaged.

Driven Disc and Pressure Plate

REMOVAL & INSTALLATION

▸ See Figures 11 and 12

Manual Transaxle

1. Remove the engine.
2. Remove the pressure plate securing bolts one turn at a time until all spring pressure is released.
3. Remove the bolts and remove the clutch assembly.

➡**Notice which side of the clutch disc faces the flywheel and install the new disc in the same direction.**

4. Before installing the new clutch, check the condition of the flywheel. It should not have excessive heat cracks and the friction surface should not be scored or warped. Check the condition of the throwout bearing. If the bearing is worn, replace it.
5. Lubricate the pilot bearing in the end of the crankshaft with grease.
6. Insert a pilot shaft, used for centering the clutch disc, through the clutch disc and place the disc against the flywheel. The pilot shaft will hold the disc in place.
7. Place the pressure plate over the disc and loosely install the bolts.

➡**Make sure the correct side of the clutch disc is facing outward. The disc will rub the flywheel if it is incorrectly positioned.**

8. After making sure that the pressure plate aligning dowels will fit into the pressure plate, gradually tighten the bolts.
9. Remove the pilot shaft and reinstall the engine.
10. Adjust the clutch pedal free-play.

DRIVE TRAIN 7-11

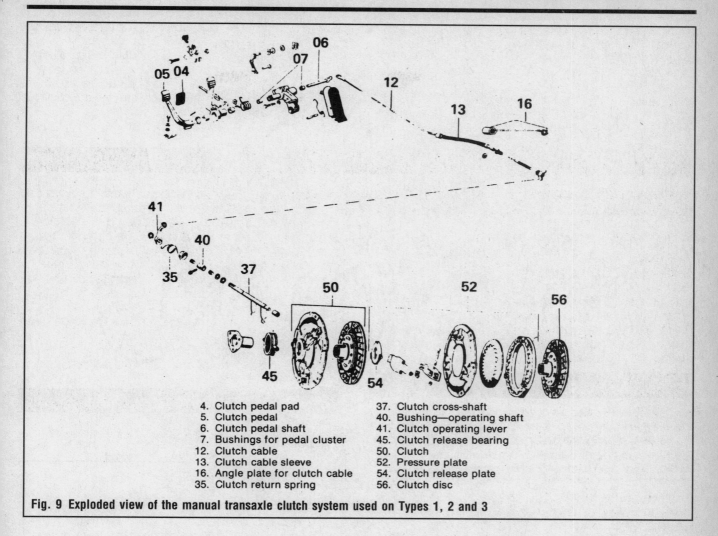

4. Clutch pedal pad
5. Clutch pedal
6. Clutch pedal shaft
7. Bushings for pedal cluster
12. Clutch cable
13. Clutch cable sleeve
16. Angle plate for clutch cable
35. Clutch return spring
37. Clutch cross-shaft
40. Bushing—operating shaft
41. Clutch operating lever
45. Clutch release bearing
50. Clutch
52. Pressure plate
54. Clutch release plate
56. Clutch disc

Fig. 9 Exploded view of the manual transaxle clutch system used on Types 1, 2 and 3

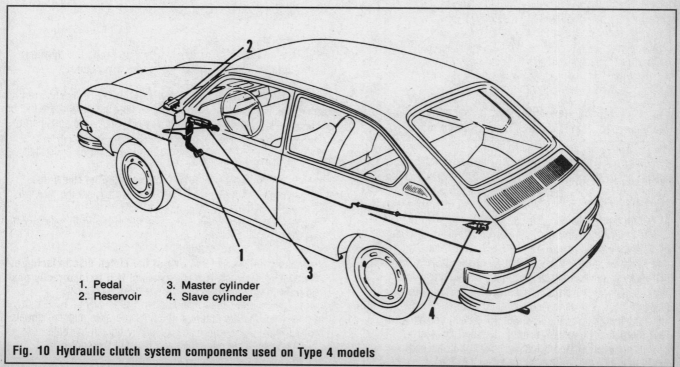

1. Pedal
2. Reservoir
3. Master cylinder
4. Slave cylinder

Fig. 10 Hydraulic clutch system components used on Type 4 models

7-12 DRIVE TRAIN

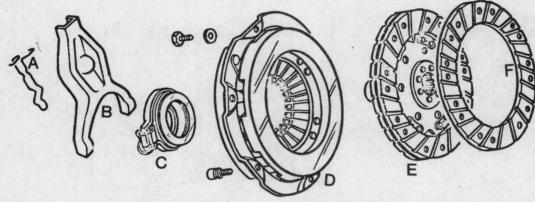

A. Retaining spring
B. Clutch operating lever
C. Release bearing
D. Pressure plate
E. Clutch plate
F. Clutch plate lining

Fig. 11 Exploded view of the clutch assembly (driven disc, pressure plate and related components) used in Type 4 models—note that the clutch face (F) is normally attached to the clutch plate (E)

Automatic Stick Shift
◆ See Figures 13 and 14

1. Disconnect the negative battery cable.
2. Remove the engine.
3. Remove the transaxle.
4. Remove the torque converter by sliding it off of the input shaft. Seal off the hub opening.
5. Mount the transaxle in a repair stand or on a suitable bench.
6. Loosen the clamp screw and pull off the clutch operating lever. Remove the transaxle cover.
7. Remove the hex nuts between the clutch housing and the transaxle case.

➡Two nuts are located inside the differential housing.

8. The oil need not be drained if the clutch is removed with the cover opening up and the gearshift housing breather blocked.
9. Pull the transaxle from the clutch housing studs.
10. Turn the clutch lever shaft to disengage the release bearing.
11. Remove both lower engine mounting bolts.
12. Loosen the clutch retaining bolts gradually and alternately to prevent distortion. Remove the bolts, pressure plate, clutch plate, and release bearing.
13. Do not wash the release bearing. Wipe it dry only.
14. Check the clutch plate, pressure plate, and release bearing

DRIVE TRAIN 7-13

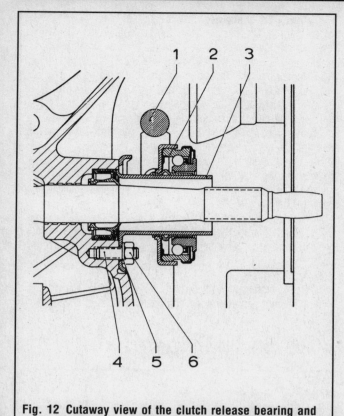

Fig. 12 Cutaway view of the clutch release bearing and guide sleeve—the guide sleeve helps with the release bearing engagement

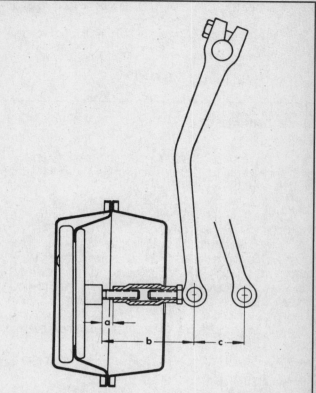

Fig. 13 Adjust the automatic stick shift clutch to the 3 dimensions shown (a, b and c)—a should equal 0.335 in., b should equal 3.03 in. and c should be adjusted to 1.6 in.

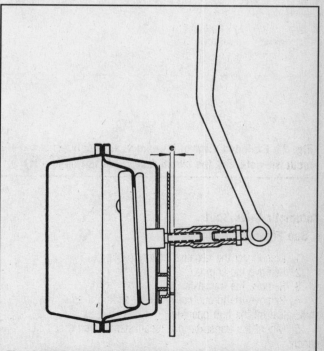

Fig. 14 When checking the clutch adjustment on automatic stick shift transaxles, e should be less than 0.16 in.

for wear and damage. Check the clutch carrier plate, needle bearing, and seat for wear. Replace the necessary parts.

15. If the clutch is wet with ATF, replace the clutch carrier plate seal and the clutch disc. If the clutch is wet with transmission oil, replace the transaxle case seal and clutch disc.

16. Coat the release bearing guide on the transaxle case neck and both lugs on the release bearing with grease. Insert the bearing into the clutch.

17. Grease the carrier plate needle bearing. Install the clutch disc and pressure plate using a pilot shaft to center the disc on the flywheel.

18. Tighten the pressure plate retaining bolts evenly and alternately. Make sure that the release bearing is correctly located in the diaphragm spring.

19. Insert the lower engine mounting bolts from the front. Replace the sealing rings if necessary. Some units have aluminum sealing rings and cap nuts.

20. Push the transaxle onto the converter housing studs. Insert the clutch lever shaft behind the release bearing lugs. Push the release bearing onto the transaxle case neck. Tighten the bolts which hold the clutch housing to the transaxle case.

21. Install the clutch operating lever.

22. It is necessary to adjust the basic clutch setting. The clutch operating lever should contact the clutch housing. Tighten the lever clamp screw slightly.

23. First adjust dimension (a) to 0.335 in. Adjust dimension (b) to 3.03 in. Finally adjust dimension (c) to 1.6 in. by repositioning the clutch lever on the clutch shaft. Tighten the lever clamp screw.

7-14 DRIVE TRAIN

24. Push the torque converter onto the support tube. Insert it into the turbine shaft by turning the converter.
25. Check the clutch play after installing the transaxle and engine.

CLUTCH CABLE ADJUSTMENT

Manual Transaxle

1. Check the clutch pedal travel by measuring the distance the pedal travels toward the floor until pressure is exerted against the clutch. The distance is 3/8 to 3/4 in.
2. To adjust the clutch, jack up the rear of the car and support it on jackstands.
3. Remove the left rear wheel.
4. Adjust the cable tension by turning the wing nut on the end of the clutch cable. Turning the wingnut counterclockwise decreases pedal free-play, turning it clockwise increases free-play.
5. When the adjustment is completed, the wings of the wingnut must be horizontal so that the lugs on the nut engage the recesses in the clutch lever.
6. Push on the clutch pedal several times and check the pedal free-play.
7. Install the wheel and lower the car.

Automatic Stick Shift

The adjustment is made on the linkage between the clutch arm and the vacuum servo unit. **To check the clutch play:**
1. Disconnect the servo vacuum hose.
2. Measure the clearance between the upper edge of the servo unit mounting bracket and the lower edge of the adjusting turnbuckle. If the clearance (e) is 0.16 in. or more, the clutch needs adjustment.
3. Reconnect the vacuum hose.

To adjust the clutch:
1. Disconnect the servo vacuum hose.
2. Loosen the turnbuckle locknut and back it off completely to the lever arm. Then turn the servo turnbuckle against the locknut. Now back off the turnbuckle 5–5½ turns. The distance between the locknut and the turnbuckle should be 0.25 in.
3. Tighten the locknut against the adjusting sleeve.
4. Reconnect the vacuum hose and road test the vehicle. The clutch is properly adjusted when Reverse gear can be engaged silently and the clutch does not slip on acceleration. If the clutch arm contacts the clutch housing, there is no more adjustment possible and the clutch plate must be replaced.

The speed of engagement of the Automatic Stick Shift clutch is regulated by the vacuum operated valve, rather than by the driver's foot. The adjusting screw is on top of the valve under a small protective cap. Adjust the valve as follows:
1. Remove the cap.
2. To slow the engagement, turn the adjusting screw ¼–½ turn clockwise. To speed engagement, turn the screw counterclockwise.
3. Replace the cap.
4. Test operation by shifting from second gear to first gear at 44 mph without depressing the accelerator. The shift should take exactly one second to occur.

CLUTCH CABLE REPLACEMENT

Types 1, 2, and 3
▶ See Figure 15

1. Jack up the car and remove the left rear wheel.
2. Disconnect the cable from the clutch operating lever.
3. Remove the rubber boot from the end of the guide tube and off the end of the cable.
4. On Type 1, unbolt the pedal cluster and remove it from the car. It will also be necessary to disconnect the brake master cylinder pushrod and throttle cable from the pedal cluster. On Type 2, remove the cover under the pedal cluster, then remove the pin from the clevis on the end of the clutch cable. On Type 3, remove

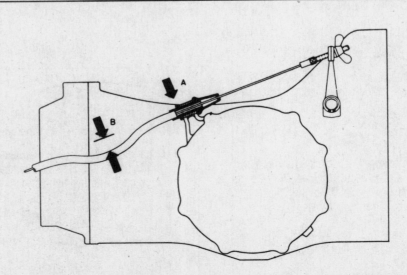

Fig. 15 For smooth clutch action on Type 1 models, dimension B should equal 1.0–1.7 in.—to adjust the cable to provide the proper amount of sag at B, install or remove washers from point A

DRIVE TRAIN

the frame head cover and remove the pin from the clevis on the end of the clutch cable.

5. Pull the cable out of its guide tube from the pedal cluster end.

6. Installation is the reverse of the removal procedure.

→ **Grease the cable before installing it and readjust the clutch pedal free-play.**

Clutch Master Cylinder

REMOVAL & INSTALLATION

Type 4

1. Siphon the hydraulic fluid from the master cylinder (clutch) reservoir.
2. Pull back the carpeting from the pedal area and lay down some absorbent rags.
3. Pull the elbow connection from the top of the master cylinder.
4. Disconnect and plug the pressure line from the rear of the master cylinder.
5. Remove the master cylinder mounting bolts and remove the cylinder to the rear.
6. Installation is the reverse of the removal procedure. Take care to bleed the system and adjust pedal free-play.

Type 2 (Vanagon)
♦ See Figure 16

1. Remove the lower front bulkhead paneling.
2. Remove the cover over the brake master cylinder by grasping the top cover to the instrument cover at the two notches at the front and pulling it off.

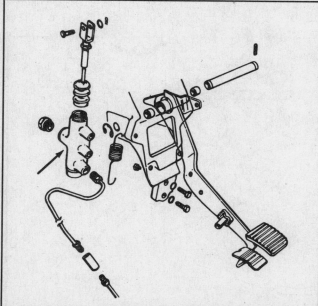

Fig. 16 Exploded view of the mounting for the hydraulic clutch master cylinder used on Vanagon models (late Type 2)

3. Disconnect and plug the reservoir hose and the clutch line from the master cylinder.
4. Disconnect the clevis pin on the master cylinder pushrod.
5. Remove the two bolts and remove the master cylinder.
6. Installation is the reverse of removal. When attaching the pushrod, play between the pushrod and the piston in the master cylinder must be 0.020 in. maximum. Adjust if necessary.

Clutch Slave Cylinder

REMOVAL & INSTALLATION

Type 4

1. Locate the slave cylinder on the bell housing.
2. Disconnect and plug the pressure line from the slave cylinder.
3. Disconnect the return spring from the pushrod.
4. Remove the retaining circlip from the boot and remove the boot.
5. Remove the circlip and slide the slave cylinder rearwards from its mount.
6. Remove the spring clip from the mount.
7. Installation is the reverse of the removal procedure. Take care to bleed the system and adjust pedal free-play.

Type 2 (Vanagon)
♦ See Figure 17

The slave cylinder is mounted on a bracket on the left side of the transaxle.

1. Jack up the vehicle and support it on stands.
2. Disconnect the clutch line and plug it to prevent fluid leakage.
3. Remove the pushrod from the clutch lever socket ball.
4. Remove the two bolts and nuts and remove the slave cylinder.

Installation is the reverse of removal. The bleeder screw is located at the top of the cylinder.

CLUTCH SYSTEM BLEEDING & ADJUSTMENT

Type 4 and Vanagon
♦ See Figures 18 and 19

→ **Perform steps 1–4 only for Vanagon models.**

Whenever air enters the clutch hydraulic system due to leakage, or if any part of the system is removed for service, the system must be bled. The hydraulic system uses high quality brake fluid meeting SAE J1703 or DOT 3 or DOT 4 specifications. Brake fluid is highly corrosive to paint finishes and care should be exercised that no spillage occurs. The procedure is as follows:

1. Top up the clutch fluid reservoir and make sure the cap vent is open.
2. Locate the slave cylinder bleed nipple and remove all dirt and grease from the valve. Attach a hose to the nipple and submerge the other end of the hose in a jar containing a few inches of clean brake fluid.
3. Find a friend to operate the clutch pedal. When your friend

7-16 DRIVE TRAIN

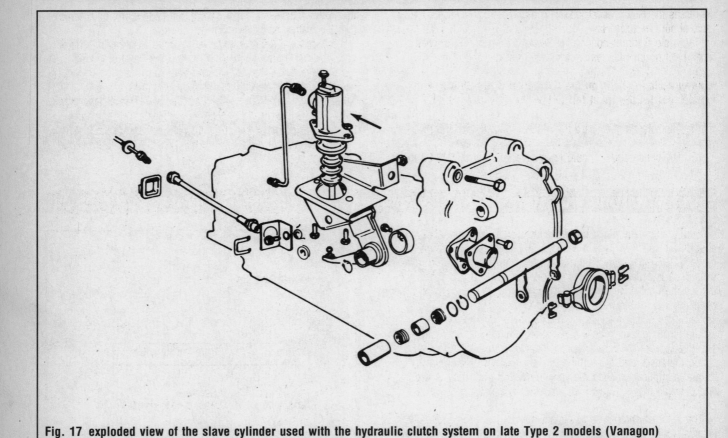

Fig. 17 exploded view of the slave cylinder used with the hydraulic clutch system on late Type 2 models (Vanagon)

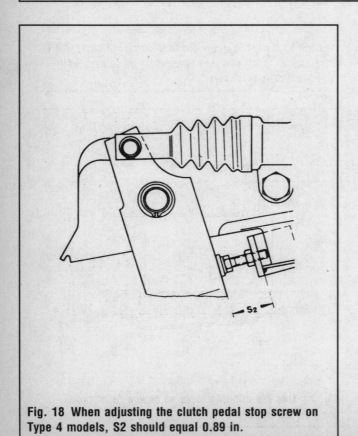

Fig. 18 When adjusting the clutch pedal stop screw on Type 4 models, S2 should equal 0.89 in.

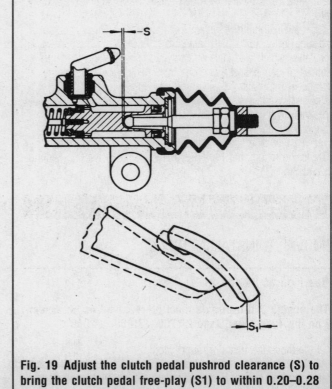

Fig. 19 Adjust the clutch pedal pushrod clearance (S) to bring the clutch pedal free-play (S1) to within 0.20–0.28 in.

DRIVE TRAIN 7-17

depresses the clutch pedal slowly to the floor, open the bleeder valve about one turn. Have your friend keep the pedal on the floor until you close the bleeder valve. Repeat this operation several times until no air bubbles are emitted from the tube.

➡ **Keep a close check on the fluid level in the fluid reservoir. Never let the level fall below the ½ full mark.**

4. After bleeding, discard the old fluid and top off the reservoir.
5. The clutch pedal should have a free-play of 0.20–0.28 in., and a 7 in. total travel. If either of the above are not to specifications, adjust the master cylinder as follows:
   a. Loosen the master cylinder pushrod locknut and shorten the pushrod length slightly.
   b. Loosen the master cylinder bolts and push the cylinder as far forward as it will go. Retighten the bolts.
   c. Remove the rubber cap from the clutch pedal stop screw and adjust distance S2 to 0.89 in. Install the rubber cap.
   d. Then lengthen the pushrod as necessary to obtain a pedal free-play of 0.20–0.28 in. Tighten the pushrod locknut.
10. Road test the car.

FULLY AUTOMATIC TRANSAXLE

A fully automatic transaxle is available on Type 3 models, available on all 1971–74 Type 4 models, and 1973 and later Type 2 models. 1976–81 Type 2's are equipped with a different transaxle which is basically the same as the one used in the VW Rabbit, although the final drive section is the same as the one used on all other models in this book. Both units consist of an automatically shifted three speed planetary transaxle and torque converter.

The torque converter is a conventional three-element design. The three elements are an impeller (driving member), a stator (reaction member), and the turbine (driven member). Maximum torque multiplication, with the vehicle starting from rest, is two and one-half to one. Maximum converter efficiency is about 96 percent.

The automatic transaxle is a planetary unit with three forward speeds, which engage automatically depending on engine loading and road speed. The converter, planetary unit, and control system are incorporated together with the final drive in a single housing. The final drive is located between the converter and the planetary gearbox.

The transaxle control system includes a gear type oil pump, a centrifugal governor, which regulates shift points, a throttle modulator valve, which evaluates engine loading according to intake manifold pressure, and numerous other regulating components assembled in the transaxle valve body.

Power flow passes through the torque converter to the turbine shaft, then to the clutch drum attached to the turbine shaft, through a clutch to a sungear. The output planet carrier then drives the rear axle shafts via the final drive.

Transaxle ranges are Park, Reverse, Neutral, Drive (3), Second (2), and First (1).

Transaxle Assembly

REMOVAL & INSTALLATION

▶ See Figures 20 and 21

➡ **The engine and transaxle must be removed as an assembly on the Type 4 and Type 2/1700, 2/1800, 2/2000.**

1. Remove the battery ground cable.
2. On the sedan, remove the cooling air intake duct with the heating fan and hoses. Remove the cooling air intake connection and bellows, then detach the hoses to the air cleaner.

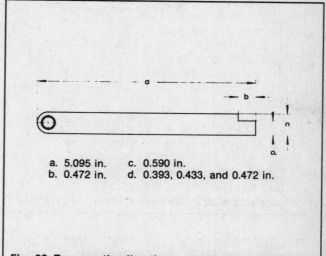

a. 5.095 in. c. 0.590 in.
b. 0.472 in. d. 0.393, 0.433, and 0.472 in.

Fig. 20 To correctly align the automatic transaxle during installation, three aligning tools must be fabricated to the specifications shown

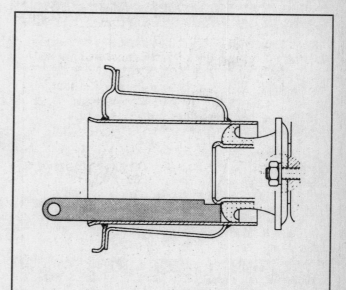

Fig. 21 Use the aligning tools as shown to correctly position the transaxle

DRIVE TRAIN

3. On station wagons, remove the warm air hoses and air cleaner. Remove the boot from between the dipstick tube and the body and the boot from between the oil filler neck and the body. Disconnect the cooling air bellows at the body.

4. Disconnect the wires at the regulator and the alternator wires at the snap connector located by the regulator. Disconnect the auxiliary air regulator and the oil pressure switch at the snap connectors located by the distributor.

5. Disconnect the fuel injection wiring on Type 3 and 4 models. There are 12 connections and they are identifiable as follows:
   a. Fuel injector cylinder 2: 2-pole, protective gray cap.
   b. Fuel injector cylinder 1: 2-pole, protective black cap.
   c. Starter: 1-pole, white.
   d. Throttle valve switch: 4-pole.
   e. Distributor: 3 pole.
   f. Thermo switch: 1-pole, white.
   g. Cold start valve: 3-pole.
   h. Temperature sensor crankcase: 2-pole.
   i. Ground connection: 3-pole, white wires.
   j. Temperature sensor for the cylinder head: 1-pole.
   k. Fuel injector cylinder 3: 2-pole, protective black cap.
   l. Fuel injector cylinder 4: 2-pole, protective gray cap.

6. Disconnect the accelerator cable.
7. Disconnect the right fuel return line.
8. Raise the car.
9. Disconnect the warm hoses from the heat exchangers.
10. Disconnect the starter wires and push the engine wiring harness through the engine cover plate.
11. Disconnect the fuel supply line and plug it.
12. Remove the heater booster exhaust pipe.
13. Remove the rear axles and cover the ends to protect them from dirt.
14. Remove the selector cable by unscrewing the cable sleeve.
15. Remove the wire from the kickdown switch.
16. Remove the bolts from the rubber transaxle mountings, taking careful note of the position, number, and thickness of the spacers that are present.

※※ CAUTION

These spacers must be reinstalled exactly as they were removed. Do not detach the transaxle carrier from the body.

17. Support the engine and transaxle assembly in such a way that it may be lowered and moved rearward at the same time.
18. Remove the engine carrier bolts and the engine and transaxle assembly from the car.

➡ **The top carrier bolts on the Type 2 are the top engine-to-transaxle bolts. Be sure to support the engine/transaxle assembly before completely removing the bolts.**

19. Matchmark the flywheel to the torque converter and remove the three attaching bolts.
20. Remove the engine-to-transaxle bolts and separate the engine and transaxle.

※※ CAUTION

Exercise care when separating the engine and transaxle, as the torque converter will easily slip off the input shaft if the transaxle is tilted downward.

To install:
21. Install and tighten the engine-to-transaxle bolts after aligning the matchmarks on the flywheel and converter.
22. Making sure the matchmarks are aligned, install the converter-to-flywheel bolts.
23. Make sure the rubber buffer is in place and the two securing studs do not project more than 0.7 in. from the transaxle case.
24. Tie a cord to the slot in the engine compartment seal. This will make positioning the seal easier.
25. Lift the assembly far enough to allow the accelerator cable to be pushed through the front engine cover.
26. Continue lifting the assembly into place. Slide the rubber buffer into the locating tube in the rear axle carrier.
27. Insert the engine carrier bolts and raise the engine until the bolts are at the top of their elongated slots. Tighten the bolts.

➡ **A set of three gauges must be obtained to check the alignment of the rubber buffer in its locating tube. The dimensions are given in the illustration as is the measuring technique. The rubber buffer is centered horizontally when the 11mm gauge can be inserted on both sides. The buffer is located vertically when the 10mm gauge can be inserted on the bottom side and the 12mm gauge can be inserted on the top side. See Steps 28 and 29 for adjustment procedures.**

28. Install the rubber transaxle mount bolts with spacers of the correct thickness. The purpose of the spacers is to center the rubber buffer vertically in its support tube. The buffer is not supposed to carry any weight; it absorbs torsional forces only.
29. To locate the buffer horizontally in its locating tube, the engine carrier must be vertical and parallel to the fan housing. It is adjusted by moving the engine carrier bolts in elongated slots. Further travel may be obtained by moving the brackets attached to the body. It may be necessary to adjust the two rear suspension wishbones with the center of the transaxle after the rubber buffer is horizontally centered. Take the car to a dealer or alignment specialist to align the rear suspension.
30. Adjust the selector level cable.
31. Connect the wire to the kickdown switch.
32. Install the rear axles. Make sure the lockwashers are placed with the convex side out.
33. Reconnect the fuel hoses and heat exchanger hoses. Install the pipe for the heater booster.
34. Lower the car and pull the engine compartment seal into place with the cord.
35. Reconnect the fuel injection and engine wiring. Push the starter wires through the engine cover plate and connect the wires to the starter.
36. Install the intake duct with the fan and hoses, as well as the cooling air intake.

Fluid Pan

REMOVAL & INSTALLATION

1. Some models have a drain plug in the pan. Remove the plug and drain the transmission oil. On models without the plug,

DRIVE TRAIN 7-19

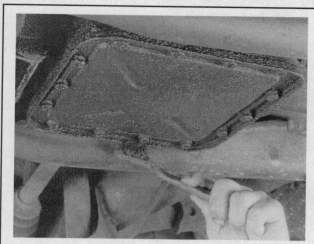

To remove the transaxle fluid pan, first clean around the mounting bolts . . .

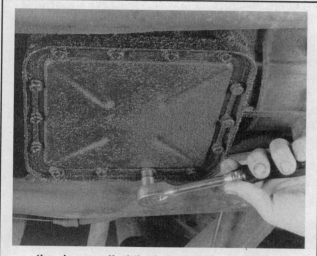

. . . then loosen all of the bolts 2 or 3 revolutions

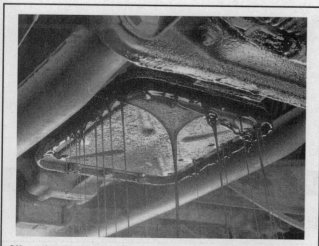

Allow the fluid to drain out of the case—make sure to have a large drain pan and lots of rags handy

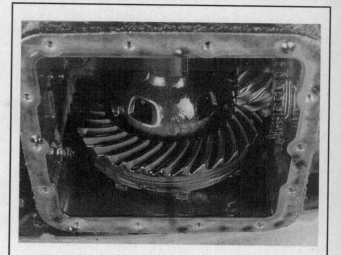

Remove the fluid pan and clean the gasket surfaces before installation—once the pan is removed, the differential can be seen

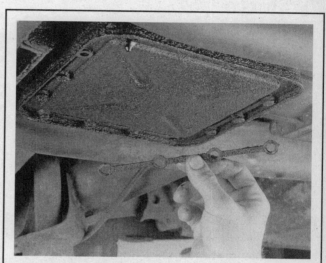

When installing the pan, make sure to install the special force-distributing hold-down to ensure proper fluid pan sealing

loosen the pan bolts 2–3 turns and lower one corner of the pan to drain the oil.

2. Remove the pan bolts and remove the pan from the transaxle.

➡ It may be necessary to tap the pan with a rubber hammer to loosen it.

3. Use a new gasket and install the pan. Tighten the bolts loosely until the pan is properly in place, then tighten the bolts fully, moving in a diagonal pattern.

➡ Do not overtighten the bolts.

4. Refill the transaxle with ATF.
5. At 5 minute intervals, retighten the pan bolts two or three times.

DRIVE TRAIN

FILTER SERVICE

The Volkswagen automatic transaxle has a filter screen secured by a screw to the bottom of the valve body. Remove the pan and remove the filter screen from the valve body.

> **✳✳ CAUTION**
>
> **Never use a cloth that will leave the slightest bit of lint in the transaxle when cleaning transaxle parts. The lint will expand when exposed to transaxle fluid and clog the valve body and filter.**

Clean the filter screen with compressed air. 1976–81 Type 2s have a non-cleanable filter which must be replaced with a new one if the ATF fluid is very dirty.

Adjustments

FRONT (SECOND) BAND

▶ See Figure 22

Except 1976–81 Type 2

Tighten the front band adjusting screw to 7 ft. lbs. Then loosen the screw and tighten it to 3.5 ft. lbs. From this position, loosen the screw exactly 1¾ to 2 turns and tighten the locknut.

1976–81 Type 2

Only the second band is adjustable on this transaxle. The adjuster is located on the passenger side of the transaxle beside the gear selector lever. It is held by a locknut. The transaxle must be horizontal when the band is adjusted or the band could jam. Tighten the adjusting screw to 7 ft. lbs., then loosen it and tighten it to 3.5 ft. lbs. From this position loosen it exactly 2½ turns and tighten the locknut.

REAR (FIRST) BAND

Except 1976–81 Type 2

▶ See Figure 23

Tighten the rear band adjusting screw to 7 ft. lbs. Then loosen the screw and retighten it to 3.5 ft. lbs. From this position, loosen the screw exactly 3¼ to 3½ turns and tighten the locknut.

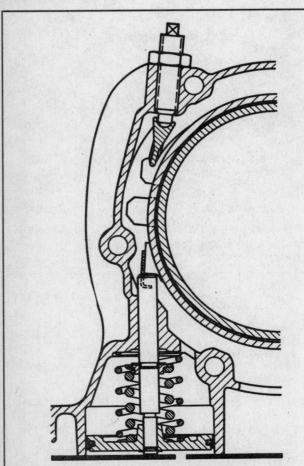

Fig. 22 Cross-sectional view of the transaxle unit, showing the front band and drum—the adjusting screw is at the top of the illustration

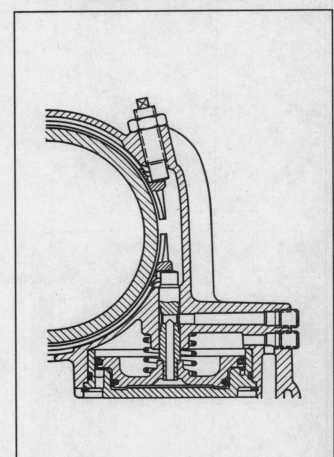

Fig. 23 the rear band is positioned in the same manner as the front band, with the adjusting screw at the top

DRIVE TRAIN 7-21

KICKDOWN SWITCH

Except 1976–81 Type 2 and All Type 4

1. Disconnect the accelerator cable return spring.
2. Move the throttle to the fully open position. Adjust the accelerator cable to give 0.02–0.04 in. clearance between the stop and the end of the throttle valve lever.
3. When the accelerator cable is adjusted and the throttle is moved to the fully open position, the kickdown switch should click. The ignition switch must be **ON** for this test.
4. To adjust the switch, loosen the switch securing screws and slide the switch back and forth until the test in Step 3 is satisfied.
5. Reconnect the accelerator cable return spring.

1976–81 Type 2
▶ See Figure 24

1. Press the accelerator pedal fully to the floor (kickdown position).
2. In this position, the play at the operating lever on the transaxle should be 0.04–0.08 in.
3. If not, adjust at the retaining screw on the accelerator pedal.

Type 4

The Type 4 switch is located in the accelerator pedal pod. On some models it can be adjusted by moving it on its mount. Adjust so that the kickdown solenoid in the transaxle can be heard to click when the accelerator is to the floor.

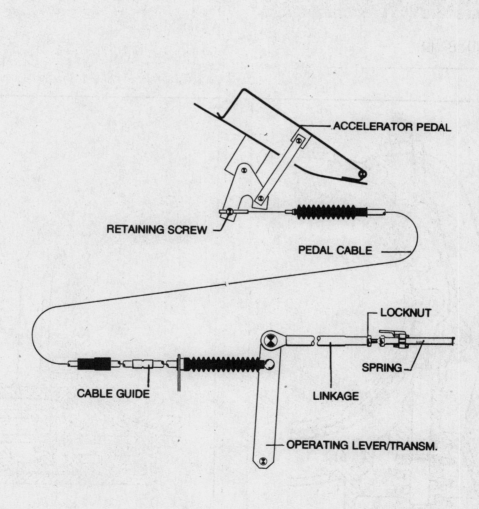

Fig. 24 1976–81 Type 2 kickdown cable and related components for the kickdown adjustment

7-22 DRIVE TRAIN

SHIFT LINKAGE

Make sure the shifting cable is not kinked or bent and that the linkage and cable are properly lubricated.

1. Move the gear shift lever to the Park position.
2. Loosen the clamp which holds the front and rear halves of the shifting rod together. Loosen the clamping bolts on the transaxle lever.
3. Press the lever on the transaxle rearward as far as possible. Spring pressure will be felt. The manual valve must be on the stop in the valve body.
4. Holding the transaxle lever against its stop, tighten the clamping bolt.
5. Holding the rear shifting rod half, push the front half forward to take up any clearance and tighten the clamp bolt.
6. Test the shift pattern.

TORSION BAR FRONT SUSPENSION 8-2
TORSION BAR 8-2
  REMOVAL & INSTALLATION 8-2
TORSION ARM 8-2
  REMOVAL & INSTALLATION 8-2
SHOCK ABSORBER 8-2
  REMOVAL & INSTALLATION 8-2
BALL JOINT 8-2
  INSPECTION 8-2
  REPLACEMENT 8-4
STRUT FRONT SUSPENSION 8-5
STRUT 8-5
  REMOVAL & INSTALLATION 8-5
TRACK CONTROL ARM 8-6
  REMOVAL & INSTALLATION 8-6
SHOCK ABSORBER 8-7
  REMOVAL & INSTALLATION 8-7
BALL JOINT 8-7
  INSPECTION 8-7
  REPLACEMENT 8-7
COIL SPRING FRONT SUSPENSION 8-7
COIL SPRING 8-7
  REMOVAL & INSTALLATION 8-7
SHOCK ABSORBER 8-9
  REMOVAL & INSTALLATION 8-9
BALL JOINT 8-9
  REMOVAL & INSTALLATION 8-9
UPPER CONTROL ARM 8-9
  REMOVAL & INSTALLATION 8-9
LOWER CONTROL ARM 8-9
  REMOVAL & INSTALLATION 8-9
FRONT END ALIGNMENT 8-9
  CASTER ADJUSTMENT 8-9
  CAMBER ADJUSTMENT 8-9
  TOE-IN ADJUSTMENT 8-10
DIAGONAL ARM REAR SUSPENSION 8-12
SHOCK ABSORBER 8-12
  REMOVAL & INSTALLATION 8-12
DIAGONAL ARM 8-12
  REMOVAL & INSTALLATION 8-12
ADJUSTMENTS 8-13
  TYPE 1 8-13
  TYPES 2 (THROUGH 1979) AND 3 8-13
COIL SPRING/TRAILING ARM REAR SUSPENSION 8-13
SHOCK ABSORBER 8-13
  REMOVAL & INSTALLATION 8-13
TRAILING ARM 8-13
  REMOVAL & INSTALLATION 8-13
ADJUSTMENTS 8-15
STEERING 8-15
STEERING WHEEL 8-15
  REMOVAL & INSTALLATION 8-15
TURN SIGNAL SWITCH 8-17
  REMOVAL & INSTALLATION 8-17
IGNITION SWITCH 8-17
  REMOVAL & INSTALLATION 8-17
IGNITION LOCK CYLINDER 8-17
  REMOVAL & INSTALLATION 8-17
STEERING LINKAGE 8-18
  REMOVAL & INSTALLATION 8-18
MANUAL STEERING GEAR 8-19
  ADJUSTMENT 8-19
SPECIFICATION CHART
WHEEL ALIGNMENT
  SPECIFICATIONS 8-11
UNLOADED REAR TENSION BAR
  SETTINGS 8-12

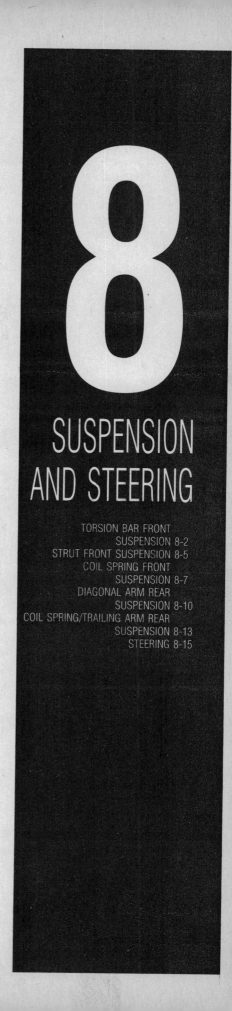

8
SUSPENSION AND STEERING

TORSION BAR FRONT SUSPENSION 8-2
STRUT FRONT SUSPENSION 8-5
COIL SPRING FRONT SUSPENSION 8-7
DIAGONAL ARM REAR SUSPENSION 8-10
COIL SPRING/TRAILING ARM REAR SUSPENSION 8-13
STEERING 8-15

8-2 SUSPENSION AND STEERING

TORSION BAR FRONT SUSPENSION

♦ See Figures 1 and 2

➡ The torsion bar front suspension system is found on Type 1 Beetles, Beetle Convertibles, Karmann Ghias, all Type 3 models and 1970–79 Type 2 models. All other models are equipped with either the strut front suspension system or the coil spring front suspension system, both of which are covered later in this section.

Each front wheel rotates on a ball joint mounted spindle. The spindle is suspended independently by a pair of torsion bars.

The principle of torsion bars is that of springing action taking place via twisting of the bars. When a front wheel goes up or down, the torsion bars are twisted, causing a downward or upward force in the opposite direction.

The supporting part of the Volkswagen front axle is the axle beam, which is two rigidly joined tubes attached to the frame with four screws. At each end of the tubes there is a side plate designed to provide additional strength and serve as the upper mounting point for the shock absorbers. Because the front axle is all-welded, it is replaced as a unit whenever damaged.

Torsion Bar

REMOVAL & INSTALLATION

1. Jack up the car and remove both wheels and brake drums.
2. Remove the ball joint nuts and remove the left and right steering knuckles. A forked ball joint removing tool is available at an auto parts store.

✳✳ CAUTION

Never strike the ball joint stud.

3. Remove the arms attached to the torsion bars on one side only. To remove the arms, loosen and remove the arm setscrew and pull the arm off the end of the torsion bar.
4. Loosen and remove the setscrew which secures the torsion bar to the torsion bar housing.
5. Pull the torsion bar out of its housing.

To install:

6. Carefully note the number of leaves and the position of the countersink marks for the torsion bar and the torsion arm.
7. Align the countersink mark in the center of the bar with the hole for the setscrew and insert the torsion bar into its housing. Install the setscrew. Install the torsion arm.
8. Reverse Steps 1–3 to complete.

Torsion Arm

REMOVAL & INSTALLATION

1. Jack up the car and remove the wheel and tire.
2. Remove the brake drum and the steering knuckle.
3. If the lower torsion arm is being removed, disconnect the stabilizer bar. To remove the stabilizer bar clamp, tap the wedge-shaped keeper toward the outside of the car or in the direction the narrow end of the keeper is pointing.
4. On Types 1 and 2, back off on the setscrew locknut and remove the setscrew. On Type 3's, remove the bolt and keeper from the end of the torsion bar.
5. Slide the torsion arm off the end of the torsion bar.
6. Installation is the reverse of the removal procedure. Check the camber and toe-in settings.

Shock Absorber

REMOVAL & INSTALLATION

1. Remove the wheel and tire.
2. Remove the nut from the torsion arm stud and slide the lower end of the shock off of the stud.
3. Remove the nut from the shock absorber shaft at the upper mounting and remove the shock from the vehicle.
4. The shock is tested by operating it by hand. As the shock is extended and compressed, it should operate smoothly over its entire stroke with an even pressure. Its damping action should be clearly felt at the end of each stroke. If the shock is leaking slightly, the shock need not be replaced. A shock that has had an excessive loss of fluid will have flat spots in the stroke as the shock is compressed and extended. That is, the pressure will feel as though it has been suddenly released for a short distance during the stroke.
5. Installation is the reverse of Steps 1–3.

Ball Joint

INSPECTION

A quick initial inspection can be made with the vehicle on the ground. Grasp the top of the tire and vigorously pull the top of the tire in and out. Test both sides in this manner. If the ball joints are excessively worn, there will be an audible tap as the ball moves around in its socket. Excess play can sometimes be felt through the tire.

A more rigorous test may be performed by jacking the car under the lower torsion arm and inserting a lever under the tire. Lift up gently on the lever so as to pry the tire upward. If the ball joints are worn, the tire will move upward 1/8–1/4 in. or more. If the tire displays excessive movement, have an assistant inspect each joint, as the tire is pried upward, to determine which ball joint is defective.

SUSPENSION AND STEERING 8-3

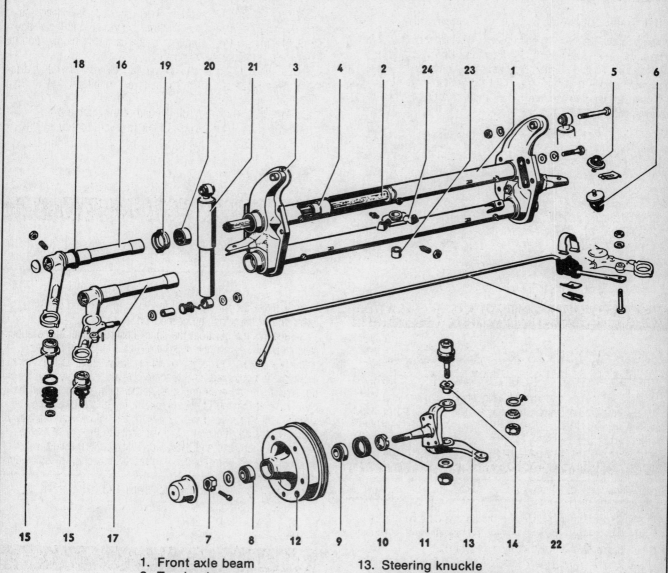

1. Front axle beam
2. Torsion bar
3. Side plate
4. Torsion arm bush
5. Upper rubber buffer
6. Lower rubber buffer
7. Clamp nut for wheel bearing adjustment
8. Outer front wheel bearing
9. Inner front wheel bearing
10. Front wheel bearing seal
11. Spacer ring
12. Brake drum
13. Steering knuckle
14. Eccentric bush for camber adjustment
15. Ball joint
16. Upper torsion arm
17. Lower torsion arm
18. Seal for torsion arm
19. Seal retainer
20. Torsion arm needle bearing
21. Shock absorber
22. Stabilizer
23. Swing lever shaft bush
24. Swing lever stop

Fig. 1 Exploded view of the front torsion bar suspension system used on 1970–79 Type 2 models

8-4 SUSPENSION AND STEERING

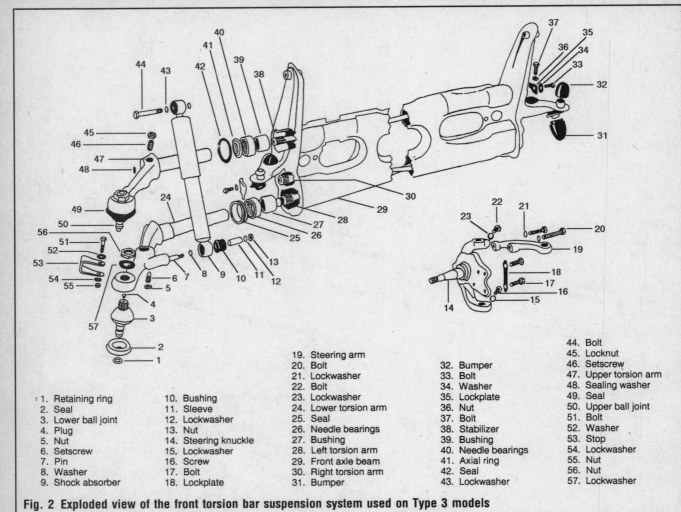

1. Retaining ring
2. Seal
3. Lower ball joint
4. Plug
5. Nut
6. Setscrew
7. Pin
8. Washer
9. Shock absorber
10. Bushing
11. Sleeve
12. Lockwasher
13. Nut
14. Steering knuckle
15. Lockwasher
16. Screw
17. Bolt
18. Lockplate
19. Steering arm
20. Bolt
21. Lockwasher
22. Bolt
23. Lockwasher
24. Lower torsion arm
25. Seal
26. Needle bearings
27. Bushing
28. Left torsion arm
29. Front axle beam
30. Right torsion arm
31. Bumper
32. Bumper
33. Bolt
34. Washer
35. Lockplate
36. Nut
37. Bolt
38. Stabilizer
39. Bushing
40. Needle bearings
41. Axial ring
42. Seal
43. Lockwasher
44. Bolt
45. Locknut
46. Setscrew
47. Upper torsion arm
48. Sealing washer
49. Seal
50. Upper ball joint
51. Bolt
52. Washer
53. Stop
54. Lockwasher
55. Nut
56. Nut
57. Lockwasher

Fig. 2 Exploded view of the front torsion bar suspension system used on Type 3 models

REPLACEMENT

♦ See Figure 3

1. Jack up the car and remove the wheel and tire.
2. Remove the brake drum and disconnect the brake line from the backing plate.
3. Remove the nut from each ball joint stud and remove the ball joint stud from the steering knuckle. Remove the steering knuckle from the car. A ball joint removal tool is available at an auto parts store. Do not strike the ball joint stud.
4. Remove the torsion arm from the torsion bar.
5. Remove the ball joint from the torsion arm by pressing it out.

To install:

6. Press a new ball joint in, making sure that the square notch in the joint is in line with the notch in the torsion arm eye.

➥Ball joints are supplied in different sizes designated by V-notches in the ring around the side of the joint. When replacing a ball joint, make sure that the new part has the same number of V-notches. If it has no notches, the replacement joint should have no notches.

7. Reverse Steps 1–4 to complete the installation.

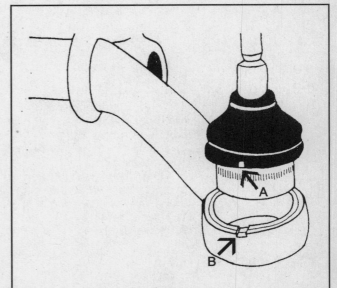

Fig. 3 When installing a new ball joint, make certain to align the notch in the ball joint (A) with the notch in the torsion arm (B)

SUSPENSION AND STEERING 8-5

STRUT FRONT SUSPENSION

♦ See Figure 4

➡ The strut front suspension system is found on Type 1 Super Beetles and Super Beetle Convertibles, and all Type 4 models. All other models are either equipped with the torsion bar front suspension system, which is covered earlier in this section, or the coil spring front suspension system, which is covered later in this section.

Each wheel is suspended independently on a shock absorber strut surrounded by a coil spring. The strut is located at the bottom by a track control arm and a ball joint, and at the top by a ball bearing which is rubber mounted to the body. The benefits of this type of suspension include a wider track, a very small amount of toe-in and camber change during suspension travel, and a reduced turning circle. The strut front suspension requires no lubrication. It is recommended, however, that the ball joint dust seals be checked every 6,000 miles and the ball joint play every 30,000 miles.

Strut

REMOVAL & INSTALLATION

1. Jack up the car and remove the wheel and tire.
2. If the left strut is to be removed, remove the speedometer cable from the steering knuckle.
3. Disconnect the brake line from the bracket on the strut.
4. At the base of the strut, bend down the locking tabs for the three bolts and remove the bolts.
5. Push down on the steering knuckle and pull the strut out of the knuckle.
6. Remove the three nuts which secure the top of the strut to the body. Before removing the last nut, support the strut so that it does not fall out of the car.
7. Installation is the reverse of the removal procedure. Always use new nuts and locking tabs during installation.

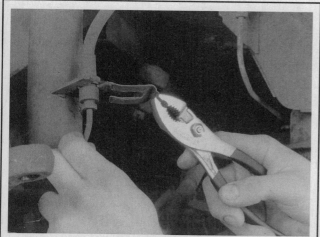

To remove the front strut, first remove the front brake line retaining clip . . .

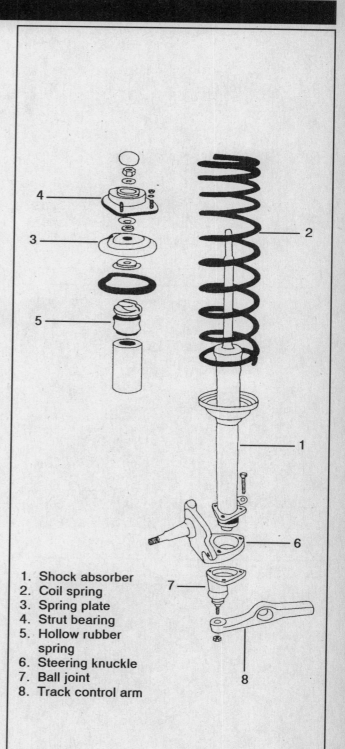

1. Shock absorber
2. Coil spring
3. Spring plate
4. Strut bearing
5. Hollow rubber spring
6. Steering knuckle
7. Ball joint
8. Track control arm

Fig. 4 Exploded view of the front strut and identification of the specific strut components used on Type 1 Super Beetles and Super Beetle Convertibles

8-6 SUSPENSION AND STEERING

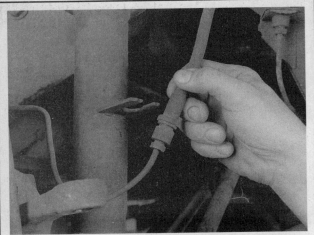

. . . then remove the brake line from the retaining bracket

. . . then pry the lower end of the strut out of the spindle

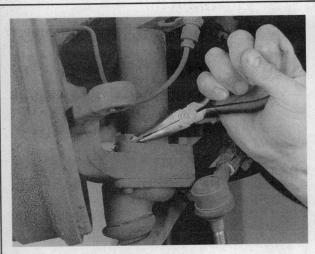

Bend down the attaching bolt tabs . . .

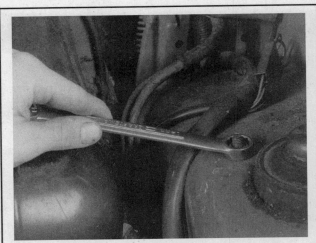

After the lower end of the strut is free of the spindle, remove the 3 upper strut bolts

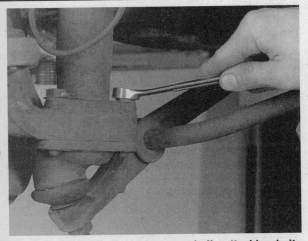

. . . remove the 3 lower strut-to-spindle attaching bolts and . . .

Track Control Arm

REMOVAL & INSTALLATION

1. Remove the ball joint stud nut and remove the stud from the control arm.
2. Disconnect the stabilizer bar from the control arm.
3. Remove the nut and eccentric bolt at the frame. This is the pivot bolt for the control arm and is used to adjust camber.
4. Pull the arm downward and remove it from the vehicle.
5. Installation is the reverse of the removal procedure. Make sure the groove in the stabilizer bar bushing is horizontal.
6. Realign the front end.

SUSPENSION AND STEERING 8-7

Whenever disconnecting a suspension component which utilizes eccentric washers for proper positioning, make certain to matchmark the washers to the bracket or component for reassembly

Shock Absorber

REMOVAL & INSTALLATION

In this type suspension system, the shock absorber is actually the supporting vertical member.
1. Remove the strut as outlined earlier.
2. It is necessary to disassemble the strut to replace the shock absorber. To remove the spring, it must be compressed. The proper type compressor is available at an auto parts store.
3. Remove the nut from the end of the shock absorber shaft and slowly release the spring. The strut can now be disassembled. Testing is the same as the torsion bar shock absorber.
4. Reverse the removal procedure for installation.

COIL SPRING FRONT SUSPENSION

♦ See Figure 5

➡The coil spring front suspension system is found only on 1980–81 Type 2 (Vanagon) models. All other models are equipped with either the strut front suspension system or the torsion bar front suspension system, both of which are covered earlier in this section.

The front suspension consists of upper and lower control arms, a separate upper coil spring/shock absorber mount, steering knuckle and attaching ball joints and a strut arm mounted on the lower control arm for stability.

Coil Spring

REMOVAL & INSTALLATION

1. Jack up the front of the vehicle and support it on stands, then remove the wheel. Remove the shock absorber.

Ball Joint

INSPECTION

Vehicles with strut suspensions have only one ball joint on each side located at the base of the strut in the track control arm.
Raise the car and support it under the frame. The wheel must be clear of the ground.
With a lever, apply upward pressure to the track control arm. Apply the pressure gently and slowly; it is important that only enough pressure is exerted to check the play in the ball joint and not compress the suspension.
Using a vernier caliper, measure the distance between the control arm and the lower edge of the ball joint flange. Record the reading. Release the pressure on the track control arm and again measure the distance between the control arm and the lower edge of the ball joint flange. Record the reading. Subtract the higher reading from the lower reading. If the difference is more than 0.10 in., the ball joint should be replaced.

➡**Remember that even in a new joint there will be measurable play because the ball in the ball joint is spring loaded.**

REPLACEMENT

1. Jack up the car and remove the wheel and tire.
2. Remove the nut from the ball joint stud and remove the stud from the track control arm.
3. Bend back the locking tab and remove the three ball joint securing screws.
4. Pull the track control arm downward and remove the ball joint from the strut.
5. Installation is the reverse of the removal procedure.

2. Compress the coil spring using a spring compressor.
3. Disconnect the stabilizer bar from the strut.
4. Measure the distance from the end of the outer nut on the strut to the tip of the strut itself, then remove the strut. This measurement is later used during installation to align the strut.
5. After the strut is removed from the lower control arm, you should be able to pull the lower ball joint-to-control arm attachment out of the control arm.
6. Remove the coil spring.

To install:
7. Seat the compressed coil spring on its cushions, making sure the grooves for the spring ends are in the correct positions.
8. Reverse remaining procedures to install. When installing strut, adjust the outside nut so that it conforms to the measurement made in Step 4.

8-8 SUSPENSION AND STEERING

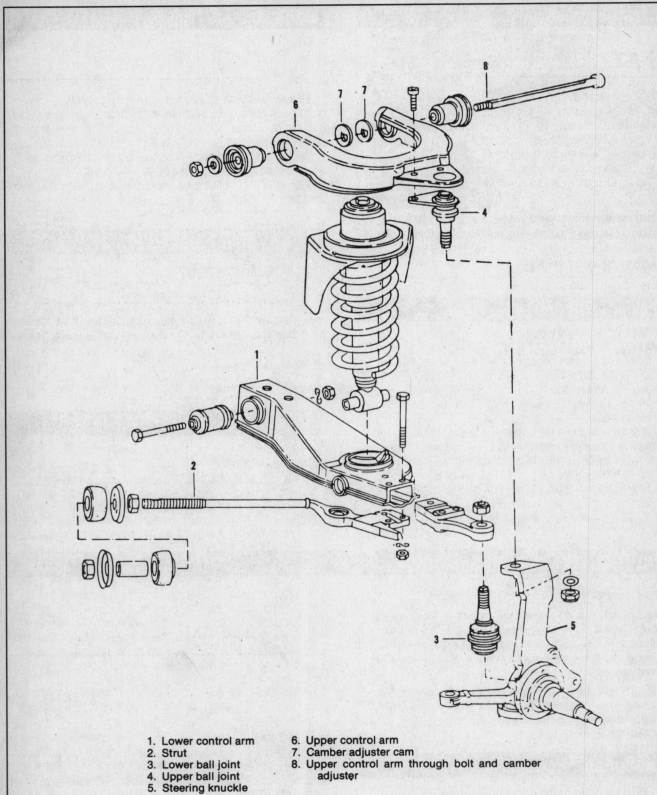

1. Lower control arm
2. Strut
3. Lower ball joint
4. Upper ball joint
5. Steering knuckle
6. Upper control arm
7. Camber adjuster cam
8. Upper control arm through bolt and camber adjuster

Fig. 5 Exploded view of the coil spring front suspension used on 1980–81 Type 2 (Vanagon) models—the stabilizer bar is not shown in this illustration

SUSPENSION AND STEERING 8-9

Shock Absorber

REMOVAL & INSTALLATION

1. Jack the front of the vehicle and remove the front wheel.
2. Loosen and remove the single retaining nut at the top of the coil spring/shock absorber upper mount.
3. Remove the through-bolt, which retains the bottom of the shock absorber to the lower control arm, and pull the shock absorber out through the bottom of the lower control arm.
4. Reverse the procedure to install.

Ball Joint

REMOVAL & INSTALLATION

Upper Ball Joint

1. Raise the vehicle and support it on jack stands, then remove the wheel.
2. Place a jack under the lower control arm as close to the steering knuckle as possible and jack up just enough to put a slight load on the coil spring.
3. Loosen the steering knuckle to ball joint nut but do not remove completely.
4. Free the ball joint from the steering knuckle using a ball joint removal tool, then remove the nut.
5. Remove the two upper ball joint-to-upper control arm bolts and remove the ball joint.
6. Reverse the procedure to install. Check wheel alignment.

Lower Ball Joint

1. Jack up the front of the vehicle and support it on stands, then remove the wheel.
2. Place a jack under the lower control arm as close to the steering knuckle as possible and put a slight load on the coil spring by raising the jack.
3. Disconnect the brake caliper hose from the caliper, and remove the brake caliper and rotor if they are in the way.
4. Loosen the upper ball joint-to-steering knuckle nut, but do not remove it. Free the upper ball joint from the steering knuckle and remove the nut.
5. Remove the lower ball joint-to-lower control arm nut and free the ball joint from the control arm using a ball joint removal tool. Remove the steering knuckle.
6. Press the ball joint off the steering knuckle.

To install:

7. Press a new ball joint in place on knuckle, observing any alignment marks on the ball joint and the knuckle.
8. Reverse the procedure to install. Bleed the brakes.

Upper Control Arm

REMOVAL & INSTALLATION

1. Jack up the vehicle and remove the front wheel.
2. Place a jack under the lower control arm and raise it to put a slight load on the coil spring.
3. Free the upper ball joint from the steering knuckle.
4. Remove the upper control arm-to-frame mounting bolt and remove the control arm.
5. Reverse the procedure to install. Check and adjust the wheel alignment.

Lower Control Arm

REMOVAL & INSTALLATION

1. Jack up the vehicle and remove the wheel.
2. Remove the coil spring, as described earlier in this section.
3. Remove the lower control arm-to-frame mounting bolt and remove the control arm.
4. Reverse the procedures to install.

Front End Alignment

CASTER ADJUSTMENT

▶ See Figure 6

Caster is the forward or backward tilt of the spindle. Forward tilt is negative caster and backward tilt is positive caster. Caster is not adjustable on the torsion bar or the strut suspensions. A slight caster adjustment can be made on the 1980–81 Type 2 Vanagon by moving the nuts on the end of the front suspension strut.

CAMBER ADJUSTMENT

▶ See Figure 7

Camber is the tilt of the top of the wheel, inward or outward, from true vertical. Outward tilt is positive, inward tilt is negative.

Torsion Bar Suspension

The upper ball joint on each side is mounted in an eccentric bushing. The bushing has a hex head and it may be rotated in either direction using a wrench.

8-10 SUSPENSION AND STEERING

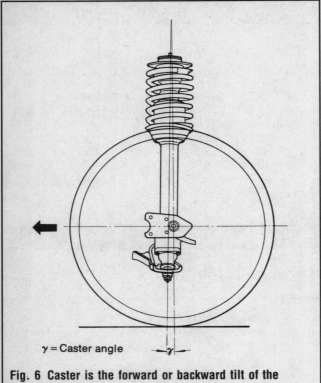

Fig. 6 Caster is the forward or backward tilt of the spindle in degrees

Strut Suspension

The track control arm pivots on an eccentric bolt. Camber is adjusted by loosening the nut and rotating the bolt.

Spring on Lower Arm

Camber is adjusted by turning the upper control arm mounting bolt, which causes the two cams on the upper control arm to force the arm in or out.

TOE-IN ADJUSTMENT

♦ See Figure 8

Toe-in is the adjustment made to make the front wheels point slightly into the front. Toe-in is adjusted on all types of front suspensions by adjusting the length of the tie-rod sleeves.

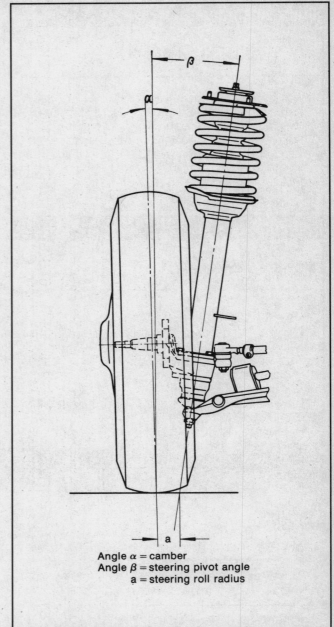

Angle α = camber
Angle β = steering pivot angle
a = steering roll radius

Fig. 7 Camber angle is the distance the top of the wheel leans out or in from the vertical plane

SUSPENSION AND STEERING

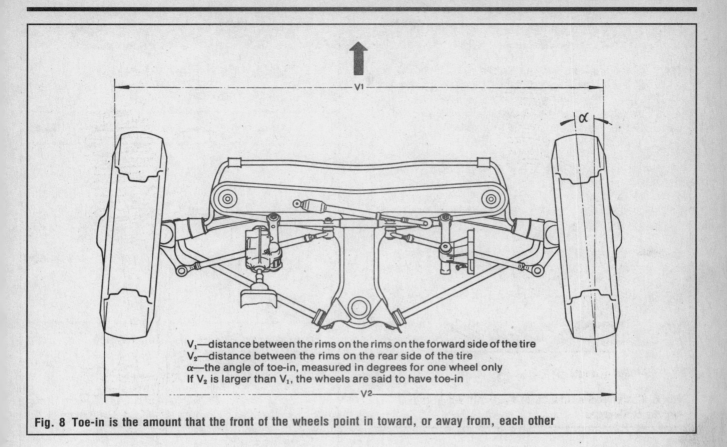

V_1—distance between the rims on the rims on the forward side of the tire
V_2—distance between the rims on the rear side of the tire
α—the angle of toe-in, measured in degrees for one wheel only
If V_2 is larger than V_1, the wheels are said to have toe-in

Fig. 8 Toe-in is the amount that the front of the wheels point in toward, or away from, each other

Wheel Alignment Specifications

| | | Front Axle | | | | | Rear Axle | | |
| | | Caster | | Camber | | | Camber | | |
Year	Model	Range (deg)	Pref Setting (deg)	Range (deg)	Pref Setting (deg)	Toe-in (in. or degrees)	Range (deg)	Pref Setting (deg)	Toe-in (deg)
1970–77	Type 1	±1°	+3° 20'	±20'	+30'	+0.071– +0.213	±40'	−1°	0' ±15'
1971–80	Type 1 ①	±35'	+2°	+20'	+1°	+0.071– +0.213	±40'	−1°	0' ±15'
1970–79	Type 2	±40'	+3°	±20'	+40'	0.0– +0.136	±30'	−50'	+10' ±20'
1980–81	Type 2 ③	±15'	7° 15'	±30'	0	+40'	±30'	−50'	0° ±10'
1970–73	Type 3	±40'	+4°	±20'	+1° 20'	+0.118	±40'	−1° 20'	0' ±15' ②
1971–74	Type 4	±35'	+1° 45'	+25' −30'	+1° 10'	+0.024– +0.165	±30'	−1°	+10' ±15'

① Super Beetle and Convertible
② Squareback given; Sedan 5' ±15'
③ Vehicle empty

DIAGONAL ARM REAR SUSPENSION

➡ The diagonal arm rear suspension system is found on Type 1, 1970–79 Type 2 and all Type 3 models. Type 4 and 1980–81 Type 2 (Vanagon) models are equipped with the coil spring/trailing arm rear suspension system, which is covered later in this section.

The rear wheels of Types 1, 2, and 3 models are independently sprung by means of torsion bars. The inside ends of the torsion bars are anchored to a body crossmember via a splined tube which is welded to the frame. The torsion bar at each side of the rear suspension has a different number of splines at each end. This makes possible the adjustment of the rear suspension.

On Type 3 models, an equalizer bar, located above the rear axle, is used to aid the handling qualities and lateral stability of the rear axle. This bar also acts progressively to soften bumps in proportion to their size.

Unloaded Rear Tension Bar Settings

Type	Model	Transmission	Setting	Range
1	all	all	20° 30'	+50'
2	221, 223, 226	Manual	21° 10'	+50'
2	222	Manual	23°	+50'
2 ③	221, 223	all	20°	+50'
2 ④	222	all	23°	+50'
3	311	Manual	23°	+50'
3	311	Automatic	24°	+50'
3	361	all	21° 30'	+50'

③ From chassis 212 2 000 001 (1972-up)
④ From chassis 212 2 000 001 (1972-up)

Shock Absorber

REMOVAL & INSTALLATION

The shock absorber is secured at the top and bottom by a through-bolt. Raise the car and remove the bolts. Remove the shock absorber from the car.

Diagonal Arm

REMOVAL & INSTALLATION

1. Remove the wheel shaft nuts.

✴✴ CAUTION

Do not raise the car to remove the nuts. They can be safely removed only if the weight of the car is on its wheels.

2. Disconnect the driveshaft on the side to be removed.
3. Remove the lower shock absorber mount. Raise the car and remove the wheel and tire.
4. Remove the brake drum, disconnect the brake lines and emergency brake cable, and remove the backing plate.
5. Matchmark the torsion bar plate and the diagonal arm with a cold chisel.
6. Remove the four bolts and nuts which secure the plate to the diagonal arm.
7. Remove the pivot bolts for the diagonal arm and remove the arm from the car.

➡ Take careful note of the washers at the pivot bolts. These washers are used to determine alignment and they must be put back in the same place.

8. Remove the spring plate hub cover.
9. Using a steel bar, lift the spring plate off of the lower suspension stop.
10. On Type 1 models, remove the five bolts at the front of the fender. On all others, remove the cover in the side of the fender.
11. Remove the spring plate and pull the torsion bar out of its housing.

➡ There are left and right torsion bars designated by an (L) or (R) on the end face. (Coat any rubber bushings with talcum powder upon installation. Do not use graphite, silicon, or grease.

To install:
12. Insert the torsion bar, outer bushing, and spring plate. Make sure that the marks you made earlier line up.
13. Using two bolts, loosely secure the spring plate hub cover. Place a thick nut between the leaves of the spring plate.
14. Lift the spring plate up to the lower suspension stop and install the remaining bolts into the hub cover. Tighten the hub cover bolts.
15. Install the diagonal arm pivot bolt and washers and peen it with a chisel. There must always be at least one washer on the outside end of the bolt.
16. Align the chisel marks and attach the diagonal arm to the spring plate.

SUSPENSION AND STEERING 8-13

17. Install the backing plate, parking brake cable, and brake lines.
18. Reconnect the shock absorber. Install the brake drum and wheel shaft nuts.
19. Reconnect the driveshaft. Bleed the brakes.
20. Install the wheel and tire.
21. Check the suspension alignment.

Adjustments

TYPE 1

The only adjustment possible is the toe-in adjustment. The adjustment is performed by varying the number of washers at the diagonal arm pivot. There must always be one washer located on the outboard side of the pivot.

TYPES 2 (THROUGH 1979) AND 3

The transaxle and engine assembly position in the vehicle is adjustable. It is necessary that the assembly be correctly centered before the suspension is aligned. It may be adjusted by moving the engine and transaxle brackets in their elongated slots.

The distance between the diagonal arms may be adjusted by moving the washers at the A-arm pivots. The washers may be positioned only two ways. Either both washers on the outboard side of the pivot or a single washer on each side of the pivot. To adjust the distance, position the diagonal arms and move the washers in the same manner at both pivots.

The wheel track angle may be adjusted by moving the diagonal arm flange in the elongated slot in the spring plate.

The toe-in is adjusted by positioning the washers and the diagonal arm pivot.

COIL SPRING/TRAILING ARM REAR SUSPENSION

▶ See Figure 9

➡The coil spring/trailing arm rear suspension system is found on Type 4 and 1980–81 Type 2 (Vanagon) models. Type 1, 1970–79 Type 2 and all Type 3 models are equipped with the diagonal arm rear suspension system, which is covered earlier in this section.

The rear wheels of Type 4 models are independently sprung by means of coil springs and trailing arms. The shock absorbers mount inside the coil springs. Each coil spring and shock absorber mounts between the trailing arm and a sheet metal shock tower. Each trailing arm pivots on a body crossmember.

The rear suspension on 1980–81 Type 2 (Vanagon) models is basically the same as the suspension system used on Type 4 models, except that the shock absorber is not mounted inside the coil spring on the 1980–81 Type 2 models.

Shock Absorber

REMOVAL & INSTALLATION

Type 4

The shock absorber is the lower stop for the suspension.

✲✲ CAUTION

The A-arm must be securely supported when the shock absorber is disconnected to prevent the spring tension from being released suddenly.

Leaving the car on the ground or raising the car and securely supporting the A-arm, remove the lower shock absorber through-bolt. To gain access to the upper shock mount, remove the access panel for each shock located at the sides of the rear luggage shelf. Remove the self-locking nut from the shock absorber shaft and remove the shock. Installation is the reverse of removal.

1980–81 Type 2

Refer to Type 4 shock absorber removal and installation for procedures; observe the CAUTION. The shock absorber is not mounted inside the spring, but behind it. Unfasten the shock absorber retaining bolts and remove the shock absorber.

Trailing Arm

REMOVAL & INSTALLATION

Type 4

1. Raise the car and place it on jackstands. Securely block up the A-arm.

✲✲ CAUTION

The A-arm must be securely supported when the shock absorber is disconnected to prevent the spring tension from being released suddenly. The shock absorber is the lower stop for the suspension.

8-14 SUSPENSION AND STEERING

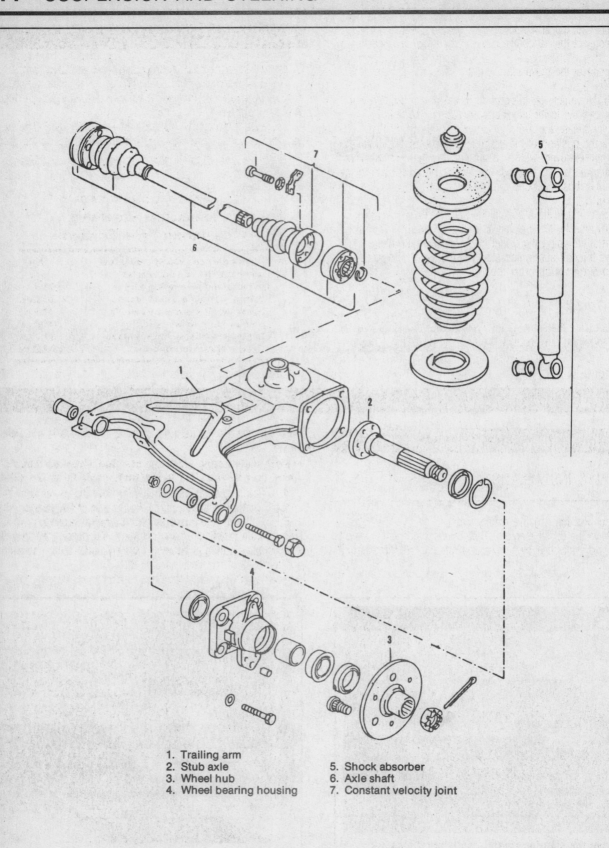

1. Trailing arm
2. Stub axle
3. Wheel hub
4. Wheel bearing housing
5. Shock absorber
6. Axle shaft
7. Constant velocity joint

Fig. 9 Exploded view of the rear suspension used on Type 2 models

SUSPENSION AND STEERING 8-15

2. Disconnect the driveshaft.
3. Disconnect the handbrake cable at the brake lever and remove it.
4. Disconnect the brake lines and the stabilizer bar, if equipped.
5. With the vehicle on the ground or the A-arm securely supported, remove the lower shock absorber mounting bolt.
6. Slowly release the A-arm and remove the coil springs.
7. Mark the position of the brackets or the eccentric bolts, whichever are removed, with a chisel. Remove the nuts which secure the brackets in the rear axle carrier, or the pivot bolts in the bonded rubber bushings, and remove the A-arm.

To install:
8. Loosely install the A-arm. If the pivot bolts were removed, install them loosely. If the eccentric bolts and brackets were removed, install, aligning the chisel marks, and tighten them.
9. Insert the coil spring and slowly compress it into place. Install the lower shock absorber mount.

1980–81 Type 2

Procedures are the same as for Type 4 models, except when removing the trailing arm, remove the trailing arm at the bushing bolts.

Adjustments

The toe-in is adjusted by the eccentric A-arm pivot bolts.
The rubber buffer centralization procedure is given in the Type 4 automatic transaxle removal & installation procedure, in Section 7 of this manual.
The track width can be adjusted by loosening the A-arm mounting bracket bolts and moving the brackets in or out to the proper position.

Type 4 Engine and Transaxle Assembly centering Specifications

Offset between vehicle center and engine/transaxle unit center	1.0 in.
Center of left measuring hole to center of right measuring hole	44.3 ± 0.04 in.
Center of left measuring hole to center of rib on transaxle	23.1 ± 0.02 in.
Center of right measuring hole to center of rib on transaxle	21.2 ± 0.02 in.

STEERING

Steering Wheel

REMOVAL & INSTALLATION

1. Disconnect the negative battery cable.
2. Remove the center emblem. This emblem will gently pry off the wheel, or is attached by screws from the back of the steering wheel.
3. Remove the nut from the steering shaft. This is a right-hand thread.

➡ **Mark the steering shaft and steering wheel so that the wheel may be installed in the same position on the shaft.**

4. Using a steering wheel puller, remove the wheel from the splined steering shaft. Do not strike the end of the steering shaft.
5. Installation is the reverse of the removal procedure. Make sure to align the match marks made on the steering wheel and steering shaft. The gap between the turn signal switch housing and the back of the wheel is 0.08–0.12 in.

To remove the steering wheel, gently pry the center cover loose from the steering wheel . . .

. . . then remove the center cover

8-16 SUSPENSION AND STEERING

Remove the electrical wire hold-down screw then . . .

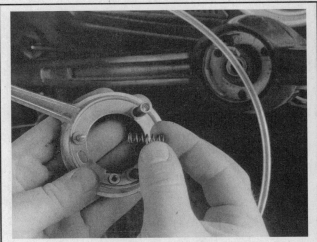
Make sure not to lose the little springs on the backside of the horn handle hold-down screws

. . . and remove the 3 horn handle hold-down screws then . . .

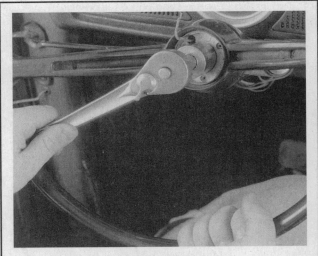
Loosen the steering wheel center nut . . .

. . . pull the horn handle off of the steering wheel

. . . then remove the nut and the large flat washer from the steering shaft

SUSPENSION AND STEERING 8-17

Matchmark the steering wheel to the steering shaft for reassembly . . .

. . . then remove the steering wheel from the shaft — often times it is not necessary to use a puller

Turn Signal Switch

REMOVAL & INSTALLATION

1. Disconnect the negative battery cable.
2. Remove the steering wheel.
3. Remove the four turn signal switch securing screws.
4. Disconnect the turn signal switch wiring plug under the steering column.
5. Pull the switch and wiring guide rail up and out of the steering column.
6. Installation is the reverse of the removal procedure. Make sure the spacers located behind the switch, if installed originally, are in position. The distance between the steering wheel and the steering column housing is 0.08–0.12 in. Install the switch with the lever in the central position.

Ignition Switch

REMOVAL & INSTALLATION

1. Disconnect the steering column wiring at the block located behind the instrument panel and pull the column wiring harness into the passenger compartment.
2. Remove the steering wheel.
3. Remove the circlip on the steering shaft.
4. Disconnect the negative battery cable.
5. Insert the key and turn the switch to the **ON** position. On Type 3 vehicles it is necessary to remove the fuse box.
6. Remove the three securing screws and slide the switch assembly from the steering column tube.

➡ It is not necessary to remove the turn signal switch at this time. If it is necessary to remove the switch from the housing, continue with the disassembly procedure.

7. Remove the turn signal switch.
8. After removing the wiring retainer, press the ignition switch wiring block upward and out of the housing and disconnect the wiring.
9. Remove the lock cylinder and the steering lock mechanism.
10. Remove the ignition switch screw and pull the ignition switch rearward.
11. Installation is the reverse of the removal procedure. When reinstalling the turn signal switch, make sure the lever is in the center position.

➡ The distance between the steering wheel and the ignition switch housing is 0.08–0.12 in. (2–3mm).

Ignition Lock Cylinder

REMOVAL & INSTALLATION

1. Proceed with Steps 1–8 in the Ignition Switch procedure.
2. With the key in the cylinder and turned to the **ON** position, pull the lock cylinder out far enough so the securing pin can be depressed through a hole in the side of the lock cylinder housing. Use a steel wire to depress the pin.
3. As the pin is depressed, pull the lock cylinder out of its housing.

To install:

4. Gently push the cylinder into its housing. Make sure the pin engages correctly and that the retainer fits easily in place. Do not force any parts together; when they are correctly aligned, they will fit easily together.

➡ On many 1976 and later Type 1's, and many Type 2's, the release hole was not drilled by the factory. Using the illustrations, drill your own release hole with a ⅛ in. drill bit.

8-18 SUSPENSION AND STEERING

Steering Linkage

REMOVAL & INSTALLATION

▶ See Figure 10

All tie-rod ends are secured by a nut which holds the tapered tie-rod end stud into a matching tapered hole. There are several ways to remove the tapered stud from its hole after the nut has been removed.

First, there are several types of removal tools available from auto parts stores. These tools include directions for their use. One of the most commonly available tools is the fork-shaped tool, which is a wedge that is forced under the tie-rod end. This tool should be used with caution because instead of removing the tie-rod end from its hole it may pull the ball out of its socket, ruining the tie-rod end.

It is also possible to remove the tie-rod end by holding a

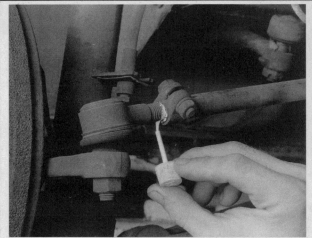

Before removing the old tie rod end, matchmark the tie rod end to the tie rod

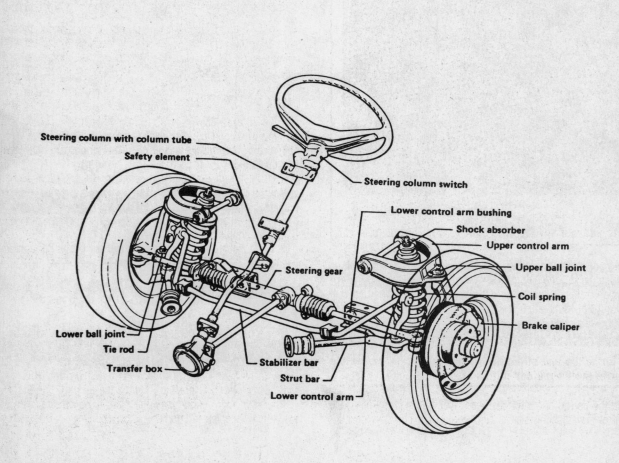

Fig. 10 Identification of the front steering components equipped on 1980–81 Type 2 (Vanagon) models

SUSPENSION AND STEERING 8-19

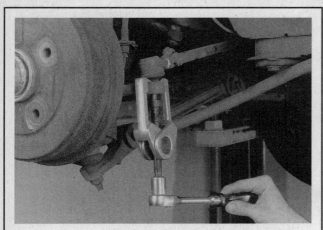

Remove the tie rod end retaining nut, then use a tie rod puller to loosen the tie rod from the steering knuckle arm . . .

. . . and remove the tie rod end from the knuckle—the tie rod end can now be unthreaded from the tie rod

heavy hammer on one side of the tapered hole and striking the opposite side of the hole sharply with another hammer. The stud will pop out of its hole.

※※ CAUTION

Never strike the end of the tie-rod end stud. It is impossible to remove the tie-rod end in this manner.

Once the tie-rod end stud has been removed, turn the tie-rod end out of the adjusting sleeve. On the pieces of the steering linkage that are not used to adjust the toe-in, the tie-rod end is welded in place and it will be necessary to replace the whole assembly.

When reassembling the steering linkage, never put lubricant in the tapered hole.

Manual Steering Gear

ADJUSTMENT

There are three types of adjustable steering gear units. The first type is the roller type, identified by the square housing cover secured by four screws, one at each corner. The second type is the worm and peg type, identified by an assymetric housing cover with the adjusting screw located at one side of the housing cover. The third type is the rack and pinion type used on 1975 Super Beetles and 1975 and later Convertibles.

Worm and Roller Type

Disconnect the steering linkage from the pitman arm and make sure the gearbox mounting bolts are tight. Have an assistant rotate the steering wheel so that the pitman arm moves alternately 10° to the left and then 10° to the right of the straight ahead position. Turn the adjusting screw in until no further play can be felt while moving the pitman arm. Tighten the adjusting screw locknut and recheck the adjustment.

Worm and Peg Type

Have an assistant turn the steering wheel back and forth through the center position several times. The steering wheel should turn through the center position without any noticeable binding.

To adjust, turn the adjusting screw inward while the assistant is turning the steering wheel. Turn the screw in until the steering begins to tighten up. Back out the adjusting screw until the steering no longer binds while turning through the center point, then tighten the adjusting screw locknut.

The adjustment is correct when there is no binding and no perceptible play.

Rack and Pinion Type
▶ See Figure 11

The steering gear requires adjustment if it begins to rattle noticeably. First, remove the access cover in the spare tire well. Then, with the car standing on all four wheels, turn the adjusting screw in by hand until it contacts the thrust washer. While holding the screw in this position, tighten the locknut.

The adjustment is correct when there is no binding and the steering self-centers properly.

8-20 SUSPENSION AND STEERING

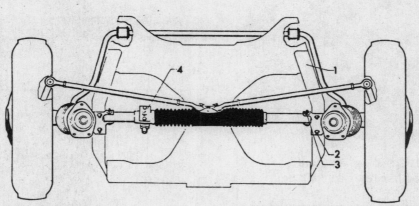

1. Side member
2. Bracket-to-side member bolts 4.5 mkg (32 ft. lbs.)
3. Steering gear-to-bracket bolts 2.5 mkg (18 ft. lbs.)
4. Adjusting screw

Fig. 11 1975 Super Beetle and 1975–79 Super Beetle Convertible steering rack and pinion details

BRAKE SYSTEM 9-2
BASIC OPERATING PRINCIPLES 9-2
    DISC BRAKES 9-2
    DRUM BRAKES 9-3
    POWER BOOSTERS 9-3
BRAKE ADJUSTMENTS 9-3
HYDRAULIC SYSTEM 9-4
MASTER CYLINDER 9-4
    REMOVAL & INSTALLATION 9-4
    OVERHAUL 9-4
HYDRAULIC SYSTEM BLEEDING 9-5
    PRESSURE BLEEDING 9-5
    MANUAL BLEEDING 9-6
FRONT DRUM BRAKES 9-6
BRAKE DRUM 9-6
    REMOVAL & INSTALLATION 9-6
    INSPECTION 9-6
BRAKE SHOES 9-8
    REMOVAL & INSTALLATION 9-8
WHEEL CYLINDER 9-8
    REMOVAL & INSTALLATION 9-8
    OVERHAUL 9-10
FRONT DISC BRAKES 9-10
BRAKE PAD 9-10
    REMOVAL & INSTALLATION 9-10
BRAKE CALIPER 9-11
    REMOVAL & INSTALLATION 9-11
    OVERHAUL 9-11
BRAKE ROTOR (DISC) 9-13
    REMOVAL & INSTALLATION 9-13
    INSPECTION 9-13
FRONT WHEEL BEARINGS 9-13
    REMOVAL & INSTALLATION 9-13
    ADJUSTMENT 9-14
REAR DRUM BRAKES 9-14
BRAKE DRUM 9-14
    REMOVAL & INSTALLATION 9-14
    INSPECTION 9-14
BRAKE SHOES 9-14
    REMOVAL & INSTALLATION 9-14
WHEEL CYLINDER 9-15
    REMOVAL & INSTALLATION 9-15
    OVERHAUL 9-17
PARKING BRAKE 9-19
CABLE 9-19
    ADJUSTMENT 9-19
    REMOVAL & INSTALLATION 9-19
SPECIFICATION CHART
  BRAKE SPECIFICATIONS 9-20

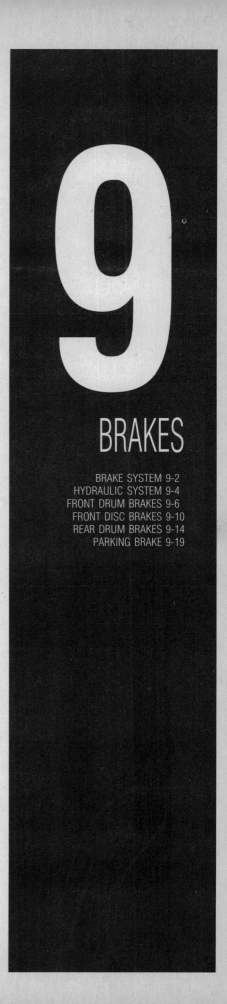

9
BRAKES

BRAKE SYSTEM 9-2
HYDRAULIC SYSTEM 9-4
FRONT DRUM BRAKES 9-6
FRONT DISC BRAKES 9-10
REAR DRUM BRAKES 9-14
PARKING BRAKE 9-19

BRAKES

BRAKE SYSTEM

Basic Operating Principles

Hydraulic systems are used to actuate the brakes of all modern automobiles. The system transports the power required to force the frictional surfaces of the braking system together from the pedal to the individual brake units at each wheel. A hydraulic system is used for two reasons.

First, fluid under pressure can be carried to all parts of an automobile by small pipes and flexible hoses without taking up a significant amount of room or posing routing problems.

Second, a great mechanical advantage can be given to the brake pedal end of the system, and the foot pressure required to actuate the brakes can be reduced by making the surface area of the master cylinder pistons smaller than that of any of the pistons in the wheel cylinders or calipers.

The master cylinder consists of a fluid reservoir along with a double cylinder and piston assembly. Double type master cylinders are designed to separate the front and rear braking systems hydraulically in case of a leak. The master cylinder coverts mechanical motion from the pedal into hydraulic pressure within the lines. This pressure is translated back into mechanical motion at the wheels by either the wheel cylinder (drum brakes) or the caliper (disc brakes).

Steel lines carry the brake fluid to a point on the vehicle's frame near each of the vehicle's wheels. The fluid is then carried to the calipers and wheel cylinders by flexible tubes in order to allow for suspension and steering movements.

In drum brake systems, each wheel cylinder contains two pistons, one at either end, which push outward in opposite directions and force the brake shoe into contact with the drum.

In disc brake systems, the cylinders are part of the calipers. At least one cylinder in each caliper is used to force the brake pads against the disc.

All pistons employ some type of seal, usually made of rubber, to minimize fluid leakage. A rubber dust boot seals the outer end of the cylinder against dust and dirt. The boot fits around the outer end of the piston on disc brake calipers, and around the brake actuating rod on wheel cylinders.

The hydraulic system operates as follows: When at rest, the entire system, from the piston(s) in the master cylinder to those in the wheel cylinders or calipers, is full of brake fluid. Upon application of the brake pedal, fluid trapped in front of the master cylinder piston(s) is forced through the lines to the wheel cylinders. Here, it forces the pistons outward, in the case of drum brakes, and inward toward the disc, in the case of disc brakes. The motion of the pistons is opposed by return springs mounted outside the cylinders in drum brakes, and by spring seals, in disc brakes.

Upon release of the brake pedal, a spring located inside the master cylinder immediately returns the master cylinder pistons to the normal position. The pistons contain check valves and the master cylinder has compensating ports drilled in it. These are uncovered as the pistons reach their normal position. The piston check valves allow fluid to flow toward the wheel cylinders or calipers as the pistons withdraw. Then, as the return springs force the brake pads or shoes into the released position, the excess fluid reservoir through the compensating ports. It is during the time the pedal is in the released position that any fluid that has leaked out of the system will be replaced through the compensating ports.

Dual circuit master cylinders employ two pistons, located one behind the other, in the same cylinder. The primary piston is actuated directly by mechanical linkage from the brake pedal through the power booster. The secondary piston is actuated by fluid trapped between the two pistons. If a leak develops in front of the secondary piston, it moves forward until it bottoms against the front of the master cylinder, and the fluid trapped between the pistons will operate the rear brakes. If the rear brakes develop a leak, the primary piston will move forward until direct contact with the secondary piston takes place, and it will force the secondary piston to actuate the front brakes. In either case, the brake pedal moves farther when the brakes are applied, and less braking power is available.

All dual circuit systems use a switch to warn the driver when only half of the brake system is operational. This switch is usually located in a valve body which is mounted on the firewall or the frame below the master cylinder. A hydraulic piston receives pressure from both circuits, each circuit's pressure being applied to one end of the piston. When the pressures are in balance, the piston remains stationary. When one circuit has a leak, however, the greater pressure in that circuit during application of the brakes will push the piston to one side, closing the switch and activating the brake warning light.

In disc brake systems, this valve body also contains a metering valve and, in some cases, a proportioning valve. The metering valve keeps pressure from traveling to the disc brakes on the front wheels until the brake shoes on the rear wheels have contacted the drums, ensuring that the front brakes will never be used alone. The proportioning valve controls the pressure to the rear brakes to lessen the chance of rear wheel lock-up during very hard braking.

Warning lights may be tested by depressing the brake pedal and holding it while opening one of the wheel cylinder bleeder screws. If this does not cause the light to go on, substitute a new lamp, make continuity checks, and, finally, replace the switch as necessary.

The hydraulic system may be checked for leaks by applying pressure to the pedal gradually and steadily. If the pedal sinks very slowly to the floor, the system has a leak. This is not to be confused with a springy or spongy feel due to the compression of air within the lines. If the system leaks, there will be a gradual change in the position of the pedal with a constant pressure.

Check for leaks along all lines and at wheel cylinders. If no external leaks are apparent, the problem is inside the master cylinder.

DISC BRAKES

Instead of the traditional expanding brakes that press outward against a circular drum, disc brake systems utilize a disc (rotor) with brake pads positioned on either side of it. An easily-seen analogy is the hand brake arrangement on a bicycle. The pads squeeze onto the rim of the bike wheel, slowing its motion. Automobile disc brakes use the identical principle but apply the braking effort to a separate disc instead of the wheel.

BRAKES

The disc (rotor) is a casting, usually equipped with cooling fins between the two braking surfaces. This enables air to circulate between the braking surfaces making them less sensitive to heat buildup and more resistant to fade. Dirt and water do not drastically affect braking action since contaminants are thrown off by the centrifugal action of the rotor or scraped off the by the pads. Also, the equal clamping action of the two brake pads tends to ensure uniform, straight line stops. Disc brakes are inherently self-adjusting. There are three general types of disc brake:

1. A fixed caliper.
2. A floating caliper.
3. A sliding caliper.

The fixed caliper design uses two pistons mounted on either side of the rotor (in each side of the caliper). The caliper is mounted rigidly and does not move.

The sliding and floating designs are quite similar. In fact, these two types are often lumped together. In both designs, the pad on the inside of the rotor is moved into contact with the rotor by hydraulic force. The caliper, which is not held in a fixed position, moves slightly, bringing the outside pad into contact with the rotor. There are various methods of attaching floating calipers. Some pivot at the bottom or top, and some slide on mounting bolts. In any event, the end result is the same.

DRUM BRAKES

Drum brakes employ two brake shoes mounted on a stationary backing plate. These shoes are positioned inside a circular drum which rotates with the wheel assembly. The shoes are held in place by springs. This allows them to slide toward the drums (when they are applied) while keeping the linings and drums in alignment. The shoes are actuated by a wheel cylinder which is mounted at the top of the backing plate. When the brakes are applied, hydraulic pressure forces the wheel cylinder's actuating links outward. Since these links bear directly against the top of the brake shoes, the tops of the shoes are then forced against the inner side of the drum. This action forces the bottoms of the two shoes to contact the brake drum by rotating the entire assembly slightly (known as servo action). When pressure within the wheel cylinder is relaxed, return springs pull the shoes back away from the drum.

Most modern drum brakes are designed to self-adjust themselves during application when the vehicle is moving in reverse. This motion causes both shoes to rotate very slightly with the drum, rocking an adjusting lever, thereby causing rotation of the adjusting screw. Some drum brake systems are designed to self-adjust during application whenever the brakes are applied. This on-board adjustment system reduces the need for maintenance adjustments and keeps both the brake function and pedal feel satisfactory.

POWER BOOSTERS

Virtually all modern vehicles use a vacuum assisted power brake system to multiply the braking force and reduce pedal effort. Since vacuum is always available when the engine is operating, the system is simple and efficient. A vacuum diaphragm is located on the front of the master cylinder and assists the driver in applying the brakes, reducing both the effort and travel he must put into moving the brake pedal.

The vacuum diaphragm housing is normally connected to the intake manifold by a vacuum hose. A check valve is placed at the point where the hose enters the diaphragm housing, so that during periods of low manifold vacuum brakes assist will not be lost.

Depressing the brake pedal closes off the vacuum source and allows atmospheric pressure to enter on one side of the diaphragm. This causes the master cylinder pistons to move and apply the brakes. When the brake pedal is released, vacuum is applied to both sides of the diaphragm and springs return the diaphragm and master cylinder pistons to the released position.

If the vacuum supply fails, the brake pedal rod will contact the end of the master cylinder actuator rod and the system will apply the brakes without any power assistance. The driver will notice that much higher pedal effort is needed to stop the car and that the pedal feels harder than usual.

Vacuum Leak Test

1. Operate the engine at idle without touching the brake pedal for at least one minute.
2. Turn off the engine and wait one minute.
3. Test for the presence of assist vacuum by depressing the brake pedal and releasing it several times. If vacuum is present in the system, light application will produce less and less pedal travel. If there is no vacuum, air is leaking into the system.

System Operation Test

1. With the engine **OFF**, pump the brake pedal until the supply vacuum is entirely gone.
2. Put light, steady pressure on the brake pedal.
3. Start the engine and let it idle. If the system is operating correctly, the brake pedal should fall toward the floor if the constant pressure is maintained.

Power brake systems may be tested for hydraulic leaks just as ordinary systems are tested.

Brake Adjustments

All models are equipped with dual hydraulic brake systems in accordance with federal regulations. In case of a hydraulic system failure, ½ braking efficiency will be retained.

After removing the access hole plugs, use a brake adjuster tool to turn the star adjusting wheels to properly position the brake shoes

9-4 BRAKES

All type 1 models (except the Karmann Ghia), and 1970 Type 2 models are equipped with front drum brakes. Discs are used at the front of all Type 1 Karmann Ghias, 1971–81 Type 2 models, and all Type 3 and 4 models. All models use rear drum brakes.

Disc brakes are self-adjusting and cannot be adjusted by hand. As the pads wear, they will automatically compensate for the wear by moving closer to the disc, maintaining the proper operating clearance.

Drum brakes, however, must be manually adjusted to take up excess clearance as the shoes wear. To adjust drum brakes, both front and rear, it is necessary to jack up the car and support it on jackstands. The wheel must spin freely. On the backing plate there are four inspection holes with a rubber plug in each hole. Two of the holes are for checking the thickness of the brake lining and the other two are used for adjustment.

➡ **There is an adjustment for each brake shoe. That means that on each wheel it is necessary to make two adjustments: one for each shoe on that wheel.**

Remove the adjustment hole plugs and, using a brake adjusting tool, insert the tool into the hole. Turn the star wheel until a slight drag is noticed as the wheel is rotated by hand. Back off on the star wheel 3–4 notches so that the wheel turns freely. Perform the same adjustment on the other shoe.

➡ **One of the star wheels in each wheel has left-hand threads and the other star wheel has right-hand threads.**

Repeat the above procedure on each wheel equipped with drum brakes.

HYDRAULIC SYSTEM

Master Cylinder

REMOVAL & INSTALLATION

✳✳ WARNING

Clean, high quality brake fluid is essential to the safe and proper operation of the brake system. You should always buy the highest quality brake fluid that is available. If the brake fluid becomes contaminated, drain and flush the system, then refill the master cylinder with new fluid. Never reuse any brake fluid. Any brake fluid that is removed from the system should be discarded.

➡ **The brake master cylinder on the 1980 Type 2 (Vanagon) is located under the dash. Remove the bottom and top dash covers for access.**

1. Drain the brake fluid from the master cylinder reservoir.

✳✳ CAUTION

Do not get any brake fluid on the paint, as it will dissolve the paint.

2. On Type 3 models, remove the master cylinder cover plate.
3. Pull the plastic elbows out of the rubber sealing rings on the top of the master cylinder, if equipped, or unfasten the fluid pipe seat nuts.
4. Remove the two bolts which secure the master cylinder to the frame and remove the cylinder. Note the spacers on Type 1 models between the frame and the master cylinder.

To install:

5. Bolt the master cylinder to the frame. Do not forget the spacers on Type 1 models.
6. Lubricate the elbows with brake fluid and insert them into the rubber seals.
7. If necessary, adjust the brake pedal free travel. On Type 1, 3, and 4 models, adjust the length of the master cylinder pushrod so that there is 5–7mm of brake pedal free-play before the pushrod contacts the master cylinder piston. On Type 2 models, the free-play is properly adjusted when the length of the pushrod, measured between the ball end and the center of the clevis pin hole, is 4.17 in.
8. Refill the master cylinder reservoir and bleed the brakes.

OVERHAUL

▸ See Figures 1 and 2

1. Remove the master cylinder from the car.
2. Remove the rubber sealing boot.
3. Remove the stop screw and sealing ring on the top of the unit.
4. Insert a small prytool in the master cylinder piston, exert inward pressure, and remove the snapring from its groove in the end of the unit. The internal parts are spring loaded and must be kept from flying out when the snapring is removed.
5. Carefully remove the internal parts of the unit and make

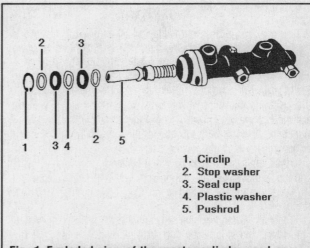

1. Circlip
2. Stop washer
3. Seal cup
4. Plastic washer
5. Pushrod

Fig. 1 Exploded view of the master cylinder used on vacuum booster equipped models

Brakes 9-5

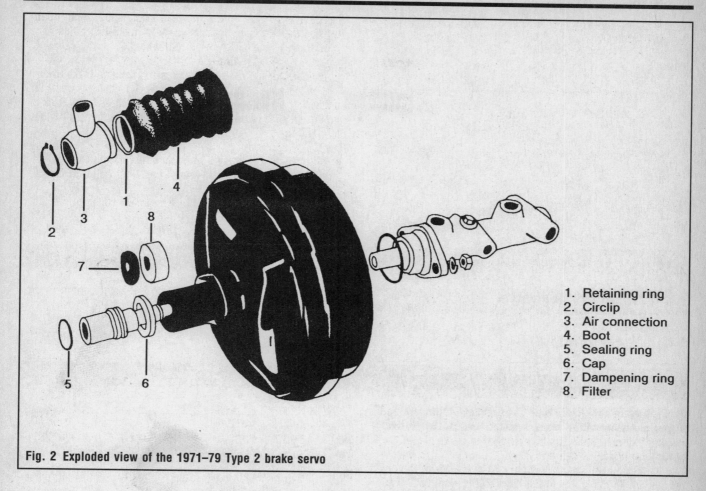

Fig. 2 Exploded view of the 1971–79 Type 2 brake servo

1. Retaining ring
2. Circlip
3. Air connection
4. Boot
5. Sealing ring
6. Cap
7. Dampening ring
8. Filter

note of their order and the orientation of the internal parts. If parts remain in the cylinder bore, they may be removed with a wire hook or very gentle application of low pressure air to the stop screw hole. Cover the end of the cylinder bore with a rag and stand away from the open end of the bore when using compressed air.

6. Use alcohol or brake fluid to clean the master cylinder and its parts.
7. It may be necessary to hone the cylinder bore, or clean it by lightly sanding it with emery cloth. Clean thoroughly after honing or sanding. Lubricate the bore with brake fluid before reassembly.

To assemble:

8. Holding the master cylinder with the open end downward, place the cup washer, primary cup, support washer, spring retainer, and spring onto the front brake circuit piston and insert the piston vertically into the master cylinder bore.
9. Assemble the rear brake circuit piston, cup washer, primary cup, support washer, spring retainer, stop sleeve, spring, and stroke limiting screw and insert the assembly into the master cylinder.
10. Install the stop washer and snapring.
11. Install the stop screw and seal, making sure the hole for the screw is not blocked by the piston. If the hole is blocked, it will be necessary to push the piston further in until the screw can be turned in.

➡ 1971–79 Type 2 vehicles have a brake servo and the order of assembly of the additional seals is illustrated.

12. Install the master cylinder and bleed the brakes.

Hydraulic System Bleeding

The hydraulic brake system must be bled any time one of the lines is disconnected or air enters the system. This may be done manually or by the pressure method.

PRESSURE BLEEDING

1. Clean the top of the master cylinder, remove the caps, and attach the pressure bleeding adapter.
2. Check the pressure bleeder reservoir for correct pressure and fluid level, then open the release valve.
3. Fasten a bleeder hose to the wheel cylinder bleeder nipple and submerge the free end of the hose in a transparent receptacle. The receptacle should contain enough brake fluid to cover the open end of the hose.
4. Open the wheel cylinder bleeder nipple and allow the fluid to flow until all bubbles disappear and an uncontaminated flow of fluid exists.

9-6 BRAKES

5. Close the nipple, remove the bleeder hose, and repeat the procedure on the other wheel cylinders or brake calipers.

MANUAL BLEEDING

This method requires two people: one to depress the brake pedal and the other to open the bleeder nipples.

1. Remove the reservoir caps and fill the reservoir.
2. Attach a bleeder hose and a clear container as outlined in the pressure bleeding procedure.
3. Have the assistant depress the brake pedal to the floor several times and then have him hold the pedal to the floor. With the pedal to the floor, open the bleeder nipple until the fluid flow ceases and then close the nipple. Repeat this sequence until there are no more air bubbles in the fluid.

➡ As the air is gradually forced out of the system, it will no longer be possible to force the brake pedal to the floor.

Periodically check the master cylinder for an adequate supply of fluid. Keep the master cylinder reservoir full of fluid to prevent air from entering the system. If the reservoir does run dry during bleeding, it will be necessary to rebleed the entire system.

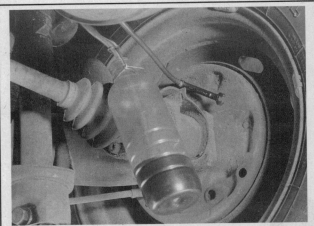

When bleeding the brake system, use a clear container full of brake fluid and a plastic hose to prevent air from re-entering the brake system

FRONT DRUM BRAKES

Brake Drum

REMOVAL & INSTALLATION

1. Jack up the car and remove the wheel and tire.
2. On the left side, remove the clip which secures the speedometer cable to the wheel bearing dust cover. Remove the dust cover.
3. Loosen the nut clamp, remove the wheel bearing adjusting nut and slide the brake drum off of the spindle. It may be necessary to back off on the brake shoe star wheels so that there is enough clearance to remove the drum.
4. Installation is the reverse of removal. Adjust the wheel bearings after installing the drum.

✱✱ CAUTION

Do not forget to readjust the brake shoes if they were disturbed during removal.

INSPECTION

If the brake drums are scored or cracked, they must be replaced or machined. If the vehicle pulls to one side or exhibits a pulsating braking action, the drum is probably out of round and

To remove the front brake drum, remove the front wheel then . . .

. . . use a small prytool to remove the dust cap from the front brake drum

BRAKES 9-7

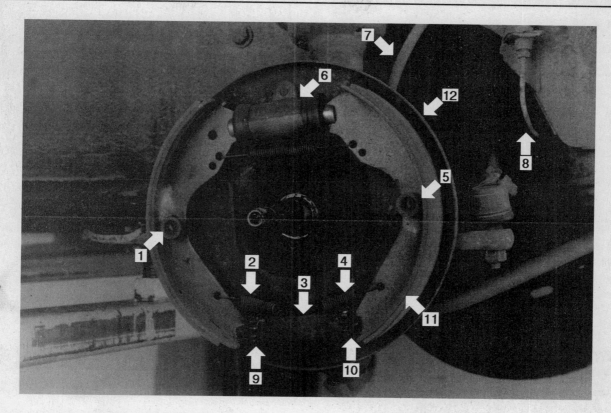

1. Hold-down spring with cup and pin (rear)
2. Rear lower return spring
3. Anchor block
4. Front lower return spring
5. Hold-down spring with cup and pin (front)
6. Wheel cylinder
7. Flexible brake hose
8. Hard brake line
9. Star adjuster wheel for rear brake shoe
10. Star adjuster wheel for front brake shoe
11. Brake shoe and lining
12. Backing plate

Identification of the front drum brake components used on most models

Loosen the nut clamp setscrew . . .

. . . then remove the nut clamp from the front spindle

9-8 BRAKES

Remove the flat washer...

...and the outer wheel bearing from the front brake drum

Pull the front brake drum off of the spindle without dragging the bearing on the spindle threads

should be checked at a machine shop. The drum may have a smooth even surface and still be out of round. The drum should be free of surface cracks and dark spots.

Brake Shoes

REMOVAL & INSTALLATION

Type 1
▶ See Figures 3 and 4

1. Jack up the car and remove the wheel and tire.
2. Remove the brake drum.
3. Remove the small disc and spring which secure each shoe to the backing plate.
4. Remove the two long springs between the two shoes.
5. Remove the shoes from the backing plate.
6. If new shoes are being installed, remove the adjusters in the end of each wheel cylinder and screw the star wheel up against the head of the adjuster. When inserting the adjusters back in the wheel cylinders, notice that the slot in the adjuster is angled and must be positioned as illustrated.
7. Position new shoes on the backing plate. The slot in the shoes and the stronger return spring must be at the wheel cylinder end.
8. Install the disc and spring which secure the shoe to the backing plate.
9. Install the brake drum and adjust the wheel bearing.

Type 2

1. Remove the brake drum.
2. Pry the rear brake shoe out of the adjuster, as illustrated, and detach the return springs. Remove the forward shoe.
3. If new shoes are to be installed, screw the star wheel up against the head of the adjuster.
4. Install the rear brake shoe.
5. Attach the return spring to the front brake shoe and then to the rear shoe.
6. Position the front brake shoe in the slot of the adjusting screw and lever it into position in the same manner as it was removed. Make sure that the return springs do not touch the brake line between the upper and lower wheel cylinders.
7. Install the brake-drum and adjust the wheel bearings.

Wheel Cylinder

REMOVAL & INSTALLATION

✱✱ WARNING

Clean, high quality brake fluid is essential to the safe and proper operation of the brake system. You should always buy the highest quality brake fluid that is available. If the brake fluid becomes contaminated, drain and flush the system, then refill the master cylinder with new fluid. Never reuse any brake fluid. Any brake fluid that is removed from the system should be discarded.

BRAKES 9-9

Front wheel brake

Rear wheel brake

FRONT
1. Adjusting screw
2. Anchor block
3. Front return spring
4. Adjusting nut
5. Guide spring with cup and pin
6. Cylinder
7. Rear return spring

8. Back plate
9. Brake shoe with lining

REAR
1. Cylinder
2. Brake shoe with lining
3. Upper return spring
4. Spring with cup and pin

5. Lower return spring
6. Adjusting screw
7. Back plate
8. Connecting link
9. Lever
10. Brake cable
11. Adjusting nut
12. Anchor block

Fig. 3 Identification of front and rear drum brake components used on Type 1 models—Beetle shown; Super Beetle and Karmann Ghia models have the front brake components rotated 90 degrees (with the cylinder at the top)

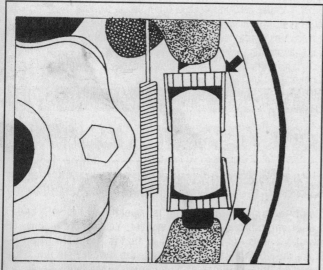

Fig. 4 During installation, the notched adjusters must be positioned as shown—Beetle

For convenience and safety, use a special brake spring tool to disengage the return springs

9-10 BRAKES

1. Remove the brake shoes.
2. On Type 1 models, disconnect the brake line from the rear of the cylinder. On Type 2 models, disconnect the brake line from the rear of the cylinder and transfer line from the front of the cylinder.
3. Remove the bolts which secure the cylinder to the backing plate and remove the cylinder from the vehicle.
4. Installation is the reverse of the removal procedure. Bleed the brakes.

OVERHAUL

♦ See Figures 5 and 6

1. Remove the wheel cylinder.
2. Remove the brake adjusters and remove the rubber boot from each end.

➥The Type 2 cylinder has only one rubber boot, piston, and cup. The rebuilding procedures are the same.

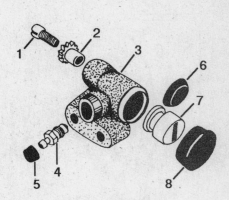

1. Adjusting screw
2. Adjusting nut
3. Housing
4. Bleeder valve
5. Dust cap
6. Cup
7. Piston
8. Boot

Fig. 6 Exploded view of the front wheel cylinder used on Type 2 models

3. On Type 1 models, push in on one of the pistons to force out the opposite piston and rubber cup. On Type 2 models, remove the piston and cup by blowing compressed air into the brake hose hole.
4. Wash the pistons and cylinder in clean brake fluid or alcohol.
5. Inspect the cylinder bore for signs of pitting, scoring, and excessive wear. If it is badly scored or pitted, the whole cylinder should be replaced. It is possible to remove the glaze and light scores with crocus cloth or a brake cylinder hone. Before rebuilding the cylinder, make sure the bleeder screw is free. If the bleeder is rusted shut or broken off, replace the entire cylinder.
6. Dip the new pistons and rubber cups in brake fluid. Place the spring in the bore and insert the rubber cups into the bore against the spring. The concave side of the rubber cup should face inward.
7. Place the pistons in the bore and install the rubber boot.
8. Install the cylinder and bleed the brakes after the shoes and drum are in place. Make sure that the brakes are adjusted.

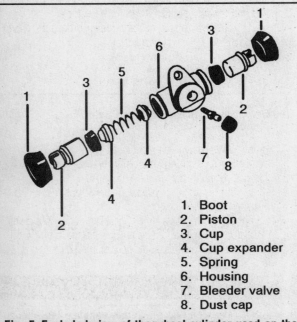

1. Boot
2. Piston
3. Cup
4. Cup expander
5. Spring
6. Housing
7. Bleeder valve
8. Dust cap

Fig. 5 Exploded view of the wheel cylinder used on the front of Type 1 models and on the rear of Types 1, 3 and 4 models

FRONT DISC BRAKES

Brake Pad

REMOVAL & INSTALLATION

♦ See Figure 7

1. Loosen, but do not remove the reservoir cover.
2. Jack up the car and remove the wheel and tire.
3. Using a punch, remove the two pins which retain the disc brake pads in the caliper.

➥If the pads are to be reused, mark the pads to insure that they are reinstalled in the same caliper and on the same side of the disc. Do not invert the pads. Changing pads from one location to another can cause uneven braking.

4. If the pads are not going to be reused, force a wedge between the disc and the pad and pry the piston back into the caliper as far as possible.

BRAKES 9-11

Fig. 7 When installing the retaining plate, make sure that the circular part (a) fits in the piston center and that the retaining plate fits down behind the relieved parts (b) of the piston—some models have a flat retainer plate with indents which fit at b

5. Using a suitable brake parts cleaner, rinse away the brake dust. Pull the old pad out of the caliper and insert a new one, taking care to note the position of the retaining plate.

→On ATE and many Teves disc brakes, before inserting the brake pads, make sure the relieved part of the piston is in the correct position to accept the retaining plate. In the illustration, the sides of the retaining plate "b" fit into the relieved area while the circular part of the plate "a" fits in the piston's center. On some Type 2's and other models, the retaining plate is simply a flat piece of metal with two notches which fit into the relieved areas in the piston. The relieved side of the piston should face against (away from) the rotation direction of the wheel when the car is going forward.

6. Now insert the wedge between the disc and pad on the opposite side and force that piston into the caliper. Remove the old pad and insert a new one.
7. If the old pads are to be reused, it is not necessary to push the piston into the caliper. Pull the pads from the caliper and reinstall the pads when necessary.
8. Install a new brake pad spreader spring and insert the retaining pins. Be careful not to shear the split clamping bushing from the pin. Insert the pin from the inside of the caliper and drive it to the outside.
9. Pump the brake pedal several times to take up the clearance between the pads and the disc before driving the car.
10. Install the wheel and tire and carefully road test the car. Apply the brakes gently for first 500 to 1,000 miles to properly break in the pads and prevent glazing.

Brake Caliper

REMOVAL & INSTALLATION

♦ See Figures 8 and 9

※※ WARNING

Clean, high quality brake fluid is essential to the safe and proper operation of the brake system. You should always buy the highest quality brake fluid that is available. If the brake fluid becomes contaminated, drain and flush the system, then refill the master cylinder with new fluid. Never reuse any brake fluid. Any brake fluid that is removed from the system should be discarded.

1. Jack up the car and remove the wheel and tire.
2. Remove the brake pads.
3. Disconnect and plug the brake line from the caliper.
4. Remove the two bolts which secure the caliper to the steering knuckle and remove the caliper from the vehicle.
5. Installation is the reverse of the removal procedure. Bleed the brakes after the caliper is installed.

OVERHAUL

→Clean all parts in alcohol or brake fluid.

1. Remove the caliper from the vehicle.
2. Remove the piston retaining plates.
3. Pry out the seal spring ring using a small prytool. Do not damage the seal beneath the ring.
4. Remove the seal with a plastic or hard rubber rod. Do not use sharp-edged or metal tools.
5. Rebuild one piston at a time. Securely clamp one piston in place so that it cannot come out of its bore. Place a block of wood between the two pistons and apply air pressure to the brake fluid port.

※※ CAUTION

Use extreme care with this technique because the piston can fly out of the caliper with tremendous force.

6. Remove the rubber seal at the bottom of the piston bore using a rubber or plastic tool.
7. Check the bore and piston for wear, rust, and pitting.
8. Install a new seal in the bottom of the bore and lubricate the bore and seal with brake fluid.
9. Gently insert the piston, making sure it does not cock and jamb in the bore.

→When installing pistons on ATE and Teves calipers, make sure the relieved part of the piston is in the correct position to accept the retaining plate for the brake pads. See brake pad removal and installation for more information and illustration.

10. Install the new outer seal and new spring ring.
11. Install the piston retaining plate.

9-12 BRAKES

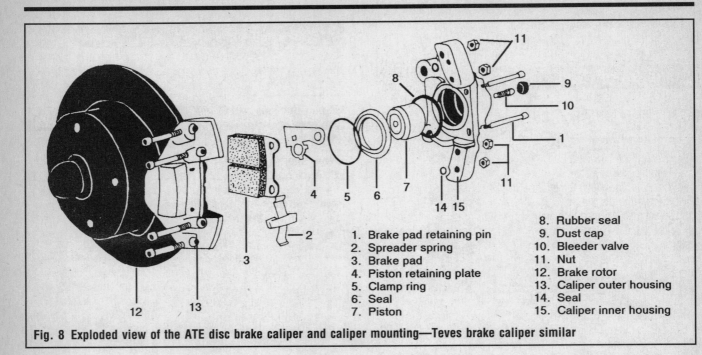

1. Brake pad retaining pin
2. Spreader spring
3. Brake pad
4. Piston retaining plate
5. Clamp ring
6. Seal
7. Piston
8. Rubber seal
9. Dust cap
10. Bleeder valve
11. Nut
12. Brake rotor
13. Caliper outer housing
14. Seal
15. Caliper inner housing

Fig. 8 Exploded view of the ATE disc brake caliper and caliper mounting—Teves brake caliper similar

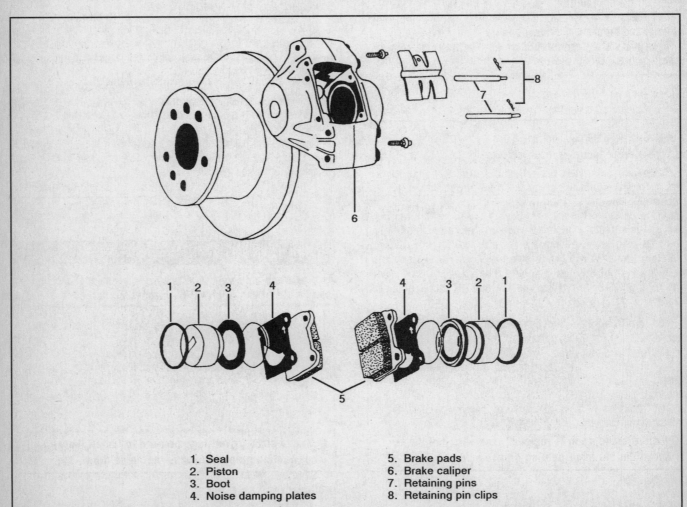

1. Seal
2. Piston
3. Boot
4. Noise damping plates
5. Brake pads
6. Brake caliper
7. Retaining pins
8. Retaining pin clips

Fig. 9 Exploded view of the Girling disc brake caliper and caliper mounting—1975 and later Type 2 models

BRAKES 9-13

12. Repeat the procedure on the other piston. Never rebuild only one side of a caliper.

Brake Rotor (Disc)

REMOVAL & INSTALLATION

1. Jack up the car and remove the wheel and tire.
2. Remove the caliper.
3. On Type 2 models, remove the three socket head bolts which secure the disc to the hub, and remove the disc from the hub. Sometimes the disc is rusted to the hub. Spray penetrating oil on the seam and tap the disc with a lead or brass hammer. If it still does not come off, screw three 8×40mm screws into the socket head holes. Tighten the screws evenly and pull the disc from the hub.
4. Type 1, 3, and 4 models, remove the wheel bearing cover. On the left side it will be necessary to remove the small clip which secures the end of the speedometer cable to the cover.
5. Loosen the nut clamp, unscrew the wheel bearing nut and remove the nut and outer wheel bearing.
6. Pull the disc off of the spindle.
7. To remove the wheel bearing races, see the "Wheel Bearing Removal and Installation" procedure.
8. Installation is the reverse of the removal procedure. Make sure the wheel bearing is properly adjusted.

INSPECTION

Visually check the rotor for excessive scoring. Minor scores will not affect performance; however, if the scores are over 1/32 in., it is necessary to replace the disc or have it resurfaced. The disc must be 0.02 in. over the wear limit to be resurfaced. The disc must be free of surface cracks and discoloration (heat bluing). Hand spin the disc and make sure that it does not wobble from side to side.

Front Wheel Bearings

REMOVAL & INSTALLATION

1. Jack up the car and remove the wheel and tire.
2. Remove the caliper and disc (if equipped with disc brakes) or brake drum.
3. To remove the inside wheel bearing, pry the dust seal out of the hub with a prytool. Lift out the bearing and its inner race.
4. To remove the outer race for either the inner or outer wheel bearing, insert a long punch into the hub opposite the end from which the race is to be removed. The race rests against a shoulder in the hub. The shoulder has two notches cut into it so that it is possible to place the end of the punch directly against the back side of the race and drive it out of the hub.
5. Carefully clean the hub.
6. Install new races in the hub. Drive them in with a soft faced hammer or a large piece of pipe of the proper diameter. Lubricate the races with a light coating of wheel bearing grease.
7. Force wheel bearing grease into the sides of the tapered roller bearings so that all spaces are filled.
8. Place a small amount of grease inside the hub.
9. Place the inner wheel bearing into its race in the hub and tap a new seal into the hub. Lubricate the sealing surface of the seal with grease.
10. Install the hub on the spindle and install the outer wheel bearing.

Once the brake drum is removed, use a seal puller to remove the old inner bearing grease seal

A correct-sized driver must be used to install the new inner bearing grease seal into the brake drum

9-14 BRAKES

11. Adjust the wheel bearing and install the dust cover.
12. Install the caliper (if equipped with disc brakes).

ADJUSTMENT

The bearing may be adjusted by feel or by a dial indicator.

To adjust the bearing by feel, tighten the adjusting nut so that all the play is taken up in the bearing. There will be a slight amount of drag on the wheel if it is hand spun. Back off fully on the adjusting nut and retighten very lightly. There should be no drag when the wheel is hand spun and there should be no perceptible play in the bearing when the wheel is grasped and wiggled from side to side.

To use a dial indicator, remove the dust cover and mount a dial indicator against the hub. Grasp the wheel at the side and pull the wheel in and out along the axis of the spindle. Read the axial play on the dial indicator. Screw the adjusting nut in or out to obtain 0.001–0.005 in. of axial play. Secure the adjusting nut and recheck the axial play.

REAR DRUM BRAKES

Brake Drum

REMOVAL & INSTALLATION

Except Type 4

1. With the wheels still on the ground, remove the cotter pin from the slotted nut on the rear axle and remove the nut from the axle. This isn't as easy as it sounds. You'll need a very long breaker bar or metal pipe for this operation.

※※ CAUTION

Make sure the emergency brake is now released.

2. Jack up the car and remove the wheel and tire.
3. The brake drum is splined to the rear axle and the drum should slip off the axle. However, the drum sometimes rusts on the splines and it is necessary to remove the drum using a puller.
4. Before installing the drum, lubricate the splines. Install the drum on the axle and tighten the nut on the axle to 217 ft. lbs. Line up a slot in the nut with a hole in the axle and insert a cotter pin. Never loosen the nut to align the slot and hole.

Type 4

The drum is held in place by the wheel lugs. Jack up the car and remove the wheel and tire. After the wheel is removed, there are two small screws that secure the drum to the hub and they must be removed before the drum will slip off the hub.

INSPECTION

Inspection is the same as given in the Front Drum Brake portion of this section.

Brake Shoes

REMOVAL & INSTALLATION

▸ See Figure 10

1. Remove the brake drum.
2. Remove both shoe retaining springs.
3. Disconnect the lower return spring.
4. Disconnect the hand brake cable from the lever attached to the rear shoe.
5. Remove the upper return spring and clip.
6. Remove the brake shoes and connecting link.
7. Remove the emergency brake lever from the rear shoe.
8. Lubricate the adjusting screws and the star wheel against the head of the adjusting screw.

After loosening the rear axle nut while the vehicle rests on the ground, raise the vehicle and remove the rear wheel and axle nut

Use a puller to draw the rear brake drum off of the rear axle shaft . . .

BRAKES 9-15

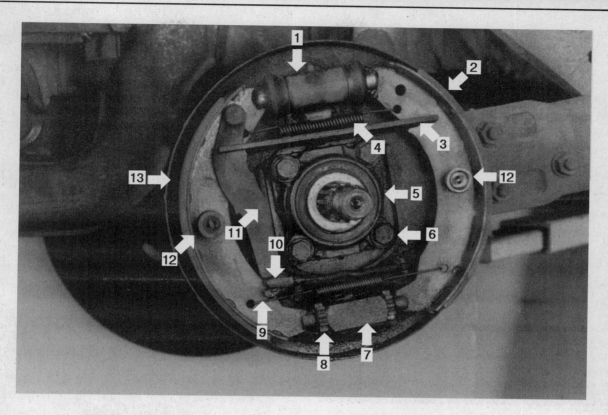

1. Wheel cylinder
2. Backing plate
3. Connecting link
4. Upper return spring
5. Rear axle shaft bearings (under cap)
6. Backing plate/Bearing cap mounting bolts
7. Anchor block
8. Star adjuster wheel (1 of 2)
9. Lower return spring
10. Parking brake cable
11. Parking brake lever
12. Hold-down spring with cup and pin

Identification of the rear brake drum components used on most models

... then lower the brake drum from the rear axle

9. Reverse Steps 1–7 to install the shoes.
10. Adjust the brakes.

Wheel Cylinder

REMOVAL & INSTALLATION

✱✱ WARNING

Clean, high quality brake fluid is essential to the safe and proper operation of the brake system. You should always buy the highest quality brake fluid that is available. If the brake fluid becomes contaminated, drain and flush the system, then refill the master cylinder with new fluid. Never reuse any brake fluid. Any brake fluid that is removed from the system should be discarded.

9-16 BRAKES

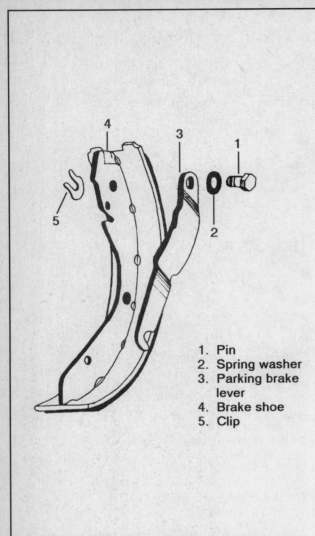

1. Pin
2. Spring washer
3. Parking brake lever
4. Brake shoe
5. Clip

Fig. 10 Details of the parking brake lever attachment to the rear brake shoes

It is also possible to remove the brake shoes by first disconnecting the parking brake cable from the brake shoe . . .

. . . then use the special tool to twist the hold-down spring cap until the slot is aligned with the tangs on the hold-down spring pin . . .

Use a suitable brake component cleaner to remove all brake dust from the components—do not use compressed air to blow the brake dust off of the components

. . . and remove the hold-down spring from both brake shoes

BRAKES 9-17

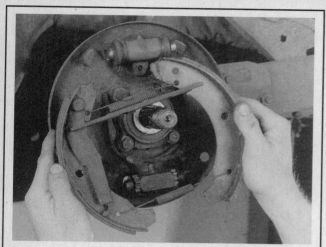

At this point, the brake shoe assembly can be removed as a whole from the backing plate . . .

. . . and remove the bearing cap

. . . and dismantled on a clean workbench

At this point, the entire backing plate/brake assembly can be removed from the rear axle shaft/bearing unit, and can be disassembled on a workbench

If, however, the backing plate needs to be removed, or if the rear bearings need servicing, simply remove the 4 bearing cap retaining bolts . . .

Remove the brake drum and brake shoes. Disconnect the brake line from the cylinder and remove the bolts which secure the cylinder to the backing plate. Remove the cylinder from the vehicle.

OVERHAUL

▶ See Figures 11 and 12

Overhaul is the same as given in the Front Drum Brake portion of this section.

9-18 BRAKES

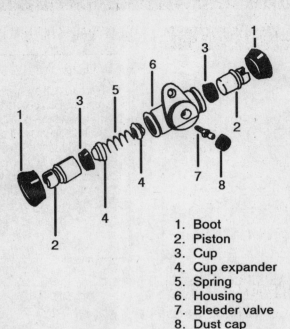

1. Boot
2. Piston
3. Cup
4. Cup expander
5. Spring
6. Housing
7. Bleeder valve
8. Dust cap

Fig. 11 Exploded view of the wheel cylinder used on the front of Type 1 models and on the rear of Types 1, 3 and 4 models

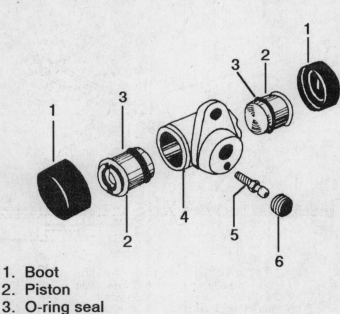

1. Boot
2. Piston
3. O-ring seal
4. Housing
5. Bleeder valve
6. Dust cap

Fig. 12 Exploded view of the rear wheel cylinder used on Type 2 models

BRAKES 9-19

PARKING BRAKE

Cable

ADJUSTMENT

Brake cable adjustment is performed at the handbrake lever in the passenger compartment on all vehicles except Type 2 models. There is a cable for each rear wheel and there are two adjusting nuts at the lever. On Type 2 models, adjust from below the vehicle.

To adjust the cable, loosen the locknut. Jack up the rear wheel to be adjusted so that it can be hand spun. Turn the adjusting nut until a very slight drag is felt as the wheel is spun. Then back off on the adjusting nut until the lever can be pulled up three notches on all models except Type 2's, six notches on Type 2 models.

※※ CAUTION

Never pull up on the handbrake lever with the cables disconnected.

REMOVAL & INSTALLATION

▶ See Figure 13

1. Disconnect the cables at the handbrake lever by removing the two nuts which secure the cables to the lever. Pull the cables rearward to remove that end from the lever bracket.
2. Remove the brake drums and detach the cable end from the lever attached to the rear brake shoe.
3. Remove the brake cable bracket from the backing plate and remove the cable from the vehicle.
4. Reverse the removal procedure for installation. Adjust the cable.

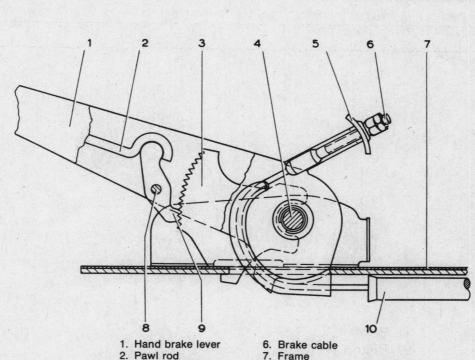

1. Hand brake lever
2. Pawl rod
3. Ratchet segment
4. Lever pin
5. Cable compensator
6. Brake cable
7. Frame
8. Pawl pin
9. Pawl
10. Cable guide tube

Fig. 13 Identification of parking brake hand lever and cable end assembly components

Brake Specifications
All measurements given are (in.) unless noted

Year	Model	Lug Nut Torque (ft. lb.)	Master Cylinder Bore	Brake Disc		Brake Drum			Minimum Lining Thickness	
				Minimum Thickness	Maximum Run-Out	Diameter	Max Machine O/S	Max Wear Limit	Front	Rear
1970–77	Type 1 (Beetle)	87–94	0.750	—	—	9.059	9.10	9.114	0.100	0.100
1970–80	Type 1 (Super Beetle)	87–94	0.750	—	—	9.768 (fr) 9.059 (rr)	9.80 (fr) 9.10 (rr)	9.823 (fr) 9.114 (rr)	0.100 0.079	0.100 0.100
1970–74	Karmann Ghia	87–94	0.938	0.335	0.0008	9.059	9.10	9.114		
1970–81	Type 2 (Bus)	87–94	0.938	0.472 ①	0.0008	9.920	9.97	9.98	0.079	0.100
1970–73	Type 3	87–94	0.750	0.393	0.0008	9.768	9.80	9.82	0.079	0.100
1970–74	Type 4	87–94	0.750	0.393	0.0008	9.768	9.80	9.82	0.079	0.100

NOTE: *Minimum lining thickness is as recommended by the manufacturer. Due to variations in state inspection regulations, the minimum allowable thickness may be different than recommended by the manufacturer.*

① 0.433 in.-Vanagon
(fr)—Front
(rr)—rear
—Not Applicable

EXTERIOR 10-2
FRONT DOORS 10-2
  REMOVAL & INSTALLATION 10-2
  DOOR STRIKER PLATE
    ALIGNING 10-3
SIDE SLIDING DOOR 10-3
  REMOVAL & INSTALLATION 10-3
  ALIGNMENT 10-4
FRONT HOOD 10-4
  REMOVAL & INSTALLATION 10-4
REAR HOOD 10-4
  REMOVAL & INSTALLATION 10-4
BUMPERS 10-6
  REMOVAL & INSTALLATION 10-6
FENDERS 10-10
  REMOVAL & INSTALLATION 10-10
AUTO BODY REPAIR 10-14
TOOLS AND SUPPLIES 10-14
RUST, UNDERCOATING, AND
  RUSTPROOFING 10-14
  RUST 10-14
AUTO BODY CARE 10-15
WASHING 10-15
CLEANERS, WAXES AND POLISHES 10-16
  CLEANERS AND
    COMPOUNDS 10-16
  WAXES AND POLISHES 10-16
    SPECIAL SURFACES 10-17

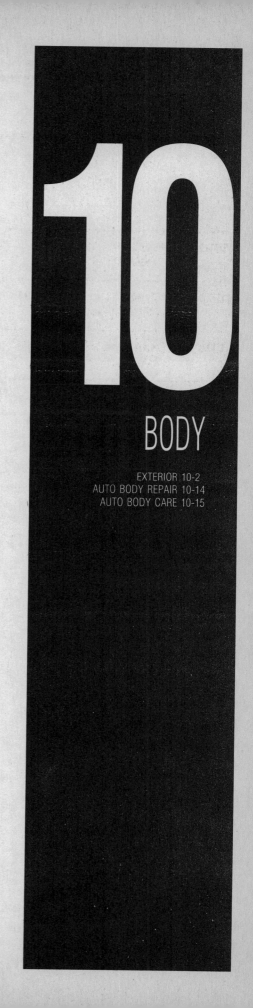

10
BODY

EXTERIOR 10-2
AUTO BODY REPAIR 10-14
AUTO BODY CARE 10-15

BODY

EXTERIOR

Front Doors

REMOVAL & INSTALLATION

▶ See Figure 1

Types 1, 3 and 4 Models

If the original door is going to be reinstalled, it is easier to press out the hinge pins rather than remove the hinges from the car body; this will avoid the need to align the door later. If you do decide to remove the hinges, carefully matchmark the hinges to the vehicle body to ease aligning the door during installation.

✳✳ WARNING

Never leave a door unsupported while one of the hinges is unscrewed, or while its hinge pin is out; doing so could bend or break the other hinge.

1. Open the door and position a padded floor jack under the door for support.
2. Remove the circlip from the lower end of the pin that holds the door check strap to the body, then remove the door check strap.
3. If the original door is to be reinstalled, perform the following:
    a. Remove the plastic caps from the tops of the hinge pins.
    b. Press out the hinge pins with a special press-out tool.

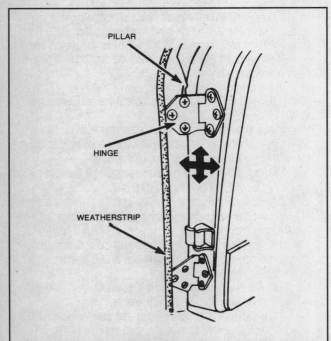

Fig. 1 With the hinge-to-body screws loose, move the door up and down or in and out until the gap around the door (when its closed) is uniform, then tighten the screws securely

✳✳ WARNING

Do not attempt to hammer the pins out of the hinges; the hinges will be damaged.

　c. With the help of an assistant, remove the door from the vehicle.
4. If the hinge pins are going to be left in place, perform the following:
    a. Pry the plastic caps (that are adjacent to the hinges) out of the body's hinge pillar (except on Karmann Ghias).
    b. Using an impact driver, loosen the Phillips head screws holding the hinges to the body.
    c. Taking care not to scratch the door, body, or fender, remove the door (together with the its hinges) from the body.

To install:

5. If the original door is being installed on the vehicle, the installation procedure is the reverse of the removal procedure. Make certain to check the weatherseal; if it is cracked, torn, or otherwise damaged, replace it.
6. If a replacement door is being installed, perform the following:
    a. Remove the lock striker plate from the lock pillar on the body.
    b. Install the new door, together with its hinges, but do not fully tighten the Phillips head screws.

➡**The hinges are screwed to movable threaded plates. This makes it possible to shift the position of the door in its opening for alignment purposes.**

　c. Align the door in the door opening so that it contacts the weatherstrip evenly all around and the door's trim molding is in line with the trim molding on the side of the body.
　d. After the door is aligned, tighten the Phillips head screws, then set them firmly with an impact driver.
　e. Install the door check strap with the pin and circlip. Install the lock striker plate and adjust it as described later in this section.

Type 2 Models

If the same door is to be reinstalled, matchmark the location of the hinges on your body. this will prevent your having to align the door later. From chassis No. 236 2126 632, new hinges (Part No. 281 831 405) are used on the front doors. When installing a new door on an earlier vehicle, you must also install the newer hinges, which are maintenance-free and do not require oiling. However, if you press out the hinge pins, you must be careful not to damage the Teflon bushing when the pin is reinstalled. Otherwise the hinge will jam and may then require periodic lubrication.

1. Remove the circlip off the pin for the door check strap. Remove the pin and disconnect the strap.
2. With the door solidly supported so that the upper hinge will not bend or break by the door's weight, remove the 2 Phillips head (or socket head) screws that hold the lower hinge to the hinge pillar. If the screws are rusted tight, loosen them with an impact driver.
3. With the door supported, remove the 2 Phillips head (or

BODY 10-3

socket head) screws from the upper hinge and remove the door together with its hinges.

To install:

4. Inspect the rubber weatherstrip around the door. If it is cracked or deformed, replace it.

➡ **Before replacing the weatherstripping, clean all of the old adhesive off of the door with solvent. Install the new weatherstripping with trim adhesive.**

5. If a new door is being installed, remove the lock striker plate from the lock pillar.

6. If the original door is being installed, mount the hinges with the matchmarks correctly aligned. If a new door is being installed, do not fully tighten the hinge mounting screws.

➡ **The hinges are screwed to movable threaded plates. This makes it possible to shift the position of the door in its opening for alignment purposes.**

7. To align a new door, position the door in the door opening so that it contacts the weatherstripping evenly all around and the door's trim molding is aligned with the trim molding on the side of the body.

8. After the door is properly aligned, tighten the hinge mounting screws. Set the Phillips head screws firmly with an impact driver.

9. Attach the door check strap to its bracket with the pin and circlip.

10. Install the door lock striker plate and adjust it as described later in this section.

DOOR STRIKER PLATE ALIGNING

All Models

After installing a replacement door, adjust the striker plate so that the rear edge of the door aligns with the body. The striker plate should also be adjusted if the door rattles or requires excessive force to close and lock.

Because of body flexion, the door lock striker plate cannot be accurately adjusted while the car is raised on jackstands or a lift.

A door rattle that persists even after all possible adjustments have been made indicates a worn rubber wedge. On 1970 and 1971 cars, this can bee corrected by placing a 0.50–1.50mm shim between the wedge and the striker plate. On 1972 and later cars, replace the wedge with a new one.

1. Remove the striker plate mounting screws and remove the striker plate from the body. Close the door and check its alignment. Adjust the door if it is out of adjustment.

2. Insert the striker plate, bottom first, in the latch. Press the latch down into its fully locked position. Turn the striker plate until it is in its normal upright (vertical) position.

3. If the striker plate exhibits vertical (up and down) free-play, the rubber wedge is worn. Either add shims or replace the wedge with a new one.

➡ **To correct either a misalignment between the door and front body, or a lack of uniformity in the gap between the door and body, adjust the hinges and not the striker plate. Only misalignment (vertical or horizontal) between the door and rear body should be corrected by adjusting the striker plate.**

4. After correcting any excessive wedge play, correct any misalignment between the door and rear body. Center the striker plat on the lock pillar by aligning the marks on the striker plate with those on the pillar.

5. Close the door. Check if the door aligns with the rear body. If necessary, adjust the position of the striker plate on the lock pillar.

➡ **The striker plate screws onto the movable threaded plate. By moving the plate up or down, you can correct any misalignment of the ridge in the body where the door and the rear body meet. Moving the plate in or out corrects the position of the door so that it will be flush with the side of the car when the door is closed.**

6. After aligning the door, feel for play between the lock and the striker plate. If there is play, or if the door will not latch, rotate the top of the striker plate toward the outside off the vehicle. If the door is too difficult or hard to close, or if the handle works stiffly, rotate the top of the striker plate toward the inside of the vehicle. Only rotate the striker plate a little at a time, then recheck the alignment. Once the proper alignment is found, tighten the striker plate screws securely.

Side Sliding Door

REMOVAL & INSTALLATION

Type 2 Models

➡ **An assistant will be necessary to safely perform this procedure.**

1. Remove the outside runner cover as follows:

   a. Remove the 3 Phillips head screws, located at the front of the outside runner cover, that hold the cover on the body.

   b. Remove the cover securing nut and bolt from inside the passenger compartment.

   c. Fully open the sliding door.

   d. Loosen the Phillips head screw on the retaining strip by about 15 revolutions.

   e. Place a punch against the head of the Phillips head screw, then, using a hammer, tap the punch sharply to drive the retaining strip toward the rear of the body.

   f. Starting at the rear of the sliding door, lift the outside runner cover up and out of the retaining strip.

   g. After taking the cover off, fully remove the Phillips head screws from the ends of the retaining strip and remove the retaining strip from the body.

2. Push the door far enough to the rear so that the guide piece and the roller on the hinge link can be lifted sideways out of the recess in the center runner.

3. Push the door completely to the rear and lift it until the upper roller can be lifted out of the top runner.

4. Swing the door slightly outward and pull the lower rollers out of the break in the bottom runner and remove the door.

To install:

5. Inspect the runners and, if necessary, straighten them. Check the guide and support rollers on the door. Replace damaged rollers. Lubricate the rollers with multipurpose grease if they turn stiffly.

6. Inspect the rubber weatherstripping around the door opening in the body. If necessary, remove the old weatherstripping, using solvent to remove any residual traces of the weatherstripping. Use a weatherstripping adhesive to glue the new weatherstripping in place.

7. Insert the door first into the bottom runner, then into the top runner.

8. Push the door forward until the roller and guide can be inserted in the break in the center runner.

9. Loosely install the retaining strip. Press the cover into the gap between the body and the retaining strip from above, then insert the beading.

➡ **Place a small amount of plastic sealing compound ("dum-dum") between the retaining strip and the body to maintain a gap while the cover and beading are being installed.**

10. Install the 2 Phillips head screws which screw in from below.

11. By turning the Phillips head screw in the front end of the retaining strip, tension the retaining strip while making sure that the beading is correctly positioned.

12. Secure the cover at the lock pillar end with the remaining Phillips head screw. Install the nut and bolt from inside the passenger compartment.

13. Align the side sliding door as described later in this section.

ALIGNMENT

The sliding door is properly adjusted if the gap between the door and the door opening is even all around. The trim or waistline on the door must align with the trim or waistline on the body and the door surface must be flush with the surface of the body.

- If the door is not aligned at the bottom, adjust the lower roller. To do this, loosen the Phillips head screw and the 2 socket head screws which hold the lower roller bracket. Insert or remove shims to adjust the height of the roller; the roller can also be moved forward or rearward.
- If the door is not flush with the body at the top, loosen the nut on the top roller shaft. Adjust the roller in or out until the top of the door is flush.
- To prevent excessive vertical door movement, loosen the 3 Phillips head screws that hold the top roller bracket. Raise the bracket until the clearance between the roller and the runner is as small as possible.
- To adjust the door gap to a uniform width, close the door. Loosen the 4 bolts on the hinge housing and adjust the angle of the hinge link.
- To check excessive latch play, press the door firmly near the hinge link. If there is any detectable play, adjust the striker plate (as described earlier in this section).
- To align the door retainer with the rear bracket, loosen the Phillips head screws and shift the position of the retainer up or down.
- To adjust the locking plate for the remote control lock, slightly loosen the 2 locking plate bolts and close the door to center the locking plate. Open the door and tighten the bolts securely. If necessary, up to 2 spacer shims can be placed behind the locking plate.

Front Hood

REMOVAL & INSTALLATION

Types 1, 3 and 4 Models

To avoid damaging the finish, cover the cowl panel with a protective cloth. To save installation time, matchmark the hinge positions before removing the hood. An assistant is needed for this procedure.

1. Matchmark the hinges to the underside of the hood.
2. Remove the 2 hood bolts from one of the hinges.
3. Have an assistant hold the unbolted side of the hood while you remove the 2 hood bolts from the other side.
4. Together, lift the hood up and off toward the front of the car.

To install:

5. Before installing the hood, check the condition of the rubber weatherstripping. If necessary, reglue or replace it.

✱✱ WARNING

If the weatherstripping is going to be replaced, make certain to remove all traces of old weatherstripping by using a suitable solvent. Otherwise the new weatherstripping may not install evenly or may fail to adhere properly to the body.

6. Attach the hood loosely. Move the hood in the elongated bolt holes until it contacts the weatherstripping evenly all around. Then tighten the bolts securely.

7. On Type 3 and 4 models, align the hood with the fenders by screwing the adjustable rubber bumpers in or out. To prevent vibrations, be sure the bumpers contact the body.

8. Check the lock operation by opening and closing the hood several times. If necessary, adjust the lock.

➡ **If it is necessary to lower the hinges after the hood has been removed, use a lever bolted to the hinge.**

To remove the hinge from the body, remove the dashboard access panel from the rear of the luggage compartment by removing the 2 knurled nuts. Also remove the front luggage compartment load liner. Pry the E-clip off the bottom pivot pin for the spring mount. Then press out the pin and swing the spring toward the front of the car until it is no longer compressed. unbolt the hinge from the bracket under the dashboard.

Type 2 Models

Type 2 models are not equipped with front hoods.

Rear Hood

REMOVAL & INSTALLATION

Type 1 Models

1. Open the rear hood, then remove the air cleaner assembly.
2. Matchmark the hinge position on the rear hood.

BODY 10-5

To remove the rear hood, first remove the air cleaner assembly (A), matchmark the hinges (B) to the hinge supports (C), then remove the 4 hinge-to-hinge support bolts (D)

10-6 BODY

3. Disconnect the negative battery cable.
4. Cover the carburetor intake to prevent the entry of dirt.
5. Disconnect the license plate light wire and unfasten it from the hood so that it will not be damaged as the hood is removed.
6. Remove the 4 bolts that hold the hood hinges to the curved hinge brackets on the body.
7. Pull the hood out and upward against spring pressure and unhook the load spring from the bracket on the vehicle's roof panel. The spring will stay with the hood when the hood is removed.
8. If necessary, remove the curved hinge brackets from the car body by removing the 3 bolts at each bracket.

To install:
9. Before installing the hood, check the condition of the rubber weatherstripping. If necessary, reglue or replace the weatherstripping.

※※ WARNING

If the weatherstripping is going to be replaced, make certain to remove all traces of old weatherstripping by using a suitable solvent. Otherwise, the new weatherstripping may not install evenly or may fail to adhere properly to the body.

10. Loosely install one hood hinge on the curved bracket on the car body.
11. Holding the hood at an angle so that you can reach behind it, engage the spring in the bracket on the car roof panel.
12. Push the hood upward and toward the car. Loosely install the remaining hinge.
13. Move the hood in the elongated bolt holes until it contacts the weatherstripping evenly all around, then tighten the hinge bolts securely.

Types 2, 3 and 4 Models

Some Type 2 models are equipped with an upper luggage compartment hood and a lower engine access hood, whereas the Type 3 and 4 models are equipped with one rear hood which supplies access to the luggage compartment and engine bay—all rear hoods on these models are removed in the same manner.

1. Before removing the rear luggage compartment hood or the rear engine access hood, matchmark its position on the hinges for easier alignment during installation.
2. Disconnect the negative battery wire.
3. Cut the license plate light wire in the middle so that it can later be rejoined with a splice or butt-end connector.
4. Remove the hinge-to-hood bolts on one side of the hood. Have an assistant support the unbolted side of the hood while you remove the hinge-to-hood bolts on the other side of the hood.
5. Lift off the rear hood together.

To install:
6. Installation is the reverse of the removal procedure. Make sure that the hinge-to-hood matchmarks are properly aligned and make certain that the latch works correctly.
7. Once the hood is properly aligned and the hinge-to-hood bolts and tightened securely, splice the two ends of the license plate light wire back together.
8. Connect the negative battery cable.

Bumpers

REMOVAL & INSTALLATION

Type 1 Models

On 1970–73 models, you can remove only the bumper itself by taking off the nuts on the back of the bumper. The brackets and reinforcement strap can stay on the car. On 1974 and later cars, the reinforcement is welded to the bumper. Remove these reinforced bumpers as a unit together with the spring-loaded damping elements that are used in place of brackets.

In addition to the welded-in reinforcement and spring-loaded damping elements, the bumpers on 1974 and later cars have plastic protectors on their ends. these protectors are intended to prevent the body work from being marred in the event that the damping elements are fully compressed.

FRONT BUMPER
♦ See Figures 2 and 3

1. Using a 13mm wrench, remove all bumper bracket-to-body bolts.
2. Pull the bumper, together with the brackets and reinforcement strap or the damping elements, out of the slots in the front fenders.
3. On 1970–73 models, remove the 3 nuts from each bracket. Separate the reinforcement strap from the bumper and the brackets.
4. Remove the 3 nuts from each bumper bracket or, on 1974 and later cars, from each damping element. remove the brackets or damping elements from the bumper.
5. Installation is the reverse of the removal procedure. Replace the rubber grommets in the front fenders if they are weathered or damaged. Lubricate the threads on the nuts and bolts before installing them.

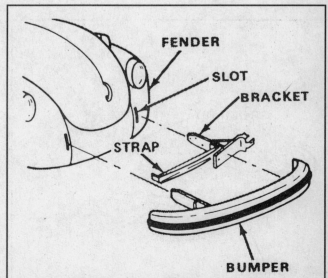

Fig. 2 The bumper and bracket should be removed together, then the bracket can be removed from the bumper

BODY 10-7

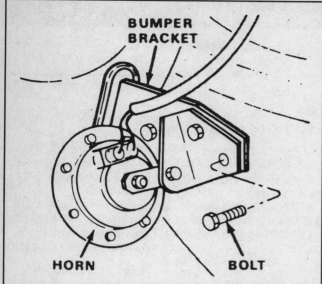

Fig. 3 The horn mounting bracket bolts to the left-hand bumper bracket and must be removed before the bumper can be removed from the car

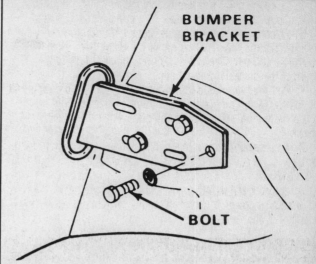

Fig. 5 A view of one of the two rear bumper brackets—note that there are 3 mounting bolts per side

When installing the bumper and the brackets on the car, install the bolts, but do not tighten them. Adjust the bumper to a uniform gap with the body before tightening the bolts to 14 ft. lbs.

REAR BUMPER

▶ See Figures 4, 5 and 6

1. Using a 13mm wrench, remove all bumper bracket-to-body bolts.
2. Pull the bumper, together with the brackets and reinforcement strap or the damping elements, out of the slots in the front fenders.
3. Remove the 3 nuts from each bumper bracket or damping

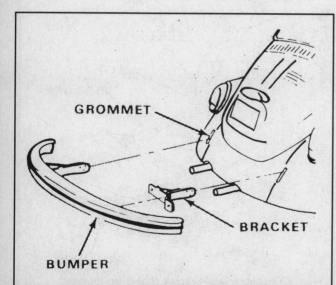

Fig. 4 The rear bumper mounts in the same manner as the front bumper—remove the 6 bracket-to-body bolts, then pull the bumper off of the car

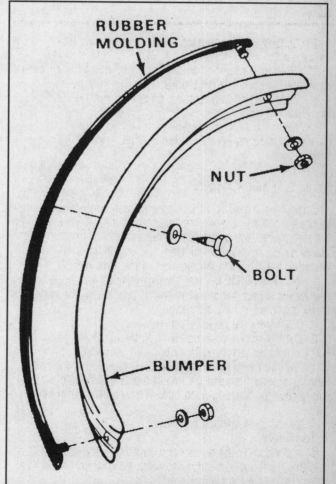

Fig. 6 If so desired, or if a new bumper is being installed and the rubber trim must be transferred, the rubber trim can be removed from the bumper by removing the 2 nuts and the center bolt

10-8 BODY

element. remove the brackets or damping elements from the bumper.

4. Installation is the reverse of the removal procedure. Replace the rubber grommets in the front fenders if they are weathered or damaged. Lubricate the threads on the nuts and bolts before installing them.

When installing the bumper and the brackets on the car, install the bolts, but do not tighten them. Adjust the bumper to a uniform gap with the body before tightening the bolts to 14 ft. lbs. On 1974 and later models, position the bumpers as far away from the body as possible.

Type 2 Models

1970–72 FRONT BUMPER

▶ See Figure 7

1. Thoroughly clean the exposed threads on all bumper mounting bolts before you attempt to remove them. If corrosion is evident, or if the nuts or bolts are difficult to turn, apply penetrating oil to the threads.
2. Remove the 2 side support bolts located just ahead of each front wheel.
3. Remove the 2 bumper bracket-to-frame rail bolts from one end of the bumper.
4. Place a support under the unbolted end of the bumper, or have someone hold it. Then remove the 2 bolts from the other end of the bumper and remove the bumper from the vehicle.
5. To remove the bumper bracket, remove the nuts and bolts attaching the bracket to the bumper.
6. To remove the end pieces, take out the 3 nuts and bolts that hold them on the bumper. Be careful not to damage the rubber seal.
7. Installation is the reverse of the removal procedure. Loosely install the assembled bumper and brackets on the vehicle. Then, by sliding the brackets and side supports on their elongated bolt holes, obtain a uniform gap between the bumper and the body. When the bumper is properly aligned, tighten the bolts to 25 ft. lbs.

1973–81 FRONT BUMPER

▶ See Figure 8

The one-piece front bumper installed on 1973–81 models is bolted on an energy-absorbing support. The bumper is removed by taking out the mounting 4 bolts. Installation is the reverse of the removal procedure. Tighten the mounting bolts to 25 ft. lbs.

1970–72 REAR BUMPER

▶ See Figures 9, 10, 11 and 12

1. Remove the bolts and nuts at both sides of the vehicle that hold the cover plate to the body.
2. Remove the 2 bolts that hold one of the bumper brackets to the frame sidemember.
3. Place a support under the unbolted end of the bumper, or have someone hold it. Then remove the 2 bolts from the other bracket and remove the bumper from the vehicle.
4. The brackets and cover plate can be unbolted at this time.

➡Thoroughly clean the exposed threads. if corrosion is evident, or if the nuts are hard to turn, apply penetrating oil.

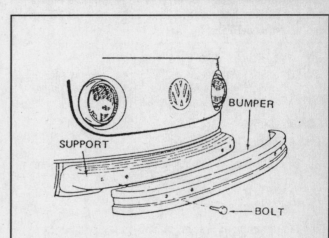

Fig. 8 On 1973–81 Type 2 models, the front bumper mounts to the support by 4 bolts, one of which is shown

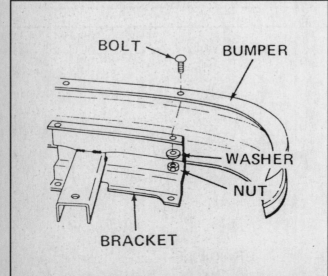

Fig. 7 Each bumper bracket is mounted to the front bumper by 4 bolts and nuts

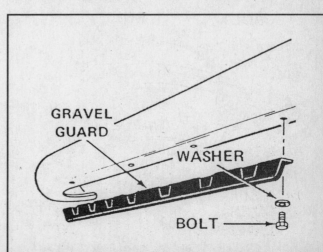

Fig. 9 The gravel guard is held onto the bottom edge of the rear bumper by bolts

BODY 10-9

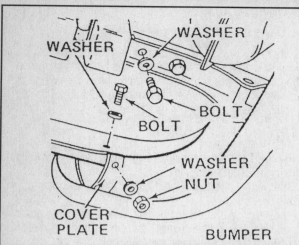

Fig. 10 Exploded view of the rear bumper and bracket mounting for 1973–81 Type 2 models

5. Installation is the reverse of the removal procedure. Be careful not to damage the rubber spacers by overtightening the bolt on the cover plate. Loosely install the assembled bumper and brackets on the vehicle. Then, by sliding the brackets and cover plate on their elongated bolt holes, obtain a uniform gap between the bumper and the body. When the bumper is properly aligned, tighten the bracket bolts to 25 ft. lbs. Do not overtighten the bolts that hold the cover plates on the body.

1973–81 REAR BUMPER
♦ See Figures 13 and 14

1. Remove the 2 bracket-securing bolts from one end off the bumper.
2. Place a support under the unbolted end of the bumper, or have someone hold it.
3. Remove the 2 bolts from the other end of the bumper, then remove the bumper from the vehicle.

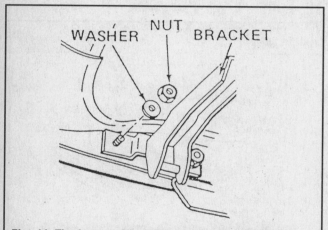

Fig. 11 The bumper bracket mounts to the bumper by two bolts and nuts—during installation, make sure to install the washers

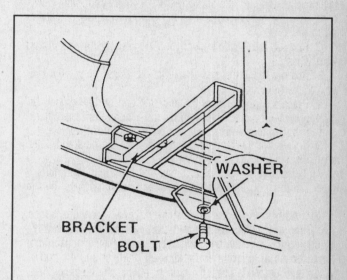

Fig. 13 Each rear bumper bracket is held onto the body with two bolts and washers

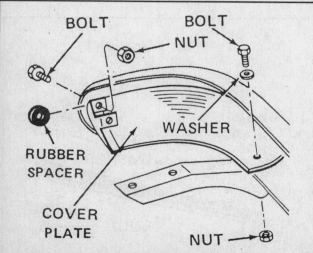

Fig. 12 Once the rear bumper is removed, the cover plate can be unbolted from the bumper

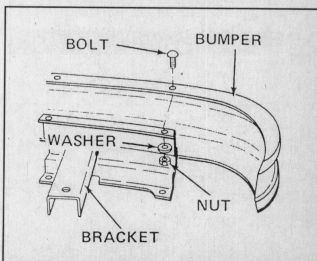

Fig. 14 To remove the bumper bracket from the bumper, simply unbolt it

10-10 BODY

To install:

4. Loosely install the assembled bumper and brackets on the vehicle.

5. By sliding the brackets on their elongated bolt holes, obtain a uniform gap between the bumper and the body.

6. When the bumper is properly aligned, tighten the bracket bolts to 25 ft. lbs.

Types 3 and 4 Models

FRONT BUMPER

Although the bumpers have undergone a major design change during the years covered by this manual, the mounting remains relatively similar.

The front bumper has a reinforcing bar between it and the front body apron.

1. Disconnect the negative battery cable.

2. Disconnect the horn wire. the horn may be left attached to the bumper until the bumper is removed from the car.

3. Remove the 2 screws securing the reinforcement bar inside the bumper.

4. Remove the 3 bolts under each front fender that hold the bumper brackets onto the body.

5. Pull the bumper forward and off the car, complete with brackets and reinforcement bar.

6. Lift out the reinforcement bar. Remove the nuts from the carriage bolts that hold the brackets to the bumper, then remove the brackets. Unbolt the horn and remove it.

7. Installation is the reverse of the removal procedure. When installing the bumper assembly, make certain that the gap between the bumper and the body is uniform. When bolting the bumper brackets to the body, remember to install the washers.

REAR BUMPER

There is no reinforcement bar behind the rear bumper. Also, the brackets used to mount the rear bumper are different from those used to mount the front bumper.

1. Remove the 2 bumper bracket-to-body bolts on each end of the bumper.

2. Pull the bumper rearward and off the vehicle, complete with brackets.

3. Unbolt the brackets and back-up lights, if applicable.

4. Installation is the reverse of the removal procedure.

Fenders

REMOVAL & INSTALLATION

Type 1 Models

Karmann Ghia fenders are welded to the body and require special tools and skills for replacement. Fenders on other models can be replaced by any car owner.

FRONT FENDERS

▸ See Figure 15

1. Disconnect the negative battery cable.

2. Use an 8mm wrench to remove the 2 nuts under the fender. Remove the turn signal/side-marker light assembly from the fender, then disconnect the light from its wires. Label the wires to ensure correct installation.

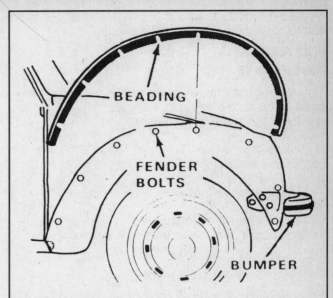

Fig. 15 When installing the front fender, make sure to install new plastic fender beading between the fender and the body

3. Pull the gasket for the turn signal/side-marker light assembly down through the fender.

4. Remove the sealed beam unit from the headlight as described in Section 6. Remove the 3 screws that hold the headlight assembly in the fender and remove the assembly.

5. Insert a small flat-bladed prytool into the small slot below each terminal in the sealed beam connector. Then pull the wire and terminal out of the rear of the connector.

6. Pull the wires and their protective hose back and out of the fender's headlight housing.

7. Remove the front bumper as described earlier in this section.

8. On left fenders, disconnect the wires from the horn, then remove the horn.

9. Remove the nut, bolt and washers that hold the fender to the running board. Then remove the 10 fender bolts and washers and remove the fender and beading.

To install:

11. Position the fender against the car body without the beading. Loosely install a bolt at the top of the fender in order to support it.

12. Start the remaining 7 bolts in their threads so that the fender hangs loosely on the car.

13. Install a new rubber spacer washer between the fender and the running board. Install, but do not tighten the nut, bolt and washers.

14. Using your hand, press the beading into the space between the fender and the body. Make sure the beading is even and that the notches have all slipped over the bolts.

15. Gradually tighten the bolts to 11 ft. lbs., checking constantly to see that the beading remains in place.

16. Tighten the bold that holds the fender to the running board to 11 ft. lbs.

17. The remainder of installation is the reverse of the removal procedure. When installing the sealed beam terminals in their connector, the yellow wire goes in the top center slot, the white wire

BODY 10-11

in the left slot and the brown wire in the right slot (as seen from the front of the car, looking into the headlight).

REAR FENDERS
♦ See Figure 16

1. Working under the fender, use an 8mm wrench to remove the tail light nuts, then remove the tail light assembly. 4 nuts secure the tail light assembly on 1973 and later cars, 2 on 1970–72 models.
2. Disconnect the wires from the tail light assembly and label them for correct reinstallation. Pull the wires and the rubber grommet down and out of the fender.
3. Remove the rear bumper, as described earlier in this section.
4. Remove the nut, bolt and washers that hold the fender to the running board. Then remove the 9 bolts and washers which hold the fender to the car body. Remove the old beading.

To install:

11. Position the fender against the car body without the beading. Loosely install a bolt at the top of the fender in order to support it.
12. Start the remaining 8 bolts in their threads so that the fender hangs loosely on the car.
13. Install a new rubber spacer washer between the fender and the running board. Install, but do not tighten the nut, bolt and washers.
14. Using your hand, press the beading into the space between the fender and the body. Make sure the beading is even and that the notches have all slipped over the bolts.
15. Gradually tighten the bolts to 11 ft. lbs., checking constantly to see that the beading remains in place.
16. Tighten the bold that holds the fender to the running board to 11 ft. lbs.
17. The remainder of installation is the reverse of the removal procedure. Make certain that the tail lights, stop lights and turn signals all work properly.

➡ Because the fenders for 1973 and later cars have larger tail light mountings, they cannot be interchanged with the fenders of earlier models.

RUNNING BOARD
♦ See Figure 17

To remove the running board, remove the nuts, bolts and washers that hold the running board to the front and rear fenders. remove the two rubber spacer washers from between the fenders and the running board. Then loosen, but do not remove, the 4 bolts that hold the running board to the body.

Installation is the reverse of the removal procedure. Use new rubber spacer washers at the front and rear of the running board. Tighten the bolts that fasten the running board to the fenders to 11 ft. lbs. Tighten the bolts which hold the running board to the body to 7 ft. lbs.

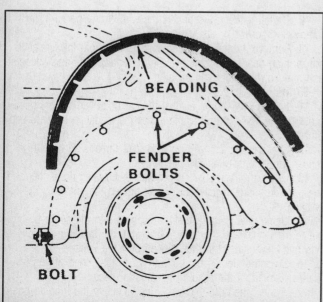

Fig. 16 The rear fender also requires plastic beading inserted between it and the vehicle body during installation

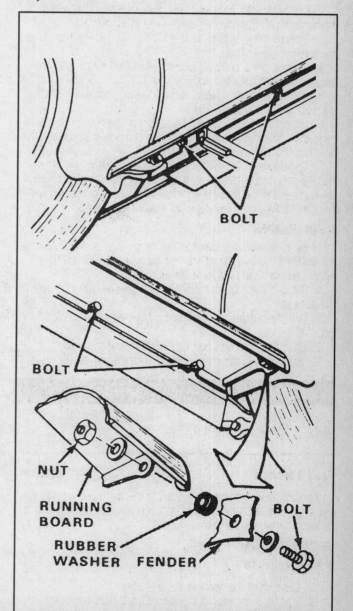

Fig. 17 Exploded view of the mounting of the running boards used on Type 1 models

10-12 BODY

Type 2 Models

The fenders on Type 2 models are not removable. If the fenders are damaged and need replacement, have the fenders repaired by someone qualified to do automotive body work.

Types 3 and 4 Models

The fenders can be removed and installed on the body without any cutting or welding. All fenders are held in place by bolts. Plastic beading forms a gasket between the individual fenders and the body itself.

FRONT FENDERS

▶ See Figure 18

The bumper brackets pass through the front fenders. The front bumper and reinforcement bar must therefore be removed prior to removing the front fender.

1. Disconnect the negative battery cable.
2. Disconnect the horn wire (left fender only).
3. Remove the tank filler flap lock and cable (right fender only).
4. Remove the hexagon head sheet metal screw, located under the edge of the fender just ahead of the door.
5. With the car door open, remove the single bolt from the door hinge pillar.
6. Carefully pull the front side panel trim off the body interior, then remove the bolts behind the front side panel trim.
7. Remove the 4th bolt underneath the instrument panel. You must take the fuse box out first when working on the left side of the car.
8. Remove the 9 sheet metal screws together with the flat washers and rubber washers.

To install:

9. Apply new **D 19** sealing tape on the hinge pillar.
10. Install the fender loosely, using all of the bolts.
11. Insert the plastic beading between the fender and the body.
12. Align the fender properly with the body and tighten all of the bolts.
13. Glue the front panel trim back with **D 12** trim cement.

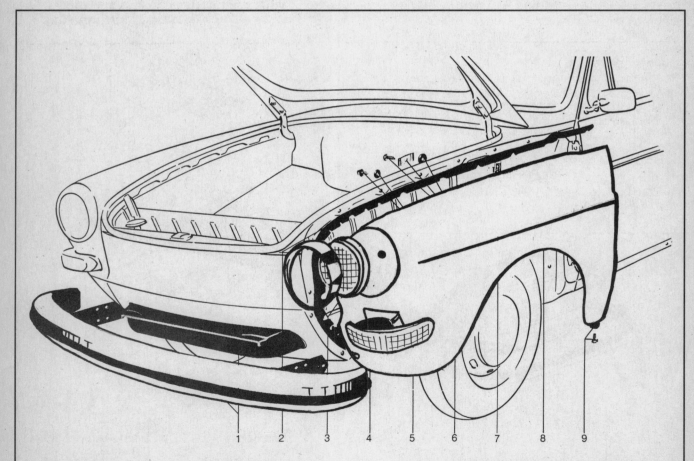

1. Bumper and reinforcement bar
2. Beading
3. Headlight
4. Turn signal
5. Sheet metal screw with flat and rubber washers (10)
6. Bolt with plate and rubber washer (2)
7. Speed nut (10)
8. Fender
9. Bolt and washer

Fig. 18 Exploded view of the front fender mounting on Type 3 models

BODY 10-13

REAR FENDERS

▶ See Figure 19

The rear fenders of the fastback and Squareback Sedans differ, but the mounting fastener locations are identical. The main difference in removal procedures is that all 8 upper mounting screws of the Squareback are removed from the rear luggage area. On the Type 3 Sedan, 5 screws are removed from inside the luggage area and 3 are removed from the car's interior. On both models, the rear bumper must be removed to get working clearance.

1. Using a wooden wedge, pry the trim molding off the lower front area of the rear fender.
2. Remove the rear seat.
3. Pull the beading off of the door pillar.
4. Remove the side trim panel from the car's interior.

➡ **The side trim panel is different in Type 3 and Squareback Sedan models. The number of spring clips that hold the side trim panel to the inner body panel also differs. However, removal is essentially the same for both models.**

5. Remove the 2 bolts that hold the fender to the side member. The bolts are located under the front forward edge of the fender.
6. Remove the rear light lens, then remove the entire rear light assembly. Disconnect the wires and pull them free of the fender, being careful not to damage the rubber grommets.
7. Remove the single bolt in the rear light assembly hole.
8. Remove the 2 sheet metal screws located under the rear lower edge of the fender.
9. Partially remove the luggage compartment lid seal from the metal band that is beside the rear light opening.
10. Remove the 2 fasteners located next to the rear light assembly opening.
11. For Squareback Sedan models, support the fender. Working from inside the rear load area, remove the 8 hexagon head sheet metal screws along the upper fender edge. Take the fender off the car.
12. For Type 3 Sedans, support the fender to avoid buckling it when the bolts are removed. Then remove the 3 hexagon head sheet metal screws from inside the car, next to the rear seat.
13. For Type 3 Sedans, remove the 5 hexagon head sheet metal screws from inside the luggage compartment, then remove the fender from the vehicle.

To install:

14. Check the fender seals and replace them if damaged.
15. Check the condition and mounting of the rubber seal

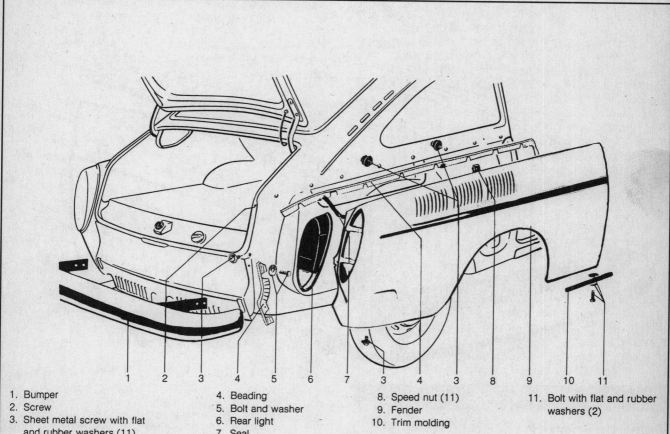

1. Bumper
2. Screw
3. Sheet metal screw with flat and rubber washers (11)
4. Beading
5. Bolt and washer
6. Rear light
7. Seal
8. Speed nut (11)
9. Fender
10. Trim molding
11. Bolt with flat and rubber washers (2)

Fig. 19 Exploded view of the rear fender mounting on Type 3 Hatchback models

around the lower part of the wheel housing. if it is damaged, replace it before installing the fender.

16. Attach the fender to the body loosely and insert the plastic beading between them.

17. Align the fender with the door and body. Tighten the bolts securely.

18. Install the exterior trim strip. glue the interior trim panels in position with **D 12** trim cement.

AUTO BODY REPAIR

You can repair most minor auto body damage yourself. Minor damage usually falls into one of several categories: (1) small scratches and dings in the paint that can be repaired without the use of body filler, (2) deep scratches and dents that require body filler, but do not require pulling, or hammering metal back into shape and (3) rust-out repairs. The repair sequences illustrated in this chapter are typical of these types of repairs. If you want to get involved in more complicated repairs including pulling or hammering sheet metal back into shape, you will probably need more detailed instructions. Chilton's *Minor Auto Body Repair, 2nd Edition* is a comprehensive guide to repairing auto body damage yourself.

Tools and Supplies

The list of tools and equipment you may need to fix minor body damage ranges from very basic hand tools to a wide assortment of specialized body tools. Most minor scratches, dings and rust holes can be fixed using an electric drill, wire wheel or grinder attachment, half-round plastic file, sanding block, various grades of sandpaper (#36, which is coarse through #600, which is fine) in both wet and dry types, auto body plastic, primer, touch-up paint, spreaders, newspaper and masking tape.

Most manufacturers of auto body repair products began supplying materials to professionals. Their knowledge of the best, most-used products has been translated into body repair kits for the do-it-yourselfer. Kits are available from a number of manufacturers and contain the necessary materials in the required amounts for the repair identified on the package.

Kits are available for a wide variety of uses, including:
- Rusted out metal
- All purpose kit for dents and holes
- Dents and deep scratches
- Fiberglass repair kit
- Epoxy kit for restyling.

Kits offer the advantage of buying what you need for the job. There is little waste and little chance of materials going bad from not being used. The same manufacturers also merchandise all of the individual products used—spreaders, dent pullers, fiberglass cloth, polyester resin, cream hardener, body filler, body files, sandpaper, sanding discs and holders, primer, spray paint, etc.

✸✸ CAUTION

Most of the products you will be using contain harmful chemicals, so be extremely careful. Always read the complete label before opening the containers. When you put them away for future use, be sure they are out of children's reach!

Most auto body repair kits contain all the materials you need to do the job right in the kit. So, if you have a small rust spot or dent you want to fix, check the contents of the kit before you run out and buy any additional tools.

Rust, Undercoating, and Rustproofing

RUST

Rust is an electrochemical process. It works on ferrous metals (iron and steel) from the inside out due to exposure of unprotected surfaces to air and moisture. The possibility of rust exists practically nationwide—anywhere humidity, industrial pollution or chemical salts are present, rust can form. In coastal areas, the problem is high humidity and salt air; in snowy areas, the problem is chemical salt (de-icer) used to keep the roads clear, and in industrial areas, sulphur dioxide is present in the air from industrial pollution and is changed to sulphuric acid when it rains. The rusting process is accelerated by high temperatures, especially in snowy areas, when vehicles are driven over slushy roads and then left overnight in a heated garage.

Automotive styling also can be a contributor to rust formation. Spot welding of panels creates small pockets that trap moisture and form an environment for rust formation. Fortunately, auto manufacturers have been working hard to increase the corrosion protection of their products. Galvanized sheet metal enjoys much wider use, along with the increased use of plastic and various rust retardant coatings. Manufacturers are also phasing out areas in automotive bodies where rust-forming moisture can collect.

To prevent rust, you must stop it before it gets started. On new vehicles, there are two ways to accomplish this.

First, the car should be treated with a commercial rustproofing compound. There are many different brands of franchised rustproofers, but most processes involve spraying a waxy "self-healing" compound under the chassis, inside rocker panels, inside doors and fender liners and similar places where rust is likely to form. Prices for a quality rustproofing job range from $100–$250, depending on the area, the brand name and the size of the vehicle.

Ideally, the vehicle should be rustproofed as soon as possible following the purchase. The surfaces of the car have begun to oxidize and deteriorate during shipping. In addition, the car may have sat on a dealer's lot or on a lot at the factory, and once the rust has progressed past the stage of light, powdery surface oxidation rustproofing is not likely to be worthwhile. Professional rustproofers feel that once rust has formed, rustproofing will simply seal in moisture already present. Most franchised rustproofing operations offer a 3–5 year warranty against rust-through, but will not support that warranty if the rustproofing is not applied within three months of the date of manufacture.

Undercoating should not be mistaken for rustproofing. Undercoating is a black, tar-like substance that is applied to the underside of a vehicle. Its basic function is to deaden noises that are transmitted from under the car. It simply cannot get into the crevices and seams where moisture tends to collect. In fact, it may clog up drainage holes and ventilation passages. Some undercoatings also tend to crack or peel with age and only create more moisture and corrosion attracting pockets.

The second thing you should do immediately after purchasing the car is apply a paint sealant. A sealant is a petroleum based product marketed under a wide variety of brand names. It has the same protective properties as a good wax, but bonds to the paint with a chemically inert layer that seals it from the air. If air can't get at the surface, oxidation cannot start.

The paint sealant kit consists of a base coat and a conditioning coat that should be applied every 6–8 months, depending on the manufacturer. The base coat must be applied before waxing, or the wax must first be removed.

Third, keep a garden hose handy for your car in winter. Use it a few times on nice days during the winter for underneath areas, and it will pay big dividends when spring arrives. Spraying under the fenders and other areas which even car washes don't reach will help remove road salt, dirt and other build-ups which help breed rust. Adjust the nozzle to a high-force spray. An old brush will help break up residue, permitting it to be washed away more easily.

It's a somewhat messy job, but worth it in the long run because rust often starts in those hidden areas.

At the same time, wash grime off the door sills and, more importantly, the under portions of the doors, plus the tailgate if you have a station wagon. Applying a coat of wax to those areas at least once before and once during winter will help fend off rust.

When applying the wax to the under parts of the doors, you will note small drain holes. These holes often are plugged with undercoating or dirt. Make sure they are cleaned out to prevent water build-up inside the doors. A small punch or penknife will do the job.

Water from the high-pressure sprays in car washes sometimes can get into the housings for parking and taillights, so take a close look. If they contain water merely loosen the retaining screws and the water should run out.

Repairing Rust Holes

One thing you have to remember about rust: even if you grind away all the rusted metal in a panel, and repair the area with any of the kits available, *eventually* the rust will return. There are two reasons for this. One, rust is a chemical reaction that causes pressure under the repair from the inside out. That's how the blisters form. Two, the back side of the panel (and the repair) is wide open to moisture, and unpainted body filler acts like a sponge. That's why the best solution to rust problems is to remove the rusted panel and install a new one or have the rusted area cut out and a new piece of sheet metal welded in its place. The trouble with welding is the expense; sometimes it will cost more than the car is worth.

One of the better solutions to do-it-yourself rust repair is the process using a fiberglass cloth repair kit (shown here). This will give a strong repair that resists cracking and moisture and is relatively easy to use. It can be used on large or small holes and also can be applied over contoured surfaces.

AUTO BODY CARE

There are hundreds—maybe thousands—of products on the market, all designed to protect or aid your car's finish in some manner. There are as many different products as there are ways to use them, but they all have one thing in common—the surface must be clean.

Washing

The primary ingredient for washing your car is water, preferably "soft" water. In many areas of the country, the local water supply is "hard" containing many minerals. The little rings or film that is left on your car's surface after it has dried is the result of "hard" water.

Since you usually can't change the local water supply, the next best thing is to dry the surface before it has a chance to dry itself.

Into the water you usually add soap. Don't use detergents or common, coarse soaps. Your car's paint never truly dries out, but is always evaporating residual oils into the air. Harsh detergents will remove these oils, causing the paint to dry faster than normal. Instead use warm water and a non-detergent soap made especially for waxed surfaces or a liquid soap made for waxed surfaces or a liquid soap made for washing dishes by hand. Other products that can be used on painted surfaces include baking soda or plain soda water for stubborn dirt.

Wash the car completely, starting at the top, and rinse it completely clean. Abrasive grit should be loaded off under water pressure; scrubbing grit off will scratch the finish. The best washing tool is a sponge, cleaning mitt or soft towel. Whichever you choose, replace it often as each tends to absorb grease and dirt.

Other ways to get a better wash include:
- Don't wash your car in the sun or when the finish is hot.
- Use water pressure to remove caked-on dirt.
- Remove tree-sap and bird effluence immediately. Such substances will eat through wax, polish and paint.

One of the best implements to dry your car is a turkish towel or an old, soft bath towel. Anything with a deep nap will hold any dirt in suspension and not grind it into the paint.

Harder cloths will only grind the grit into the paint making more scratches. Always start drying at the top, followed by the hood and trunk and sides. You'll find there's always more dirt near the rocker panels and wheelwells which will wind up on the rest of the car if you dry these areas first.

Cleaners, Waxes and Polishes

Before going any further, you should know the function of various products.

Cleaners—remove the top layer of dead pigment or paint.

Rubbing or polishing compounds—used to remove stubborn dirt, get rid of minor scratches, smooth away imperfections and partially restore badly weathered paint.

Polishes—contain no abrasives or waxes; they shine the paint by adding oils to the paint.

Waxes—are a protective coating for the polish.

CLEANERS AND COMPOUNDS

Before you apply any wax, you'll have to remove oxidation, road film and other types of pollutants that washing alone will not remove.

The paint on your car never dries completely. There are always residual oils evaporating from the paint into the air. When enough oils are present in the paint, it has a healthy shine (gloss). When too many oils evaporate the paint takes on a whitish cast known as oxidation. The idea of polishing and waxing is to keep enough oil present in the painted surface to prevent oxidation; but when it occurs, the only recourse is to remove the top layer of "dead" paint, exposing the healthy paint underneath.

Products to remove oxidation and road film are sold under a variety of generic names—polishes, cleaner, rubbing compound, cleaner/polish, polish/cleaner, self-polishing wax, pre-wax cleaner, finish restorer and many more. Regardless of name there are two types of cleaners—abrasive cleaners (sometimes called polishing or rubbing compounds) that remove oxidation by grinding away the top layer of "dead" paint, or chemical cleaners that dissolve the "dead" pigment, allowing it to be wiped away.

Abrasive cleaners, by their nature, leave thousands of minute scratches in the finish, which must be polished out later. These should only be used in extreme cases, but are usually the only thing to use on badly oxidized paint finishes. Chemical cleaners are much milder but are not strong enough for severe cases of oxidation or weathered paint.

The most popular cleaners are liquid or paste abrasive polishing and rubbing compounds. Polishing compounds have a finer abrasive grit for medium duty work. Rubbing compounds are a coarser abrasive and for heavy duty work. Unless you are familiar with how to use compounds, be very careful. Excessive rubbing with any type of compound or cleaner can grind right through the paint to primer or bare metal. Follow the directions on the container—depending on type, the cleaner may or may not be OK for your paint. For example, some cleaners are not formulated for acrylic lacquer finishes.

When a small area needs compounding or heavy polishing, it's best to do the job by hand. Some people prefer a powered buffer for large areas. Avoid cutting through the paint along styling edges on the body. Small, hand operations where the compound is applied and rubbed using cloth folded into a thick ball allow you to work in straight lines along such edges.

To avoid cutting through on the edges when using a power buffer, try masking tape. Just cover the edge with tape while using power. Then finish the job by hand with the tape removed. Even then work carefully. The paint tends to be a lot thinner along the sharp ridges stamped into the panels.

Whether compounding by machine or by hand, only work on a small area and apply the compound sparingly. If the materials are spread too thin, or allowed to sit too long, they dry out. Once dry they lose the ability to deliver a smooth, clean finish. Also, dried out polish tends to cause the buffer to stick in one spot. This in turn can burn or cut through the finish.

WAXES AND POLISHES

Your car's finish can be protected in a number of ways. A cleaner/wax or polish/cleaner followed by wax or variations of each all provide good results. The two-step approach (polish followed by wax) is probably slightly better but consumes more time and effort. Properly fed with oils, your paint should never need cleaning, but despite the best polishing job, it won't last unless it's protected with wax. Without wax, polish must be renewed at least once a month to prevent oxidation. Years ago (some still swear by it today), the best wax was made from the Brazilian palm, the Carnuba, favored for its vegetable base and high melting point. However, modern synthetic waxes are harder, which means they protect against moisture better, and chemically inert silicone is used for a long lasting protection. The only problem with silicone wax is that it penetrates all layers of paint. To repaint or touch up a panel or car protected by silicone wax, you have to completely strip the finish to avoid "fish-eyes."

Under normal conditions, silicone waxes will last 4–6 months, but you have to be careful of wax build-up from too much waxing. Too thick a coat of wax is just as bad as no wax at all; it stops the paint from breathing.

Combination cleaners/waxes have become popular lately because they remove the old layer of wax plus light oxidation, while putting on a fresh coat of wax at the same time. Some cleaners/waxes contain abrasive cleaners which require caution, although many cleaner/waxes use a chemical cleaner.

Applying Wax or Polish

You may view polishing and waxing your car as a pleasant way to spend an afternoon, or as a boring chore, but it has to be done to keep the paint on your car. Caring for the paint doesn't require special tools, but you should follow a few rules.

1. Use a good quality wax.
2. Before applying any wax or polish, be sure the surface is completely clean. Just because the car looks clean, doesn't mean it's ready for polish or wax.
3. If the finish on your car is weathered, dull, or oxidized, it will probably have to be compounded to remove the old or oxidized paint. If the paint is simply dulled from lack of care, one of the non-abrasive cleaners known as polishing compounds will do the trick. If the paint is severely scratched or really dull, you'll probably have to use a rubbing compound to prepare the finish for waxing. If you're not sure which one to use, use the polishing compound, since you can easily ruin the finish by using too strong a compound.
4. Don't apply wax, polish or compound in direct sunlight, even if the directions on the can say you can. Most waxes will not cure properly in bright sunlight and you'll probably end up with a blotchy looking finish.
5. Don't rub the wax off too soon. The result will be a wet, dull looking finish. Let the wax dry thoroughly before buffing it off.

6. A constant debate among car enthusiasts is how wax should be applied. Some maintain pastes or liquids should be applied in a circular motion, but body shop experts have long thought that this approach results in barely detectable circular abrasions, especially on cars that are waxed frequently. They advise rubbing in straight lines, especially if any kind of cleaner is involved.

7. If an applicator is not supplied with the wax, use a piece of soft cheesecloth or very soft lint-free material. The same applies to buffing the surface.

SPECIAL SURFACES

One-step combination cleaner and wax formulas shouldn't be used on many of the special surfaces which abound on cars. The one-step materials contain abrasives to achieve a clean surface under the wax top coat. The abrasives are so mild that you could clean a car every week for a couple of years without fear of rubbing through the paint. But this same level of abrasiveness might, through repeated use, damage decals used for special trim effects. This includes wide stripes, wood-grain trim and other appliques.

Painted plastics must be cleaned with care. If a cleaner is too aggressive it will cut through the paint and expose the primer. If bright trim such as polished aluminum or chrome is painted, cleaning must be performed with even greater care. If rubbing compound is being used, it will cut faster than polish.

Abrasive cleaners will dull an acrylic finish. The best way to clean these newer finishes is with a non-abrasive liquid polish. Only dirt and oxidation, not paint, will be removed.

Taking a few minutes to read the instructions on the can of polish or wax will help prevent making serious mistakes. Not all preparations will work on all surfaces. And some are intended for power application while others will only work when applied by hand.

Don't get the idea that just pouring on some polish and then hitting it with a buffer will suffice. Power equipment speeds the operation. But it also adds a measure of risk. It's very easy to damage the finish if you use the wrong methods or materials.

Caring for Chrome

Read the label on the container. Many products are formulated specifically for chrome, but others contain abrasives that will scratch the chrome finish. If it isn't recommended for chrome, don't use it.

Never use steel wool or kitchen soap pads to clean chrome. Be careful not to get chrome cleaner on paint or interior vinyl surfaces. If you do, get it off immediately.

GLOSSARY

AIR/FUEL RATIO: The ratio of air-to-gasoline by weight in the fuel mixture drawn into the engine.

AIR INJECTION: One method of reducing harmful exhaust emissions by injecting air into each of the exhaust ports of an engine. The fresh air entering the hot exhaust manifold causes any remaining fuel to be burned before it can exit the tailpipe.

ALTERNATOR: A device used for converting mechanical energy into electrical energy.

AMMETER: An instrument, calibrated in amperes, used to measure the flow of an electrical current in a circuit. Ammeters are always connected in series with the circuit being tested.

AMPERE: The rate of flow of electrical current present when one volt of electrical pressure is applied against one ohm of electrical resistance.

ANALOG COMPUTER: Any microprocessor that uses similar (analogous) electrical signals to make its calculations.

ARMATURE: A laminated, soft iron core wrapped by a wire that converts electrical energy to mechanical energy as in a motor or relay. When rotated in a magnetic field, it changes mechanical energy into electrical energy as in a generator.

ATMOSPHERIC PRESSURE: The pressure on the Earth's surface caused by the weight of the air in the atmosphere. At sea level, this pressure is 14.7 psi at 32°F (101 kPa at 0°C).

ATOMIZATION: The breaking down of a liquid into a fine mist that can be suspended in air.

AXIAL PLAY: Movement parallel to a shaft or bearing bore.

BACKFIRE: The sudden combustion of gases in the intake or exhaust system that results in a loud explosion.

BACKLASH: The clearance or play between two parts, such as meshed gears.

BACKPRESSURE: Restrictions in the exhaust system that slow the exit of exhaust gases from the combustion chamber.

BAKELITE: A heat resistant, plastic insulator material commonly used in printed circuit boards and transistorized components.

BALL BEARING: A bearing made up of hardened inner and outer races between which hardened steel balls roll.

BALLAST RESISTOR: A resistor in the primary ignition circuit that lowers voltage after the engine is started to reduce wear on ignition components.

BEARING: A friction reducing, supportive device usually located between a stationary part and a moving part.

BIMETAL TEMPERATURE SENSOR: Any sensor or switch made of two dissimilar types of metal that bend when heated or cooled due to the different expansion rates of the alloys. These types of sensors usually function as an on/off switch.

BLOWBY: Combustion gases, composed of water vapor and unburned fuel, that leak past the piston rings into the crankcase during normal engine operation. These gases are removed by the PCV system to prevent the buildup of harmful acids in the crankcase.

BRAKE PAD: A brake shoe and lining assembly used with disc brakes.

BRAKE SHOE: The backing for the brake lining. The term is, however, usually applied to the assembly of the brake backing and lining.

BUSHING: A liner, usually removable, for a bearing; an anti-friction liner used in place of a bearing.

CALIPER: A hydraulically activated device in a disc brake system, which is mounted straddling the brake rotor (disc). The caliper contains at least one piston and two brake pads. Hydraulic pressure on the piston(s) forces the pads against the rotor.

CAMSHAFT: A shaft in the engine on which are the lobes (cams) which operate the valves. The camshaft is driven by the crankshaft, via a belt, chain or gears, at one half the crankshaft speed.

CAPACITOR: A device which stores an electrical charge.

CARBON MONOXIDE (CO): A colorless, odorless gas given off as a normal byproduct of combustion. It is poisonous and extremely dangerous in confined areas, building up slowly to toxic levels without warning if adequate ventilation is not available.

CARBURETOR: A device, usually mounted on the intake manifold of an engine, which mixes the air and fuel in the proper proportion to allow even combustion.

CATALYTIC CONVERTER: A device installed in the exhaust system, like a muffler, that converts harmful byproducts of combustion into carbon dioxide and water vapor by means of a heat-producing chemical reaction.

CENTRIFUGAL ADVANCE: A mechanical method of advancing the spark timing by using flyweights in the distributor that react to centrifugal force generated by the distributor shaft rotation.

CHECK VALVE: Any one-way valve installed to permit the flow of air, fuel or vacuum in one direction only.

GLOSSARY 10-19

CHOKE: A device, usually a moveable valve, placed in the intake path of a carburetor to restrict the flow of air.

CIRCUIT: Any unbroken path through which an electrical current can flow. Also used to describe fuel flow in some instances.

CIRCUIT BREAKER: A switch which protects an electrical circuit from overload by opening the circuit when the current flow exceeds a predetermined level. Some circuit breakers must be reset manually, while most reset automatically.

COIL (IGNITION): A transformer in the ignition circuit which steps up the voltage provided to the spark plugs.

COMBINATION MANIFOLD: An assembly which includes both the intake and exhaust manifolds in one casting.

COMBINATION VALVE: A device used in some fuel systems that routes fuel vapors to a charcoal storage canister instead of venting them into the atmosphere. The valve relieves fuel tank pressure and allows fresh air into the tank as the fuel level drops to prevent a vapor lock situation.

COMPRESSION RATIO: The comparison of the total volume of the cylinder and combustion chamber with the piston at BDC and the piston at TDC.

CONDENSER: 1. An electrical device which acts to store an electrical charge, preventing voltage surges. 2. A radiator-like device in the air conditioning system in which refrigerant gas condenses into a liquid, giving off heat.

CONDUCTOR: Any material through which an electrical current can be transmitted easily.

CONTINUITY: Continuous or complete circuit. Can be checked with an ohmmeter.

COUNTERSHAFT: An intermediate shaft which is rotated by a mainshaft and transmits, in turn, that rotation to a working part.

CRANKCASE: The lower part of an engine in which the crankshaft and related parts operate.

CRANKSHAFT: The main driving shaft of an engine which receives reciprocating motion from the pistons and converts it to rotary motion.

CYLINDER: In an engine, the round hole in the engine block in which the piston(s) ride.

CYLINDER BLOCK: The main structural member of an engine in which is found the cylinders, crankshaft and other principal parts.

CYLINDER HEAD: The detachable portion of the engine, usually fastened to the top of the cylinder block and containing all or most of the combustion chambers. On overhead valve engines, it contains the valves and their operating parts. On overhead cam engines, it contains the camshaft as well.

DEAD CENTER: The extreme top or bottom of the piston stroke.

DETONATION: An unwanted explosion of the air/fuel mixture in the combustion chamber caused by excess heat and compression, advanced timing, or an overly lean mixture. Also referred to as "ping".

DIAPHRAGM: A thin, flexible wall separating two cavities, such as in a vacuum advance unit.

DIESELING: A condition in which hot spots in the combustion chamber cause the engine to run on after the key is turned off.

DIFFERENTIAL: A geared assembly which allows the transmission of motion between drive axles, giving one axle the ability to turn faster than the other.

DIODE: An electrical device that will allow current to flow in one direction only.

DISC BRAKE: A hydraulic braking assembly consisting of a brake disc, or rotor, mounted on an axle, and a caliper assembly containing, usually two brake pads which are activated by hydraulic pressure. The pads are forced against the sides of the disc, creating friction which slows the vehicle.

DISTRIBUTOR: A mechanically driven device on an engine which is responsible for electrically firing the spark plug at a predetermined point of the piston stroke.

DOWEL PIN: A pin, inserted in mating holes in two different parts allowing those parts to maintain a fixed relationship.

DRUM BRAKE: A braking system which consists of two brake shoes and one or two wheel cylinders, mounted on a fixed backing plate, and a brake drum, mounted on an axle, which revolves around the assembly.

DWELL: The rate, measured in degrees of shaft rotation, at which an electrical circuit cycles on and off.

ELECTRONIC CONTROL UNIT (ECU): Ignition module, module, amplifier or igniter. See Module for definition.

ELECTRONIC IGNITION: A system in which the timing and firing of the spark plugs is controlled by an electronic control unit, usually called a module. These systems have no points or condenser.

END-PLAY: The measured amount of axial movement in a shaft.

ENGINE: A device that converts heat into mechanical energy.

EXHAUST MANIFOLD: A set of cast passages or pipes which conduct exhaust gases from the engine.

10-20 GLOSSARY

FEELER GAUGE: A blade, usually metal, of precisely predetermined thickness, used to measure the clearance between two parts.

FIRING ORDER: The order in which combustion occurs in the cylinders of an engine. Also the order in which spark is distributed to the plugs by the distributor.

FLOODING: The presence of too much fuel in the intake manifold and combustion chamber which prevents the air/fuel mixture from firing, thereby causing a no-start situation.

FLYWHEEL: A disc shaped part bolted to the rear end of the crankshaft. Around the outer perimeter is affixed the ring gear. The starter drive engages the ring gear, turning the flywheel, which rotates the crankshaft, imparting the initial starting motion to the engine.

FOOT POUND (ft. lbs. or sometimes, ft.lb.): The amount of energy or work needed to raise an item weighing one pound, a distance of one foot.

FUSE: A protective device in a circuit which prevents circuit overload by breaking the circuit when a specific amperage is present. The device is constructed around a strip or wire of a lower amperage rating than the circuit it is designed to protect. When an amperage higher than that stamped on the fuse is present in the circuit, the strip or wire melts, opening the circuit.

GEAR RATIO: The ratio between the number of teeth on meshing gears.

GENERATOR: A device which converts mechanical energy into electrical energy.

HEAT RANGE: The measure of a spark plug's ability to dissipate heat from its firing end. The higher the heat range, the hotter the plug fires.

HUB: The center part of a wheel or gear.

HYDROCARBON (HC): Any chemical compound made up of hydrogen and carbon. A major pollutant formed by the engine as a byproduct of combustion.

HYDROMETER: An instrument used to measure the specific gravity of a solution.

INCH POUND (inch lbs.; sometimes in.lb. or in. lbs.): One twelfth of a foot pound.

INDUCTION: A means of transferring electrical energy in the form of a magnetic field. Principle used in the ignition coil to increase voltage.

INJECTOR: A device which receives metered fuel under relatively low pressure and is activated to inject the fuel into the engine under relatively high pressure at a predetermined time.

INPUT SHAFT: The shaft to which torque is applied, usually carrying the driving gear or gears.

INTAKE MANIFOLD: A casting of passages or pipes used to conduct air or a fuel/air mixture to the cylinders.

JOURNAL: The bearing surface within which a shaft operates.

KEY: A small block usually fitted in a notch between a shaft and a hub to prevent slippage of the two parts.

MANIFOLD: A casting of passages or set of pipes which connect the cylinders to an inlet or outlet source.

MANIFOLD VACUUM: Low pressure in an engine intake manifold formed just below the throttle plates. Manifold vacuum is highest at idle and drops under acceleration.

MASTER CYLINDER: The primary fluid pressurizing device in a hydraulic system. In automotive use, it is found in brake and hydraulic clutch systems and is pedal activated, either directly or, in a power brake system, through the power booster.

MODULE: Electronic control unit, amplifier or igniter of solid state or integrated design which controls the current flow in the ignition primary circuit based on input from the pick-up coil. When the module opens the primary circuit, high secondary voltage is induced in the coil.

NEEDLE BEARING: A bearing which consists of a number (usually a large number) of long, thin rollers.

OHM: (Ω) The unit used to measure the resistance of conductor-to-electrical flow. One ohm is the amount of resistance that limits current flow to one ampere in a circuit with one volt of pressure.

OHMMETER: An instrument used for measuring the resistance, in ohms, in an electrical circuit.

OUTPUT SHAFT: The shaft which transmits torque from a device, such as a transmission.

OVERDRIVE: A gear assembly which produces more shaft revolutions than that transmitted to it.

OVERHEAD CAMSHAFT (OHC): An engine configuration in which the camshaft is mounted on top of the cylinder head and operates the valve either directly or by means of rocker arms.

OVERHEAD VALVE (OHV): An engine configuration in which all of the valves are located in the cylinder head and the camshaft is located in the cylinder block. The camshaft operates the valves via lifters and pushrods.

OXIDES OF NITROGEN (NOx): Chemical compounds of nitrogen produced as a byproduct of combustion. They combine with hydrocarbons to produce smog.

GLOSSARY 10-21

OXYGEN SENSOR: Used with the feedback system to sense the presence of oxygen in the exhaust gas and signal the computer which can reference the voltage signal to an air/fuel ratio.

PINION: The smaller of two meshing gears.

PISTON RING: An open-ended ring which fits into a groove on the outer diameter of the piston. Its chief function is to form a seal between the piston and cylinder wall. Most automotive pistons have three rings: two for compression sealing; one for oil sealing.

PRELOAD: A predetermined load placed on a bearing during assembly or by adjustment.

PRIMARY CIRCUIT: The low voltage side of the ignition system which consists of the ignition switch, ballast resistor or resistance wire, bypass, coil, electronic control unit and pick-up coil as well as the connecting wires and harnesses.

PRESS FIT: The mating of two parts under pressure, due to the inner diameter of one being smaller than the outer diameter of the other, or vice versa; an interference fit.

RACE: The surface on the inner or outer ring of a bearing on which the balls, needles or rollers move.

REGULATOR: A device which maintains the amperage and/or voltage levels of a circuit at predetermined values.

RELAY: A switch which automatically opens and/or closes a circuit.

RESISTANCE: The opposition to the flow of current through a circuit or electrical device, and is measured in ohms. Resistance is equal to the voltage divided by the amperage.

RESISTOR: A device, usually made of wire, which offers a preset amount of resistance in an electrical circuit.

RING GEAR: The name given to a ring-shaped gear attached to a differential case, or affixed to a flywheel or as part of a planetary gear set.

ROLLER BEARING: A bearing made up of hardened inner and outer races between which hardened steel rollers move.

ROTOR: 1. The disc-shaped part of a disc brake assembly, upon which the brake pads bear; also called, brake disc. 2. The device mounted atop the distributor shaft, which passes current to the distributor cap tower contacts.

SECONDARY CIRCUIT: The high voltage side of the ignition system, usually above 20,000 volts. The secondary includes the ignition coil, coil wire, distributor cap and rotor, spark plug wires and spark plugs.

SENDING UNIT: A mechanical, electrical, hydraulic or electromagnetic device which transmits information to a gauge.

SENSOR: Any device designed to measure engine operating conditions or ambient pressures and temperatures. Usually electronic in nature and designed to send a voltage signal to an on-board computer, some sensors may operate as a simple on/off switch or they may provide a variable voltage signal (like a potentiometer) as conditions or measured parameters change.

SHIM: Spacers of precise, predetermined thickness used between parts to establish a proper working relationship.

SLAVE CYLINDER: In automotive use, a device in the hydraulic clutch system which is activated by hydraulic force, disengaging the clutch.

SOLENOID: A coil used to produce a magnetic field, the effect of which is to produce work.

SPARK PLUG: A device screwed into the combustion chamber of a spark ignition engine. The basic construction is a conductive core inside of a ceramic insulator, mounted in an outer conductive base. An electrical charge from the spark plug wire travels along the conductive core and jumps a preset air gap to a grounding point or points at the end of the conductive base. The resultant spark ignites the fuel/air mixture in the combustion chamber.

SPLINES: Ridges machined or cast onto the outer diameter of a shaft or inner diameter of a bore to enable parts to mate without rotation.

TACHOMETER: A device used to measure the rotary speed of an engine, shaft, gear, etc., usually in rotations per minute.

THERMOSTAT: A valve, located in the cooling system of an engine, which is closed when cold and opens gradually in response to engine heating, controlling the temperature of the coolant and rate of coolant flow.

TOP DEAD CENTER (TDC): The point at which the piston reaches the top of its travel on the compression stroke.

TORQUE: The twisting force applied to an object.

TORQUE CONVERTER: A turbine used to transmit power from a driving member to a driven member via hydraulic action, providing changes in drive ratio and torque. In automotive use, it links the driveplate at the rear of the engine to the automatic transmission.

TRANSDUCER: A device used to change a force into an electrical signal.

TRANSISTOR: A semi-conductor component which can be actuated by a small voltage to perform an electrical switching function.

TUNE-UP: A regular maintenance function, usually associated with the replacement and adjustment of parts and components in the electrical and fuel systems of a vehicle for the purpose of attaining optimum performance.

Glossary

TURBOCHARGER: An exhaust driven pump which compresses intake air and forces it into the combustion chambers at higher than atmospheric pressures. The increased air pressure allows more fuel to be burned and results in increased horsepower being produced.

VACUUM ADVANCE: A device which advances the ignition timing in response to increased engine vacuum.

VACUUM GAUGE: An instrument used to measure the presence of vacuum in a chamber.

VALVE: A device which control the pressure, direction of flow or rate of flow of a liquid or gas.

VALVE CLEARANCE: The measured gap between the end of the valve stem and the rocker arm, cam lobe or follower that activates the valve.

VISCOSITY: The rating of a liquid's internal resistance to flow.

VOLTMETER: An instrument used for measuring electrical force in units called volts. Voltmeters are always connected parallel with the circuit being tested.

WHEEL CYLINDER: Found in the automotive drum brake assembly, it is a device, actuated by hydraulic pressure, which, through internal pistons, pushes the brake shoes outward against the drums.

MASTER INDEX

ACCELERATOR CABLE 5-16
   REMOVAL & INSTALLATION 5-16
ADD-ON ELECTRICAL EQUIPMENT 6-11
ADJUSTMENTS (FULLY AUTOMATIC TRANSAXLE) 7-17
   FRONT (SECOND) BAND 7-20
   KICKDOWN SWITCH 7-21
   REAR (FIRST) BAND 7-20
   SHIFT LINKAGE 7-22
ADJUSTMENTS (DIAGONAL ARM REAR SUSPENSION) 8-13
   TYPE 1 8-13
   TYPES 2 (THROUGH 1979) AND 3 8-13
ADJUSTMENTS (COIL SPRING/TRAILING ARM REAR SUSPENSION) 8-15
ADJUSTMENT (VALVE LASH) 2-12
AIR CLEANER 1-22
   REMOVAL & INSTALLATION 1-22
AIR CONDITIONING 1-32
   DISCHARGING, EVACUATING AND CHARGING 1-34
   GENERAL SERVICING PROCEDURES 1-33
   SAFETY PRECAUTIONS 1-32
   SYSTEM INSPECTION 1-34
AIR FLAP AND THERMOSTAT 3-39
   ADJUSTMENT 3-39
AIR INJECTION SYSTEM 1973–74 4-7
   SERVICING 4-7
AIR POLLUTION 4-2
ALTERNATOR, GENERATOR, AND REGULATOR SPECIFICATIONS (SPECIFICATION CHARTS) 3-16
AUTO BODY CARE 10-15
AUTO BODY REPAIR 10-14
AUTOMATIC STICK SHIFT TRANSAXLE 7-5
AUTOMOTIVE EMISSIONS 4-3
AUTOMOTIVE POLLUTANTS 4-2
   HEAT TRANSFER 4-3
   TEMPERATURE INVERSION 4-2
AUXILIARY AIR REGULATOR 5-25
   TESTING 5-25
AVOIDING THE MOST COMMON MISTAKES 1-2
AVOIDING TROUBLE 1-2
BALL JOINT (TORSION BAR FRONT SUSPENSION) 8-2
   INSPECTION 8-2
   REPLACEMENT 8-4
BALL JOINT (STRUT FRONT SUSPENSION) 8-7
   INSPECTION 8-7
   REPLACEMENT 8-7
BALL JOINT (COIL SPRING FRONT SUSPENSION) 8-9
   REMOVAL & INSTALLATION 8-9
BASIC FUEL SYSTEM DIAGNOSIS 5-2
BASIC OPERATING PRINCIPLES 9-2
   DISC BRAKES 9-2
   DRUM BRAKES 9-3
   POWER BOOSTERS 9-3
BATTERY, STARTING AND CHARGING SYSTEMS 3-4
   BASIC OPERATING PRINCIPLES 3-4
BATTERY (ROUTINE MAINTENANCE) 1-26
   BATTERY FLUID 1-27
   CABLES 1-28
   CHARGING 1-29
   GENERAL MAINTENANCE 1-26
   REPLACEMENT 1-30
BATTERY (ENGINE ELECTRICAL) 3-14
   REMOVAL & INSTALLATION 3-14
BELTS 1-30
   ADJUSTMENT 1-31
   INSPECTION 1-30
BOLTS, NUTS AND OTHER THREADED RETAINERS 1-8
BRAKE ADJUSTMENTS 9-3
BRAKE CALIPER 9-11
   OVERHAUL 9-11
   REMOVAL & INSTALLATION 9-11
BRAKE DRUM (REAR DRUM BRAKES) 9-14

INDEX

INSPECTION 9-14
    REMOVAL & INSTALLATION 9-14
BRAKE DRUM (FRONT DRUM BRAKES) 9-6
    INSPECTION 9-6
    REMOVAL & INSTALLATION 9-6
BRAKE PAD 9-10
    REMOVAL & INSTALLATION 9-10
BRAKE ROTOR (DISC) 9-13
    INSPECTION 9-13
    REMOVAL & INSTALLATION 9-13
BRAKE SHOES (REAR DRUM BRAKES) 9-14
    REMOVAL & INSTALLATION 9-14
BRAKE SHOES (FRONT DRUM BRAKES) 9-8
    REMOVAL & INSTALLATION 9-8
BRAKE SPECIFICATIONS (SPECIFICATION CHARTS) 9-20
BRAKE SYSTEM 9-2
BREAKER POINTS AND CONDENSER 2-8
    DWELL ANGLE ADJUSTMENT 2-8
    POINT GAP ADJUSTMENT 2-8
    REMOVAL & INSTALLATION 2-8
BUMPERS 10-6
    REMOVAL & INSTALLATION 10-6
CABLE (PARKING BRAKE) 9-19
    ADJUSTMENT 9-19
    REMOVAL & INSTALLATION 9-19
CAMSHAFT AND TIMING GEARS 3-33
    REMOVAL & INSTALLATION 3-33
CAPACITIES (SPECIFICATION CHARTS) 1-55
CARBURETED FUEL SYSTEM 5-2
CARBURETED VEHICLES (IDLE SPEED AND MIXTURE ADJUSTMENTS) 2-13
    SOLEX 30 PICT-3 (1970 TYPE 1 AND TYPE 2 MODELS) 2-14
    SOLEX 34 PDSIT-2/3 TWIN CARBURETORS (1972–74) TYPE 2 MODELS 2-14
    SOLEX PICT-3 (1971–74 TYPE 1 MODELS, 1971 TYPE 2 MODELS) AND SOLEX 34 PICT-4 (1973–74 TYPE 1 CALIFORNIA MODELS) 2-14
CARBURETOR SPECIFICATIONS (SPECIFICATION CHARTS) 5-15
CARBURETORS 5-7
    ADJUSTMENTS 5-8
    OVERHAUL 5-11
    REMOVAL & INSTALLATION 5-7
CATALYTIC CONVERTER SYSTEM 4-12
CHASSIS GREASING 1-48
CHASSIS NUMBER CHART (SPECIFICATION CHARTS) 1-16
CLEANERS, WAXES AND POLISHES 10-16
    CLEANERS AND COMPOUNDS 10-16
    SPECIAL SURFACES 10-17
    WAXES AND POLISHES 10-16
CLUTCH MASTER CYLINDER 7-15
    REMOVAL & INSTALLATION 7-15
CLUTCH SLAVE CYLINDER 7-15
    CLUTCH SYSTEM BLEEDING & ADJUSTMENT 7-15
    REMOVAL & INSTALLATION 7-15
CLUTCH 7-10
COIL SPRING FRONT SUSPENSION 8-7
COIL SPRING/TRAILING ARM REAR SUSPENSION 8-13
COIL SPRING 8-7
    REMOVAL & INSTALLATION 8-7
COLD START VALVE 5-23
    CHECKING VALVE LEAKAGE 5-24
    CHECKING VALVE OPERATION 5-24
COMPONENT LOCATIONS
    UNDER HOOD ENGINE COMPONENTS 3-18
CONNECTING RODS 3-35
    REMOVAL & INSTALLATION 3-35
COOLING 1-49
    ENGINE 1-49
    TRANSMISSION 1-49
CRANKCASE EMISSIONS 4-5
CRANKCASE RECONDITIONING 3-51
    ALIGN BORE THE CRANKCASE 3-51
    ASSEMBLE THE CRANKSHAFT 3-54
    ASSEMBLING THE CRANKCASE 3-55
    CHECK CONNECTING ROD BEARING (OIL) CLEARANCE 3-53
    CHECK CONNECTING ROD SIDE CLEARANCE, AND FOR STRAIGHTNESS 3-51
    CHECK CRANKSHAFT END-PLAY 3-55
    CHECK MAIN BEARING (OIL) CLEARANCE 3-53
    CHECK THE CAMSHAFT BEARINGS 3-54
    CHECK THE LIFTERS (TAPPETS) 3-54
    CHECK TIMING GEAR BACKLASH 3-55
    CLEAN AND INSPECT THE CAMSHAFT 3-53
    DISASSEMBLE THE CRANKSHAFT 3-52
    DISASSEMBLING CRANKCASE 3-51
    HOT TANK THE CRANKCASE 3-51
    INSPECT THE CONNECTING RODS 3-52
    INSPECT THE CRANKCASE 3-51
    INSPECT THE CRANKSHAFT 3-52
    INSTALLING THE CRANKSHAFT AND CAMSHAFT 3-54
CRANKCASE VENTILATION SYSTEM (EMISSION CONTROLS) 4-5
    SERVICING 4-5
CRANKCASE VENTILATION (ROUTINE MAINTENANCE) 1-26
    SERVICING 1-26
CRANKCASE 3-33
    DISASSEMBLY & ASSEMBLY 3-33
CRANKSHAFT 3-34
    REMOVAL & INSTALLATION 3-34
CRANKSHAFT AND CONNECTING ROD SPECIFICATIONS (SPECIFICATION CHARTS) 3-21
CYLINDER HEAD RECONDITIONING 3-42
    CHECK THE VALVE SPRINGS 3-48
    CHECK THE VALVE STEM-TO-GUIDE CLEARANCE (VALVE ROCK) 3-44
    CHECKING VALVE SEAT CONCENTRICITY 3-48
    DE-CARBON THE CYLINDER HEAD 3-43
    DEGREASE THE REMAINING CYLINDER HEAD PARTS 3-43
    HOT-TANK THE CYLINDER HEAD 3-42
    IDENTIFY THE VALVES 3-42
    INSPECT THE PUSHRODS AND PUSHROD TUBES 3-49
    INSPECT THE ROCKER SHAFTS AND ROCKER ARMS 3-49
    INSTALL THE VALVES 3-49
    KNURLING THE VALVE GUIDES 3-44
    LAPPING THE VALVES 3-48
    REMOVE THE VALVES AND SPRINGS 3-42
    REPLACING THE VALVE GUIDES 3-44
    REPLACING VALVE SEAT INSERTS 3-47
    RESURFACING (GRINDING) THE VALVE FACE 3-45
    RESURFACING THE VALVE SEATS 3-47
CYLINDER HEAD 3-24
    OVERHAUL & VALVE GUIDE REPLACEMENT 3-27
    REMOVAL & INSTALLATION 3-24
    VALVE SEATS 3-27
DECELERATION CONTROL 4-12
    CHECKING THE DECELERATION VALVE 4-13
DESCRIPTION (ENGINE MECHANICAL) 3-16
DESCRIPTION (GASOLINE FUEL INJECTION SYSTEM) 5-16
    AIR FLOW CONTROLLED 5-16
    NON-AIR FLOW CONTROLLED 5-16
DIAGONAL ARM REAR SUSPENSION 8-12
DIAGONAL ARM 8-12
    REMOVAL & INSTALLATION 8-12
DISTRIBUTOR 3-7
    REMOVAL & INSTALLATION 3-7

INDEX 10-25

DRIVEN DISC AND PRESSURE PLATE 7-10
    CLUTCH CABLE ADJUSTMENT 7-14
    CLUTCH CABLE REPLACEMENT 7-14
    REMOVAL & INSTALLATION 7-10
DRIVESHAFT AND CONSTANT VELOCITY (CV) JOINT 7-8
DRIVESHAFT AND CV-JOINT ASSEMBLY 7-8
    CONSTANT VELOCITY JOINT OVERHAUL 7-9
    REMOVAL & INSTALLATION 7-8
ELECTRIC FUEL PUMP 5-17
    ADJUSTMENTS 5-21
    REMOVAL & INSTALLATION 5-21
ELECTRONIC CONTROL (BRAIN) BOX 5-17
ELECTRONIC IGNITION 2-9
EMISSION CONTROLS 4-5
ENGINE COOLING 3-39
ENGINE ELECTRICAL 3-2
ENGINE IDENTIFICATION CHART (SPECIFICATION CHARTS) 1-19
ENGINE LUBRICATION 3-36
ENGINE MECHANICAL 3-16
ENGINE OIL AND FUEL RECOMMENDATIONS 1-44
ENGINE REBUILDING 3-40
ENGINE (SERIAL NUMBER IDENTIFICATION) 1-14
ENGINE (ENGINE MECHANICAL) 3-21
    REMOVAL & INSTALLATION 3-21
ENGLISH TO METRIC CONVERSION CHARTS (SPECIFICATION CHARTS) 1-56
EVAPORATIVE EMISSION CONTROL SYSTEM 4-6
    SERVICING 4-7
EVAPORATIVE EMISSIONS 4-5
EXHAUST GAS RECIRCULATION SYSTEM 4-7
    EGR VALVE INSPECTION 4-10
    EGR VALVE SERVICE 4-10
    GENERAL DESCRIPTION 4-8
EXHAUST GASES 4-3
    CARBON MONOXIDE 4-4
    HYDROCARBONS 4-3
    NITROGEN 4-4
    OXIDES OF SULFUR 4-4
    PARTICULATE MATTER 4-4
EXTERIOR 10-2
FAN HOUSING AND FAN 3-39
    REMOVAL & INSTALLATION 3-39
FASTENERS, MEASUREMENTS AND CONVERSIONS 1-8
FENDERS 10-10
    REMOVAL & INSTALLATION 10-10
FIRING ORDERS 2-8
FLUID CHANGES 1-45
    ENGINE 1-45
    FINAL DRIVE HOUSING 1-48
    TRANSAXLE 1-47
FLUID DISPOSAL 1-43
FLUID LEVEL CHECKS 1-43
    BRAKE FLUID 1-44
    DIFFERENTIAL 1-44
    ENGINE OIL 1-43
    STEERING GEAR (EXCEPT RACK AND PINION TYPE) 1-44
    TRANSAXLE 1-43
FLUID PAN (FULLY AUTOMATIC TRANSAXLE) 7-18
    FILTER SERVICE 7-20
    REMOVAL & INSTALLATION 7-18
FLUIDS AND LUBRICANTS 1-43
FRONT DISC BRAKES 9-10
FRONT DOORS 10-2
    DOOR STRIKER PLATE ALIGNING 10-3
    REMOVAL & INSTALLATION 10-2
FRONT DRUM BRAKES 9-6
FRONT END ALIGNMENT 8-9
    CAMBER ADJUSTMENT 8-9
    CASTER ADJUSTMENT 8-9
    TOE-IN ADJUSTMENT 8-10
FRONT HOOD 10-4
    REMOVAL & INSTALLATION 10-4
FRONT TURN SIGNAL/SIDE MARKER LIGHTS 6-18
    REMOVAL & INSTALLATION 6-18
FRONT WHEEL BEARINGS 9-13
    ADJUSTMENT 9-14
    REMOVAL & INSTALLATION 9-13
FUEL EVAPORATION CONTROL SYSTEM 1-26
    SERVICING 1-26
FUEL FILTER 1-25
    SERVICING 1-25
FUEL GAUGE AND CLOCK ASSEMBLY 6-16
    REMOVAL & INSTALLATION 6-16
FUEL INJECTED VEHICLES (IDLE SPEED AND MIXTURE ADJUSTMENTS) 2-16
    ALL TYPE 3 AND TYPE 4; 1975 AND LATER TYPE 1 AND TYPE 2 2-16
FUEL INJECTORS 5-21
    CHECKING FUEL INJECTOR OPERATION 5-22
    CHECKING THE FUEL INJECTOR SIGNAL 5-22
    REMOVAL & INSTALLATION 5-22
FUEL PRESSURE REGULATOR 5-24
    ADJUSTMENT (NON-AIR FLOW CONTROLLED ONLY) 5-24
    REMOVAL & INSTALLATION 5-24
    TESTING 5-25
FUEL TANK 5-26
FULLY AUTOMATIC TRANSAXLE 7-17
FUSES 6-20
FUSES (SPECIFICATION CHARTS) 6-20
GASOLINE FUEL INJECTION SYSTEM 5-16
GENERAL ENGINE SPECIFICATIONS (SPECIFICATION CHARTS) 3-20
GENERAL RECOMMENDATIONS 1-48
GENERATOR AND ALTERNATOR 3-9
    ALTERNATOR PRECAUTIONS 3-9
    REMOVAL & INSTALLATION 3-9
HANDLING A TRAILER 1-50
HEADLIGHTS 6-16
    REMOVAL & INSTALLATION 6-16
HEATER 6-12
HEATER BLOWER 6-13
    REMOVAL & INSTALLATION 6-13
HEATER CABLES 6-13
    REMOVAL & INSTALLATION 6-13
HELI-COIL SPECIFICATIONS (SPECIFICATION CHARTS) 3-41
HISTORY 1-14
HITCH (TONGUE) WEIGHT 1-48
HOW TO BUY A USED VEHICLE 1-53
HOW TO USE THIS BOOK 1-2
HYDRAULIC SYSTEM BLEEDING 9-5
    MANUAL BLEEDING 9-6
    PRESSURE BLEEDING 9-5
HYDRAULIC SYSTEM 9-4
HYDRAULIC VALVE LIFTERS 3-38
    ADJUSTMENT 3-38
    BLEEDING 3-38
    REMOVAL & INSTALLATION 3-38
IDLE SPEED AND MIXTURE ADJUSTMENTS 2-13
IDLE SPEED REGULATOR 5-25
    TESTING 5-25
IGNITION COIL 3-6
    TESTING 3-6
IGNITION LOCK CYLINDER 8-17
    REMOVAL & INSTALLATION 8-17
IGNITION SWITCH (INSTRUMENT CLUSTER) 6-16
IGNITION SWITCH (STEERING) *8-17*

10-26 INDEX

REMOVAL & INSTALLATION 8-17
IGNITION TIMING 2-10
INDUSTRIAL POLLUTANTS 4-2
INSTRUMENT CLUSTER 6-15
INTAKE MANIFOLD 3-28
   REMOVAL & INSTALLATION 3-28
JACKING AND HOISTING 1-50
JACKING PRECAUTIONS 1-51
JUMP STARTING A DEAD BATTERY 1-51
JUMP STARTING PRECAUTIONS 1-52
JUMP STARTING PROCEDURE 1-52
LICENSE PLATE LIGHT 6-19
   REMOVAL & INSTALLATION 6-19
LIGHTING 6-16
LINKAGE 6-15
   REMOVAL & INSTALLATION 6-15
LOWER CONTROL ARM 8-9
   REMOVAL & INSTALLATION 8-9
MAINTENANCE OR REPAIR? (HOW TO USE THIS BOOK) 1-2
MAINTENANCE (ELECTRONIC IGNITION) 2-10
   TROUBLESHOOTING 2-10
MANUAL STEERING GEAR 8-19
   ADJUSTMENT 8-19
MANUAL TRANSAXLE 7-2
MASTER CYLINDER 9-4
   OVERHAUL 9-4
   REMOVAL & INSTALLATION 9-4
MECHANICAL FUEL PUMP 5-4
   REMOVAL & INSTALLATION 5-4
   TESTING & ADJUSTING 5-6
MODEL IDENTIFICATION 1-14
MUFFLERS, TAILPIPES AND HEAT EXCHANGERS 3-28
   REMOVAL & INSTALLATION 3-28
NATURAL POLLUTANTS 4-2
OIL COOLER 3-36
   REMOVAL & INSTALLATION 3-36
OIL PRESSURE RELIEF VALVE 3-38
   REMOVAL & INSTALLATION 3-38
OIL PUMP 3-36
   REMOVAL & INSTALLATION 3-36
OIL STRAINER 3-36
   REMOVAL & INSTALLATION 3-36
OIL VISCOSITY SELECTION CHART (SPECIFICATION CHARTS) 1-45
PARKING BRAKE 9-19
PISTON AND RING SPECIFICATIONS (SPECIFICATION CHARTS) 3-21
PISTONS AND CYLINDERS 3-30
   REMOVAL & INSTALLATION 3-30
POINT-TYPE IGNITION SYSTEMS 3-5
POINT TYPE IGNITION 2-8
PRECAUTIONS 2-9
PRESSURE SENSOR (NON-AIR FLOW CONTROLLED ONLY) 5-25
REAR BRAKE/TURN SIGNAL/BACK-UP LIGHTS 6-17
   REMOVAL & INSTALLATION 6-17
REAR DRUM BRAKES 9-14
REAR HOOD 10-4
   REMOVAL & INSTALLATION 10-4
ROCKER SHAFTS 3-27
   REMOVAL & INSTALLATION 3-27
ROUTINE MAINTENANCE 1-20
RUST, UNDERCOATING, AND RUSTPROOFING 10-14
   RUST 10-14
SAFETY PRECAUTIONS 6-2
SEAT BELT/STARTER INTERLOCK 6-16
SERIAL NUMBER IDENTIFICATION 1-14
SERVICING YOUR VEHICLE SAFELY 1-7
DO'S 1-7
DON'TS 1-8
SHOCK ABSORBER (DIAGONAL ARM REAR SUSPENSION) 8-12

REMOVAL & INSTALLATION 8-12
SHOCK ABSORBER (COIL SPRING/TRAILING ARM REAR SUSPENSION) 8-13
   REMOVAL & INSTALLATION 8-13
SHOCK ABSORBER (TORSION BAR FRONT SUSPENSION) 8-2
   REMOVAL & INSTALLATION 8-2
SHOCK ABSORBER (STRUT FRONT SUSPENSION) 8-7
   REMOVAL & INSTALLATION 8-7
SHOCK ABSORBER (COIL SPRING FRONT SUSPENSION) 8-9
   REMOVAL & INSTALLATION 8-9
SIDE SLIDING DOOR 10-3
   ALIGNMENT 10-4
   REMOVAL & INSTALLATION 10-3
SPARK PLUGS 2-4
   CHECKING & REPLACING SPARK PLUG CABLES 2-7
   REMOVAL & INSTALLATION 2-4
   RETHREADING SPARK PLUG HOLE 2-7
   SPARK PLUG HEAT RANGE 2-4
SPECIAL TOOLS 1-6
SPECIFICATION CHARTS
   ALTERNATOR, GENERATOR, AND REGULATOR SPECIFICATIONS 3-16
   BRAKE SPECIFICATIONS 9-2
   CAPACITIES 1-55
   CARBURETOR SPECIFICATIONS 5-15
   CHASSIS NUMBER CHART 1-16
   CRANKSHAFT AND CONNECTING ROD SPECIFICATIONS 3-21
   ENGINE IDENTIFICATION CHART 1-19
   ENGLISH TO METRIC CONVERSION CHARTS 1-56
   FUSES 6-20
   GENERAL ENGINE SPECIFICATIONS 3-20
   HELI-COIL SPECIFICATIONS 3-41
   OIL VISCOSITY SELECTION CHART 1-45
   PISTON AND RING SPECIFICATIONS 3-21
   STARTER SPECIFICATIONS 3-15
   TORQUE SPECIFICATIONS 3-20
   TUNE-UP SPECIFICATIONS 2-2
   UNLOADED REAR TENSION BAR SETTINGS 8-12
   VALVE SPECIFICATIONS 3-21
   WHEEL ALIGNMENT SPECIFICATIONS 8-11
SPEEDOMETER CABLE 6-16
   REMOVAL & INSTALLATION 6-16
SPEEDOMETER 6-15
   REMOVAL & INSTALLATION 6-15
STANDARD AND METRIC MEASUREMENTS 1-12
STARTER 3-13
   REMOVAL & INSTALLATION 3-13
   SOLENOID REPLACEMENT 3-14
   STARTER/SEAT BELT INTERLOCK 3-13
STARTER SPECIFICATIONS (SPECIFICATION CHARTS) 3-15
STEERING 8-15
STEERING LINKAGE 8-18
   REMOVAL & INSTALLATION 8-18
STEERING WHEEL 8-15
   REMOVAL & INSTALLATION 8-15
STRUT FRONT SUSPENSION 8-5
STRUT 8-5
   REMOVAL & INSTALLATION 8-5
TANK ASSEMBLY 5-26
   REMOVAL & INSTALLATION 5-26
TEMPERATURE SENSORS I AND II 5-25
   REMOVAL & INSTALLATION 5-25
THROTTLE VALVE POSITIONER 4-13
   ADJUSTMENT 4-14
THROTTLE VALVE SWITCH 5-23
   ADJUSTMENT (NON-AIR FLOW CONTROLLED ONLY) 5-23
   REMOVAL & INSTALLATION 5-23
TIPS (HOW TO BUY A USED VEHICLE) 1-53

INDEX

ROAD TEST CHECKLIST 1-54
USED VEHICLE CHECKLIST 1-53
TIRES AND WHEELS 1-39
CARE OF SPECIAL WHEELS 1-42
INFLATION & INSPECTION 1-40
TIRE DESIGN 1-40
TIRE ROTATION 1-39
TIRE STORAGE 1-40
TOOLS AND EQUIPMENT 1-3
TOOLS AND SUPPLIES 10-14
TORQUE 1-9
TORQUE ANGLE METERS 1-12
TORQUE WRENCHES 1-11
TORQUE SPECIFICATIONS (SPECIFICATION CHARTS) 3-20
TORSION ARM 8-2
REMOVAL & INSTALLATION 8-2
TORSION BAR FRONT SUSPENSION 8-2
TORSION BAR 8-2
REMOVAL & INSTALLATION 8-2
TOWING THE VEHICLE 1-50
TRACK CONTROL ARM 8-6
REMOVAL & INSTALLATION 8-6
TRAILER TOWING 1-48
TRAILER WEIGHT 1-48
TRAILER WIRING 6-21
TRAILING ARM 8-13
REMOVAL & INSTALLATION 8-13
TRANSAXLE ASSEMBLY (FULLY AUTOMATIC TRANSAXLE) 7-17
REMOVAL & INSTALLATION 7-17
TRANSAXLE ASSEMBLY (MANUAL TRANSXALE) 7-2
REMOVAL & INSTALLATION 7-2
SHIFT LINKAGE ADJUSTMENT 7-4
TRANSAXLE ASSEMBLY (AUTOMATIC STICK SHIFT TRANSAXLE) 7-6
REMOVAL & INSTALLATION 7-6
SHIFT LINKAGE ADJUSTMENT 7-7
TRANSAXLE (SERIAL NUMBER IDENTIFICATION) 1-15
TRIGGER CONTACTS (NON-AIR FLOW CONTROLLED ONLY) 5-24
REMOVAL & INSTALLATION 5-24
TROUBLESHOOTING (UNDERSTANDING AND TROUBLESHOOTING ELECTRICAL SYSTEMS) 6-3
BASIC TROUBLESHOOTING THEORY 6-4
TEST EQUIPMENT 6-4
TESTING 6-6
TUNE-UP PROCEDURES 2-3
TUNE-UP SPECIFICATIONS (SPECIFICATION CHARTS) 2-2
TURN SIGNAL SWITCH 8-17
REMOVAL & INSTALLATION 8-17
UNDERSTANDING AND TROUBLESHOOTING ELECTRICAL SYSTEMS 6-2
UNDERSTANDING BASIC ELECTRICITY 6-2

AUTOMOTIVE CIRCUITS 6-3
CIRCUITS 6-2
SHORT CIRCUITS 6-3
THE WATER ANALOGY 6-2
UNDERSTANDING ELECTRICITY 3-2
BASIC CIRCUITS 3-2
TROUBLESHOOTING 3-3
UNDERSTANDING THE AUTOMATIC STICK SHIFT TRANSAXLE 7-5
OPERATION 7-5
UNDERSTANDING THE CLUTCH 7-10
UNDERSTANDING THE FUEL SYSTEM 5-2
CARBURETOR 5-2
FUEL FILTERS 5-2
FUEL PUMP 5-2
FUEL TANK 5-2
UNDERSTANDING THE MANUAL TRANSAXLE 7-2
UNLOADED REAR TENSION BAR SETTINGS (SPECIFICATION CHARTS) 8-12
UPPER CONTROL ARM 8-9
REMOVAL & INSTALLATION 8-9
VACUUM DIAGRAMS 4-14
VALVE LASH 2-12
VALVE SPECIFICATIONS (SPECIFICATION CHARTS) 3-21
VEHICLE (CHASSIS) (SERIAL NUMBER IDENTIFICATION) 1-14
VEHICLE CERTIFICATION LABEL (SERIAL NUMBER IDENTIFICATION) 1-14
VOLTAGE REGULATOR 3-13
REMOVAL & INSTALLATION 3-13
VOLTAGE ADJUSTMENT 3-13
WASHING 10-15
WHEEL ALIGNMENT SPECIFICATIONS (SPECIFICATION CHARTS) 8-11
WHEEL BEARINGS 1-48
REMOVAL, PACKING & INSTALLATION 1-48
WHEEL CYLINDER 9-15
OVERHAUL 9-17
REMOVAL & INSTALLATION 9-15
WHEEL CYLINDER 9-8
OVERHAUL 9-10
REMOVAL & INSTALLATION 9-8
WHERE TO BEGIN 1-2
WINDSHIELD WIPERS 6-14
WINDSHIELD WIPERS 1-35
ELEMENT (REFILL) CARE AND REPLACEMENT 1-35
WIPER BLADE AND ARM 6-14
REMOVAL & INSTALLATION 6-14
WIPER MOTOR 6-14
REMOVAL & INSTALLATION 6-14
WIRING DIAGRAMS 6-22
WIRING HARNESSES 6-8
WIRING REPAIR 6-8